Annals of Mathematics Studies

Number 88

FOUNDATIONAL ESSAYS ON TOPOLOGICAL MANIFOLDS, SMOOTHINGS, AND TRIANGULATIONS

BY

ROBION C. KIRBY

AND

LAURENCE C. SIEBENMANN

PRINCETON UNIVERSITY PRESS

AND

UNIVERSITY OF TOKYO PRESS

PRINCETON, NEW JERSEY

1977

Published in Japan exclusively by
University of Tokyo Press;
In other parts of the world by
Princeton University Press

Printed in the United States of America
by Princeton University Press, Princeton, New Jersey

Library of Congress Cataloging in Publication data will
be found on the last printed page of this book

Rep!

FOREWORD

These five essays consolidate several parts of the subject of topological manifolds that were brought within reach by our two short articles [Ki_1] and [KS_1] published in 1969. Many of the theorems proved here were announced by us in [KS_3] [KS_4] [Si_{10}]. Certainly we have not labored in isolation; our bibliography bears clear testimony to this!

Preliminary versions of segments of these essays (particularly Essay I) have been in limited circulation since early 1970. Polycopied versions of all five were released in August 1972 (from Orsay), and reissued in November 1973.

The final camera-ready manuscript was prepared by the second author, Wanda Jones, and Arlene Spurlock during 1974 and 1975 using the IBM selectric composer system (hitherto exploited but little by mathematicians, cf. [Whole, p. 435]). G. Blattmann inked the drawings.

Essential moral support and criticism were contributed by our colleagues, particularly by R. D. Edwards; and a multitude of errata and points needing revision were tracked down by: R. D. Edwards, A. Fathi, L. Guillou, H. Hähl, M. Handel, A. Marin, M. Scharlemann, R. Stern, J. Väisälä, Y.-M. Visetti. (It is a statistical certainty that many errata (or worse) persist. With our readers' assistance we hope to keep a list available upon request.)

Important direct or indirect support has come during this period from several institutions, including the NSF (USA), the CNRS (France), and our home institutions.

For all these contributions, large and small, to the successful completion of this monograph, we are happy to express our heartfelt thanks.

Robion C. Kirby
University of California
Berkeley, CA 94720
USA

Laurence C. Siebenmann
Université de Paris-Sud
91405 Orsay
France

TABLE OF CONTENTS

Essay I includes (as Appendix B) a note by John Milnor on submerging the punctured torus. Annex 3 includes (tacked on at the end) a note on involutions by Michael Atiyah. Both are published here for the first time.

GUIDE

Ample introduction to these five essays is provided by our three articles [Ki_1] [KS_1] [Si_{10}] (of the combined bibliography) that have been reprinted as annexes to this volume. There the reader will find motivation and get a preliminary view of the whole subject.

The interdependence of the essays is roughly as follows:

	Essay I Deformation of smooth and piecewise linear manifold structures.	→ **Essay II** Deformation of sliced families of manifold structures
⇙	⇘	⇓
Essay III Some basic theorems about topological manifolds	**Essay IV** Stable classification of smooth and piecewise linear manifold structures	**Essay V** Classification of sliced families of manifold structures

There are two itineraries we recommend:

1) To acquire an understanding of topological manifolds as such, using a minimum of machinery, read Essay I then Essay III .

2) To learn about classification of smoothings and triangulations read Essay V , referring back to the relevant parts of Essay II .

Many readers will find itinerary 2) too steep. For them we suggest Essay I and then Essay IV ; the latter presents the basic classifications in a leisurely and old-fashioned way.

On the other hand, itinerary 2) will lead the reader to very sophisticated proofs of the results of Essay I , permitting him to pass, if he wishes, straight on to Essay III .

Some readers will be most interested in understanding how the Hauptvermutung and the triangulation conjecture fail for manifolds. They should set out on a peripheral route [KS_1] , [Si_{10}, §2] , [IV, App. B] [V, App. B] , referring also to [KiK] .

On first reading, one can safely bypass the many alternatives digressions and generalizations offered.

Some basic definitions conventions and notations are collected in §2 of Essay I .

Essay I

DEFORMATION OF SMOOTH AND PIECEWISE LINEAR MANIFOLD STRUCTURES

by

R. Kirby and L. Siebenmann

Section headings in Essay I

§1. INTRODUCTION

This essay presents proofs based on handlebody theory in the sense of Smale [Sm_2] of three theorems concerning PL (= piecewise-linear) or DIFF (= differentiable C^∞) manifold structures on topological manifolds. To state them in a simple form, let M denote a metrizable topological manifold without boundary and of (finite) dimension $\geqslant 5$.

Concordance Implies Isotopy *(see §4). Let* Γ *be a CAT (= DIFF or PL) manifold structure on* $M \times I$, *where* $I = [0,1] \subset R^1$ *and let* $\Sigma \times 0$ *be its restriction to* $M \times 0$. *Then there exists an isotopy (= path of homeomorphisms)* $h_t : M \times I \to M \times I$, $0 \leqslant t \leqslant 1$, *such that* h_0 *is the identity map* $id|M \times I$ *and* h_1 *gives a CAT manifold isomorphism* $h_1 : (M \times I)_{\Sigma \times I} \to (M \times I)_\Gamma$ *while* h_t *fixes* $M \times 0$ *pointwise for all* t *in* $[0,1]$. *Further,* h_t *can be as near to the identity as we please.*

The structure Γ is said to give a **concordance** from Σ to Σ' where $\Sigma' \times 1 = \Gamma|(M \times 1)$, and Σ, Σ' are said to be **concordant**. The theorem implies that concordant structures Σ, Σ' are isotopic, i.e. are related by an isotopy of $id|M$ to a CAT isomorphism $M_\Sigma \to M_{\Sigma'}$.

Concordance Extension *(see §4). Let* Γ *be a CAT manifold structure on* $U \times I \subset M \times I$ *where* U *is an open subset of* M *and let* Σ *be a CAT structure on* M *such that the restrictions* $\Gamma|U \times 0$ *and* $\Sigma \times 0|U \times 0$ *coincide. Then* Γ *extends to a CAT structure* Γ' *on* $M \times I$ *such that* $\Gamma'|M \times 0 = \Sigma \times 0$.

Note the resemblance to the classical homotopy extension property.

Product Structure Theorem *(see §5). Let* Θ *be a CAT manifold structure on* $M \times R^s$, $s \geqslant 1$. *There exists a concordant CAT structure* $\Sigma \times R^s$ *on* $M \times R^s$ *obtained from a CAT structure* Σ *on* M *by producting with* R^s.

† The abiding limitation to dimensions $\geqslant 5$ cannot be removed entirely (see §4, §5, [Si_8]) ; perhaps it suffices here to assume M is noncompact or of dimension $\geqslant 4$.

These three results are parallel to existing theorems about compatible DIFF structures on PL manifolds due respectively to Munkres [Mu_2] ; Hirsch [Hi_4] and Munkres [Mu_3] ; and Cairns and Hirsch [Ca] [Hi_2] [Hi_4] [Hi_7] . Our proofs, however do not much resemble existing proofs of these PL-DIFF analogues. Also we eventually manage to deduce them in dimensions $\geqslant 5$, see §5.3 , §5.4 and [II, §4] .

Historically, the above theorems were first sought after to permit a classification up to isotopy of manifold structures in the framework of Milnor's microbundle theory. Essay IV explains this application in detail.

Curiously, some of the most important applications of these theorems concern not manifold structures, but rather basic geometrical properties of topological manifolds. These include topological transversality theorems, topological handlebody theory, and topological facts about simple homotopy type. They are the subject of Essay III . We remark that the proofs in this essay, in depending mainly on DIFF (or PL) handlebody theory, thereby depend on transversality ideas and on some simple homotopy theory.

Our main technical tool is in fact the CAT (= DIFF or PL)

S-Cobordism Theorem† *(see [Ke_1] [Mi_8] [Hu_2] [St_3] [RS_3]). Let $(W; V, V')$ be a compact connected CAT cobordism‡ V to V', which is relative in the sense that $\partial W - int(V \cup V') \cong \partial V \times I$ (where $\cong$ denotes CAT isomorphism ; ∂ indicates boundary ; int indicates formal manifold interior) . Suppose that the inclusions $i : V \hookrightarrow W$, $i' : V' \hookrightarrow W$ are homotopy equivalences , dim $W \geqslant 6$, and i has zero torsion in the Whitehead group in $Wh(\pi_1 W)$. Then $(W; V, V') \cong V \times (I; 0, 1)$.*‡‡

† When $\pi_1 W = 0$, so that $Wh(\pi_1 W) = 0$, the result is called the *h-cobordism theorem* . The 'h' stands for homotopy equivalence ; the 's' stands for simple homotopy equivalence , (i.e. one with zero torsion) .

‡ This implies that V and V′ are CAT submanifolds of ∂W admitting CAT collarings in W (see Appendix A) .

‡‡ This isomorphism can clearly extend the identity isomorphism $V \cong V \times 0$; it can *also* extend the given isomorphism $\partial W - int(V \cup V') \cong \partial V \times I$ as a device exploiting a CAT collaring of ∂V in V readily shows , cf. [Si_4 , Prop. II , fig. 4] , or step 3 in proof of 4.1 below .

In addition, we use the non-compact version of the s-cobordism theorem [Si_7] (see 3.1.1) with $V = M \times R$. It can be reduced by Stallings' engulfing methods to the above compact version with $V = M \times S^1$.

To apply these s-cobordism theorems, we need the algebraic theorem of Grothendieck and Bass-Heller-Swan [BHS] [Bas] to the effect that $\widetilde{K}_0 Z[A] = 0 = Wh(A)$ for any finitely generated free abelian group A.

Finally one must know that the annulus conjecture is true in dimensions ≥ 6 in order to prove the Product Structure Theorem (not its companions though). This in inevitably so, for the latter converts an annulus of dimension ≥ 6 into a smooth h-cobordism. Unfortunately the proof of the annulus conjecture in dimensions ≥ 5 given in [Ki_1] relies on the principal theorems of nonsimply-connected surgery. As we shall note in §5, this unfortunate dependence of the Product Structure Theorem and its corollaries on sophisticated surgery can be artificially eliminated by restating it (and its corollaries) so as to replace topological manifolds everywhere by STABLE† topological manifolds, i.e. manifolds equipped with atlases of charts related by homeomorphisms that are STABLE in the sense of Brown and Gluck [BrnG].

One can give a competitive alternative proof of the Product Structure Theorem, if one is willing to use the stability theorem for TOP_m/CAT_m ($m \geq 5$) [KS_1] [LR_2] [IV, §9.4] [V, §5.2] and a (relative) homotopy theoretic classification theorem for CAT structures derived by immersion theoretic methods.‡ A significant advantage our present proof has is the greater simplicity of the prerequisites. The alternative proof necessarily uses the same handlebody theory to establish the stability theorem. In addition it uses immersion theory. And in the immersion theory are hidden two notable further prerequisites, namely Milnor's microbundle theory and the isotopy extension theorem for the topological category.

† The capital letters should serve to avoid confusion with other uses of the word 'stable'.

‡ Similar remarks apply to the Concordance Implies Isotopy Theorem and the Concordance Extension Theorem.

§2. SOME DEFINITIONS, CONVENTIONS, AND NOTATIONS

These are adopted for all five essays (although some inconsistencies persist). Experienced readers should manage to guess them all; less experienced readers will need to use this section as a glossary.

A set consisting of a single element x is often denoted by x itself rather than by $\{x\}$, provided no confusion will result. The empty set is denoted by ϕ (the Greek 'phi' will be of another shape: φ).

We agree that a **space** is a topological space, until Essay [V] where some semi-simplicial spaces and quasi-spaces appear as well (cf. [V, Appendix A]). Of course any space may have much additional structure.

Given a subset S of a topological space X we write $\mathring{S}$ for the **interior** of S in X and write δS for its **frontier** in X (although by rights X should appear in both notations). Avoid confusion with $\mathrm{int}\, S$ and ∂S defined (below) when S is a manifold. If Y is a subspace of X (perhaps not even meeting S), a **neighborhood** of S in Y is defined to be a set $V \subset Y$ of the form $V = U \cap Y$ where U is a neighborhood (classically defined) of S in X. (This usage agrees with the usual definition of a neighborhood of ∞ in euclidean n-space R^n.)

Maps of topological spaces are understood to be continuous maps. A map $f: X \to Y$ is said to have a certain property **over** a subset $S \subset Y$ of the target in case the restricted mapping $f^{-1}S \to S$ enjoys this property. Thus 'over' means 'on the preimage of'.

An **imbedding** (also spelled embedding) is a map $f: X \to Y$ that gives a homeomorphism $X \to f(X)$ onto its image.

Manifolds and polyhedra are assumed to be metrizable and in fact to have a given metric, denoted most frequently by d. Products may be assumed to have a standard product metric (your favorite).

Our manifolds are, unless the contrary is stated, finite dimensional manifolds of one of the three classical sorts: topological (= TOP), piecewise linear (= PL), or smooth C^∞ (= DIFF).

A topological (= TOP) **manifold** M (possibly with boundary) of dimension $n < \infty$ is a metrizable topological space M such that each point x in M admits an open neighborhood U and an homeomorphism $\kappa : U \to \kappa(U)$ onto an open subset of the euclidean half-space $R^n_+ = \{(x_1, \ldots, x_n) \in R^n \mid x_1 \geq 0\}$. The open set U is called a (coordinate) **chart**; but so is the homeomorphism κ and also the composed imbedding $U \xrightarrow{\kappa} R^n_+$. The points x in M that correspond to points $\kappa(x)$ in the hyperplane $\{(x_1, \ldots, x_n) \in R^n_+ \mid x_1 = 0\}$ form the **boundary** ∂M of M which by invariance of domain [ES, p.303] forms a closed subset of M that is a $(n-1)$-manifold with empty boundary. The (formal) **interior** of M is the set $M - \partial M$ denoted $\operatorname{int} M$; it is a manifold with empty boundary.

A TOP manifold is Hausdorff and paracompact since it is supposed metric; each component is separable and is sigma-compact (the union of countably many compact sets).

A piecewise linear (= PL) **manifold structure** Σ on a topological n-manifold M is a complete (= maximal) piecewise-linearly compatible atlas † of coordinate charts κ to R^n_+. Piecewise-linearly compatible charts are charts that are related on their overlaps by piecewise linear homeomorphisms of open subsets of R^n_+. Such homeomorphisms form what is called a pseudo-group ‡. For *any* pseudo-group G of homeomorphisms on R^n_+ there is a corresponding notion of G-structure, see [KN, §1] . Piecewise linearity and piecewise linear techniques are explained in the textbooks [Ze_1] [Hu_2] [St_3] [Gla] [RS_3] .

A **smooth** (= differentiable C^∞ ; or DIFF) manifold structure is similarly defined using the pseudo-group of C^∞ diffeomorphisms of open subsets of R^n_+ . See [Mi_7] .

A DIFF structure with (cubical) **corners** is defined in terms of the pseudo-group of diffeomorphisms of open subsets of the positive cone

$$R^n_\square = \{(x_1, \ldots, x_n) \in R^n \mid x_1 \geq 0, \ldots, x_n \geq 0\} .$$

† An atlas is a collection $\{\kappa_i\}$ of charts $\kappa_i : U_i \to \kappa_i(U)$ such that $\cup_i U_i = M$.

‡ A pseudo-group G on a space X is a subcategory of the category of all homeomorphisms between open subsets of X , obeying certain axioms: (a) All morphisms of G are isomorphisms, (b) the identity map of X is in G , (c) every restriction of a homeomorphism in G (to an open subset) is also in G , (d) a homeomorphism $h : U \to V$ belongs to G whenever, for some open cover $\{U_\alpha\}$ of U , all the restrictions $U_\alpha \to hU_\alpha$ of h belong to G .

A **CAT manifold** where CAT = PL , or DIFF , or DIFF with corners, is now defined to be a topological manifold M equipped with a CAT manifold structure Σ . The symbol M_Σ is the usual notation for it; M alone is often used if this is unambiguous, as when Σ is in some sense standard or fixed. For example R^n , $I = [0,1]$, and R^n_+ are CAT manifolds.

More about (cubical) corners.

Since $R^n_\square$ is piecewise-linearly homeomorphic to R^n_+ , every PL or TOP 'manifold with corners' would be, in a natural way, simply a manifold with boundary. But there is no diffeomorphism $R^n_\square \to R^n_+$, $n \geqslant 2$.

The **corner set** of a DIFF manifold M with corners is by definition the set of points in ∂M mapped by charts to points in the subset of R^n where $\geqslant 2$ coordinates are zero. Note that this subset is respected by diffeomorphisms between open subsets of $R^n_\square$ (but not by PL homemorphisms). The corner set is thus a diffeomorphism invariant. It may be empty; even the boundary may be empty.

The inevitability of DIFF manifolds with cubical corners appears in the fact that the product $M \times N$ of two DIFF manifolds with nonempty boundaries is a DIFF manifold whose corner set is $\partial M \times \partial N$. The cartesian product of two DIFF manifolds with corners is in a natural way a DIFF manifold with corners, because $R^m_\square \times R^n_\square = R^{m+n}_\square$.

There is no denying that corners can be a nuisance. We shall occasionally have to pause to eliminate some by *'unbending'* , see [III, §4.3] . Compare [$Do_{1,2}$] , [Ce_5] .

We now adopt some conventions for these essays to specify three exceptional cases when a DIFF manifold (with no mention of corners) should be understood to have certain corners. These conventions will spare us almost all further explicit mention of corners.

(a) Let N be a given DIFF manifold with corners such as the s-fold product $I^s = I \times I \times \ldots \times I$, $s \geqslant 0$, or $R^s_\square$. If M is a DIFF manifold (with boundary) we agree to speak of the DIFF manifold $M \times N$ (or $N \times M$) , without mentioning the corners, since the notation gives sufficient warning.

(b) Again, if M is a TOP manifold with boundary, when we speak of a DIFF structure Σ on $M \times N$, we mean a DIFF structure with corner set C exactly what it would be if M were a DIFF manifold with boundary, namely $C = \partial M \times \partial N \cup M \times D$, where D is the corner set of N .

(c) Thirdly, if W^n gives a DIFF cobordism V to V' of DIFF (n–1)-submanifolds with boundary of ∂W (disjoint and closed as subsets), we agree that $\partial V \cup \partial V'$ is the corner set of W .

Every open subset M_0 of a CAT manifold M has an induced CAT manifold structure; with this structure, M_0 is called an **open CAT submanifold .**

If Σ is a CAT structure on M and $h : M \to M'$ is a homeomorphism, the **image CAT structure** $h\Sigma$ on N has as typical chart a composed homeomorphism

$$hU \xrightarrow{h^{-1}} U \xrightarrow{\kappa} \kappa U$$

where $\kappa : U \to \kappa U$ is a chart of Σ .

A homeomorphism of CAT manifolds $h : M_\Sigma \to N_{\Sigma'}$ is a **CAT isomorphism** if $\Sigma' = h\Sigma$, or equivalently $\Sigma = h^{-1}\Sigma'$. We then write $M_\Sigma \cong N_{\Sigma'}$; and in case $N_{\Sigma'}$ is in addition an open submanifold of a CAT manifold N^+ , we call the composition $h_+ : M_\Sigma \to N^+$ of h with the inclusion $N_{\Sigma'} \hookrightarrow N^+$ a **CAT imbedding** .

One can define a **CAT map** $f : M \to N$ of CAT manifolds to be a map of the underlying sets which, when expressed locally in terms of coordinate charts, is respectively continuous if CAT = TOP , piecewise-linear if CAT = PL , or differentiable C^∞ if CAT = DIFF (possibly with corners). There results a **category** of CAT manifolds which is occasionally itself denoted CAT . Note that its isomorphisms are as described above. The sign $\cong$ indicates isomorphism in whatever category we happen to be working.

In the above spirit one can define a **polyhedron** to be a metric space equipped with a maximal piecewise-linearly compatible atlas of charts to (locally finite) simplicial complexes – the atlas can be called a **polyhedral structure** . Every simplicial complex, and every PL manifold, is naturally a polyhedron. Piecewise-linearity of maps of polyhedra makes good sense. Thus we obtain an enlarged PL category. It is an elementary fact that every polyhedron is PL homeomorphic to a simplicial complex.

A PL map that is a homeomorphism is a PL isomorphism. This is clearly quite untrue for DIFF in place of PL , as the map $x \mapsto x^3$ on R^1 shows. Cf. [SchS] .

A space X is said to be **triangulated** if it is a simplicial complex; the term is particularly convenient if space carries additional structure such as a DIFF (manifold) structure. A triangulation of a polyhedron is said to be PL if it agrees with the polyhedral structure. A triangulation of a DIFF manifold M_Σ is said to be a **Whitehead or DIFF triangulation** if for each closed simplex σ the inclusion $\sigma \hookrightarrow M_\Sigma$ is a smooth nonsingular imbedding (into the DIFF manifold) .

A PL structure **S** and a DIFF structure Σ (possibly with corners) on a TOP manifold M are **Whitehead compatible** if some PL triangulation of $M_{\mathbf{S}}$ is a DIFF triangulation of M_Σ . It is then sometimes convenient to say that **S** *is a Whitehead* (or DIFF) *triangulation of* Σ ; although strictly speaking **S** is merely *determined by* some Whitehead triangulation of Σ . It is not difficult to see that Whitehead compatibility of **S** and Σ can be tested *locally* as below [I, § 5.3] .

A simplicial complex PL isomorphic to a PL manifold is known as a **combinatorial manifold** ; it is characterized by the fact that the link of each simplex is PL isomorphic to a PL sphere or ball.

Warning: For us PL structure means PL manifold structure. In 1974, R. D. Edwards constructed some triangulations of R^n , $n \geq 5$, that are not combinatorial ; in other words, he constructed some polyhedral structures on R^n that are not PL manifold structures. See [Ed_6] .

It is still not known (in 1975) whether every TOP manifold can be triangulated, i.e. admits a polyhedral structure. See [Si_9] [Matu] [GalS] .

A map $f : X \to Y$ in any category CAT we have met (from the category of topological spaces and continuous maps on) is called a CAT *trivial bundle* if there exists a CAT object F and a CAT isomorphism $\theta : F \times Y \to X$ such that $f\theta = p_2$ where p_2 is the projection to Y . More generally, f is called a CAT (locally trivial) **bundle** or bundle projection if each point y in Y has an open neighborhood over which f is a CAT trivial bundle. More generally still, f is called a CAT **submersion** if each point x in X has an open neighborhood U such that f(U) is open in Y and the restriction $U \to f(U)$ of f is a CAT trivial bundle. These ideas are the subject of Essay II . (See [II , § 1] for related definitions.)

A CAT **microbundle** is most succinctly defined to be a CAT retraction that is a CAT submersion; we will always assume that its fibers (= point preimages) are CAT manifolds without boundary. With due warning, we occasionally deviate slightly from this definition. See [III , § 1] [IV , § 1] [V ; § 1, § 2] where many related definitions appear.

A map $f : X \times P \to Y \times P$ is said to be a **product along P near a point** $(x,p) \in X \times P$ if there exists a neighborhood $X_0 \times P_0$ of (x,p) and a map $\varphi : X_0 \to Y$ such that $f = \varphi \times (id \mid P)$ on $X_0 \times P_0$. Also f is said to be a product along P **near a subset** $S \subset X \times P$ if f is a product along P near each point (x,p) in S .

Similarly, given a CAT structure Θ on a product $M \times P$ of a TOP manifold M with a CAT manifold $P = P_\Sigma$; we shall say that Θ is a **product along** P **near a point** z of $M \times P$ in case there exists an open neighborhood $M_o \times P_o$ of z and a CAT structure Σ_0 on M_o such that $\Theta = \Sigma_0 \times \Sigma$ on $M_o \times P_o$. Also we shall say that Θ is a product along P **near a subset** $S \subset M \times P$ if it is one near each point z in S .

Homotopies isotopies and concordances .

A **homotopy** is a map $F : I \times X \to I \times Y$ such that $p_1 = p_1 F$, where p_1 denotes projection to the first factor I . In case F is an isomorphism it is called an **isotopy** . When F is an open imbedding it is called an **isotopy through open imbeddings** .

This language applies in the several categories CAT we have met, not just the category of continuous maps of topological spaces.

Note that F is uniquely determined by the family $f_t : X \to Y$, $0 \leqslant t \leqslant 1$, of maps such that $F(t,x) = (t, f_t(x))$ for all (t,x) in $I \times X$. We often write f_t , $0 \leqslant t \leqslant 1$, instead of F ; and we often say that F is a homotopy from f_0 to f_1 , writing $f_0 \simeq f_1$. Note that f_t will inherit from F the property of being an CAT isomorphism or a CAT open imbedding.

The homotopy f_t , $0 \leqslant t \leqslant 1$, is said to **respect** a subset $S \subset X$ if $f_t(S) \subset S$ for $0 \leqslant t \leqslant 1$. It is said to **fix** S (pointwise) if $f_t(x) = f_0(x)$ for all x in S and all t in I . It is said to be **rel** (or 'relative to') S in case, for some open neighborhood U of S in X , the homotopy F fixes U . The **support** of the homotopy F is the closure in X of the set of all $x \in X$ such that $f_t(x) \neq f_0(x)$ for some t in I .

[Similarly, a self-map $f : X \to X$ **respects** $S \subset X$ if $f(S) \subset S$; **fixes** S (pointwise) if $f \mid S = id$; and is **rel** S if $f = id$ near S . Also the **support** of f is the closure in X of the set of all $x \in X$ such that $f(x) \neq x$.]

We say that the homotopy F is **conditioned** if F is a product along I near $0 \times X$ and near $1 \times X$. Any homotopy can be replaced by a conditioned one by a simple change of parameter.

Occasionally homotopies and isotopies are defined with another 'parameter' space in place of I , such as a simplex or a closed interval [a,b] , $-\infty \leqslant a < b \leqslant \infty$. Also the parameter space may be placed first (as above) or last, at pleasure – we take the same liberty with all similar notions, e.g. concordances.

Two conditioned CAT homotopies defined for adjacent intervals [a,b] and [b,c] compose naturally to give a unique conditioned homotopy defined for [a,c] . The conditioning is quite superfluous for CAT = PL ; but not for DIFF . So for CAT = DIFF we agree to compose only conditioned homotopies.

Strongly analogous conventions will be adopted concerning CAT (= DIFF or PL) concordances (sometimes without special explanation). A CAT **concordance** Γ is a CAT structure on a product $I \times M$ where M is a TOP manifold, so that Γ gives by restriction CAT structures $\Sigma_i \times \{i\}$ on $M \times \{i\}$ for $i = 0,1$; and we say that it gives a CAT concordance $\Gamma : \Sigma_0 \simeq \Sigma_1$.

A **sliced** concordance is a special sort of concordance discussed in Essay II .

The concordance Γ is **rel** $S \subset M$ if $\Gamma = I \times \Sigma_0$ near $I \times S$. Its **support** is the closure in M of the set of all $x \in M$ such that Γ fails to be rel $I \times \{x\}$.

It is **conditioned** if Γ is a product along I near $0 \times M$ and $1 \times M$. A concordance $\Gamma : \Sigma_0 \simeq \Sigma_1$ rel C where C is a closed subset of M can be replaced by a conditioned one by using the (local) collaring theorems of Appendix A (this essay) .

Our use of the term **submanifold** is restricted to submanifolds that are **clean** in the sense that they meet the (possibly empty) boundary transversally (see precise definition below) .

Examples: In the closed right half x-y plane R^2_+ , the positive x-axis is a clean 1-submanifold; the 1st quadrant is a clean 2-submanifold; but the y-axis is not a clean submanifold. A nonempty boundary ∂M of a manifold M is never a clean submanifold of M .

Define a subset M of a CAT manifold W^w (CAT = TOP , PL or DIFF) to be a **clean CAT m-submanifold** of W if M is a closed subset of W and each point of M lies in a chart

$$\kappa : U \to \kappa(U) \subset R^w_+ = R^1_+ \times R^{w-1}$$

such that $\kappa(U \cap M)$ is the intersection of $\kappa(U)$ with $R^m_{++} = R^2_\square \times R^{m-2} \subset R^w_+$. Thus M is a CAT m-manifold (with model R^m_{++}), and the pair (W^w, M^m) is locally CAT isomorphic near M^m to (R^w_+, R^m_{++}). The **codimension** of M is $w-m$. By convention, $R^1_{++} = R^1_+$ and $R^0_{++} = R^0_+ = R^0$.

To define clean submanifolds of DIFF manifolds with corners, one should use, in place of (R^w_+, R^m_{++}), models of the form

$$(R^a_\square \times R^b \times R^c\,,\ R^a_\square \times R^b_\square \times R^{c'}) ,$$

with $c' \leqslant c$.

A **locally flat** submanifold is similarly defined using in place of (R^w_+, R^m_{++}) the model

$$R^m_\square \times (R^{w-m}, 0)$$

Finally we mention some standard objects.

B^k has one of two meanings. When PL structures are being discussed,

$$B^k = \{(x_1, \ldots, x_k) \in R^k \ ;\ |x_1| \leqslant 1, \ldots, |x_k| \leqslant 1\} = [-1,1]^k$$

is a standard PL k-**ball** (or k-disc or k-disk or k-cube) in euclidean space. When DIFF structures are being discussed.

$$B^k = \{(x_1, \ldots, x_k) \in R^k \ ;\ x_1^2 + x_2^2 + \ldots + x_k^2 \leqslant 1\}$$

the standard smooth unit k-ball in R^k. In each case $S^{k-1} = \partial B^k$ is a standard (k−1)-**sphere**.

For any subset S of a vector space V and any real number λ we define

$$\lambda S = \{\lambda x \ ;\ x \in S \subset V\} .$$

The n-**torus** T^n and the hieroglyph ⊐, both defined in the next section, will reappear in later essays, as will the **collarings** of Appendix A.

§3. HANDLES THAT CAN ALWAYS BE STRAIGHTENED

By a CAT **handle problem** (CAT = DIFF or PL) we mean a topological imbedding $h : B^k \times R^n \to V^{k+n}$ into a CAT manifold which is a CAT imbedding near $\partial B^k \times R^n = S^{k-1} \times R^n$. Think of $B^k \times B^n$ as a model k-handle with core $B^k \times 0$. We say that the problem h can be **solved** in the event that there exists an isotopy h_t, $0 \leqslant t \leqslant 1$, of h through imbeddings such that, 1) h_1 is a CAT imbedding near $B^k \times B^n$, and 2) outside some compact set, and near $\partial B^k \times R^n$, one has $h_t = h$ for all t. Then we sometimes say that the handle h can be **straightened** (if CAT = PL), or **smoothed** (if CAT = DIFF).

The incentive to study handle problems came as follows. Let $f : M \to N$ be a homeomorphism of PL manifolds where M has a handle decomposition. Suppose one wants to isotop f to a piecewise-linear homeomorphism. One observes that it would suffice to solve a sequence of handle problems. But the solution of the annulus conjecture in [Ki_1] shows how to straighten 0-handles in dimension $\geqslant 5$. We attempted to straighten k-handles for $k > 0$ in [KS_1] and arrived at the conclusion that, while this is possible for $k + n \geqslant 5$ and $k \neq 3$, some 3-handles cannot be straightened. An obstruction in Z_2 is involved, analogous to one in Milnor's $\Gamma_7 = Z_{28}$ to *smoothing* a seven handle. We remark in passing that if $k + n \leqslant 3$, all handles *can* be straightened (or smoothed)†. This is a result of Moise for $k + n = 3$.

Our basic method for solving handle problems serves in a wide variety of situations. This essay is based upon one key application of this method in a theorem *designed to involve no obstruction in its statement and to require no techniques beyond handlebody theory in its proof.*

The following convention will save many words. In this section, when a manifold of the form $I \times M$ is under consideration, where M is some manifold with boundary ∂M, then the symbol $\sqsupset(I \times M)$ or simply the hieroglyph $\sqsupset$ will stand for $I \times \partial M \cup 1 \times M$ by reason of a similarity when $M = I$.

† The method of [KS_1] also proves this; see [Si_8, §5] or [Si_{11}] for some assistance.

THEOREM 3.1.

Suppose $h : (I,0) \times B^k \times R^n \to (X,V)$ *is a homeomorphism onto a CAT manifold pair such that* h *is a CAT imbedding near* $\sqsupset$. *If* $m = n + k \geqslant 5$, *there exists an isotopy* $h_t : (I,0) \times B^k \times R^n \to (X,V)$, $0 \leqslant t \leqslant 1$, *of* $h = h_0$ *such that*

1) h_1 *is a CAT imbedding near* $I \times B^k \times B^n$ *and*
2) $h_t = h$ *near* $\sqsupset$ *and outside* $I \times B^k \times rB^n$ *for some* $r > 0$.

This states that a handle problem $0 \times B^k \times R^n \xrightarrow{h} V$ is solvable if it is 'concordant' to a solved handle problem. Better, one can in a sense solve the whole concordance of handle problems. One can regard it as a 'concordance implies isotopy' theorem. For brevity we shall often write B_+ for $I \times B^k$.

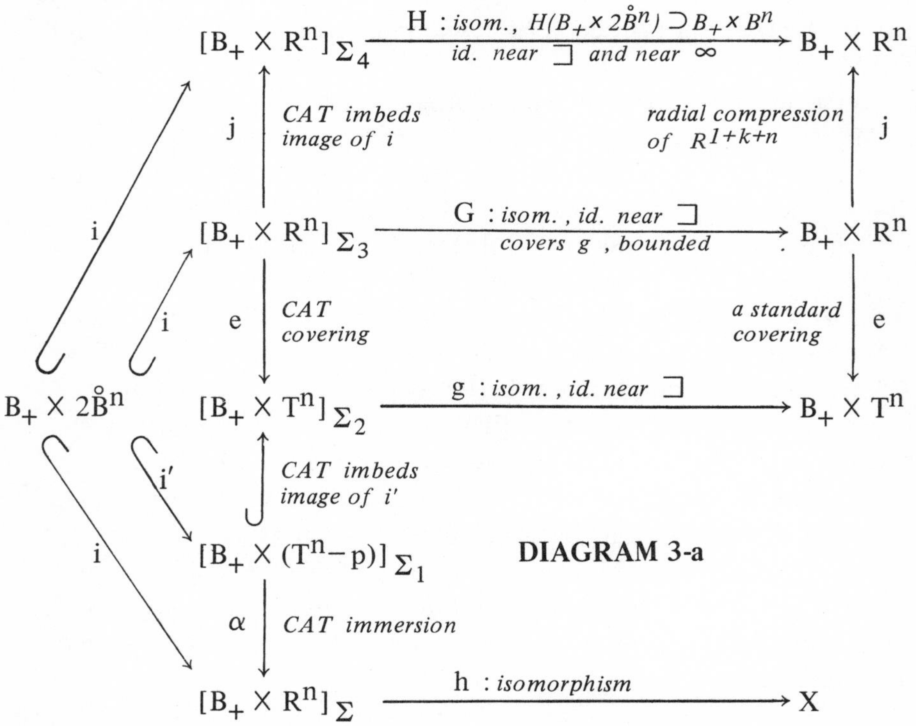

DIAGRAM 3-a

Notes: *The diagram commutes, except in the top rectangle. By restriction one gets a similar diagram with* $0 \times B^k$ *in place of* $B_+ = I \times B^k$, *and* V *in place of* X. *Each CAT structure* $\Sigma, \Sigma_1, \Sigma_2, \Sigma_3, \Sigma_4$ *is standard near* $\sqsupset$. *The pieces are constructed in the order* $e, p, i', i, \alpha, j, \Sigma, \Sigma_1, \Sigma_2, g, G, \Sigma_3, H, \Sigma_4$.

PROOF OF 3.1.

We begin with a degenerate case: $n = 0$, $k \geqslant 5$. The h-cobordism theorem (see § 1) provides a CAT isomorphism $h' : (I,0) \times B^k \to (X,V)$ equal h near $\sqsupset$. Now $f = h'^{-1}h$ is a self-homeomorphism of $I \times B^k$ equal to the identity near $\sqsupset$. Thus we can extend it by the identity to a self-homeomorphism of $[0,\infty) \times R^k \supset I \times B^k$. Now the Alexander isotopy [A$_1$] defined by $f_t(x) = (1-t)\, f(\frac{x}{1-t})$ for $0 \leqslant t < 1$, and by $f_1(x) = x$ determines the wanted isotopy of h via the formula $h_t = h'f_t$, $0 \leqslant t \leqslant 1$. ∎

The skeleton of the proof of 3.1 for $n \geqslant 1$ *is the Diagram 3 - a*. It is a variant of the Main Diagram of [KS$_1$] mentioned in [KS$_1$, Proposition 1]. To motivate this relatively intricate construction we begin by showing how its end-products H and Σ_4 lead to the proof of 3.1. The vertical maps on the left of the diagram are all CAT imbeddings on $B_+ \times 2\mathring{B}^n$ or its image by i'; so from commutativity of the triangles at the left we see that $\Sigma_4 = \Sigma$ on $B_+ \times 2\mathring{B}^n$. Let $H_t : B_+ \times R^n \to B_+ \times R^n$ be the Alexander isotopy of H to the identity fixing points near $\sqsupset$ and outside some compactum (independent of t). To define it, one extends H by the identity to a self-homeomorphism of $[0,\infty) \times R^{n+k}$. Then one sets $H_t(x) = tH(x/t)$ for $0 < t \leqslant 1$, and sets H_0 = identity.

The asserted isotopy h_t of $h : B_+ \times R^n \to X$ is defined by

$$h_t = hH_t^{-1}, \qquad 0 \leqslant t \leqslant 1 .$$

This isotopy equals h outside some compactum and near $\sqsupset$ since the Alexander isotopy H_t is the identity there. Visibly $h_0 = h$. Finally $h_1 = hH^{-1}$ is the composition

$$B_+ \times R^n \xrightarrow[\cong]{H^{-1}} [B_+ \times R^n]_{\Sigma_4} \xrightarrow{id} [B_+ \times R^n]_{\Sigma} \xrightarrow[\cong]{h} X .$$

Since $H^{-1}(B_+ \times B^n) \subset [B_+ \times 2\mathring{B}^n]_{\Sigma_4}$ and $\Sigma_4 = \Sigma$ on $B_+ \times 2\mathring{B}^n$ we see that h_1 is a CAT imbedding near $B_+ \times B^n = I \times B^k \times B^n$. This completes the proof of 3.1 given Diagram 3-a . ■

We proceed now to construct the pieces of Diagram 3-a in the following order : $e, p, i', i, \alpha, j, \Sigma, \Sigma_1, \Sigma_2, g, G, \Sigma_3, H, \Sigma_4$. The first six do not depend on h .

Construction of e , p , i′ , i . Recall that the n-torus T^n is the quotient of R^n by the integer subgroup Z^n . Let $\rho : R^n \to T^n$ be the quotient map taking residues of the coordinates mod 1 . T^n has the CAT structure that makes ρ a CAT covering map . Set $\bar{e}(y) = \rho(y/8)$ and define $e(t, x, y) = (t, x, \bar{e}(y))$ for $(t, x, y) \in I \times B^k \times R^n$. Then e imbeds $I \times B^k \times 2B^n$ and we choose the point $p = (½, ½, \ldots, ½) \in T^n$ to be disjoint from $\bar{e}(2B^n)$, so that $i'(t, x, y) = e(t, x, y)$ defines an imbedding $i' : I \times B^k \times 2\mathring{B}^n \to I \times B^k \times (T^n - p)$. The inclusion i in Diagram 3-a is the natural one .

Construction of α . We use a CAT submersion $\bar{\alpha} : T^n - p \to R^n$;† then α will be $(\mathrm{id} \mid B_+) \times \bar{\alpha}$. J. Milnor has provided us with an explicit construction of $\bar{\alpha}$ which appears in Appendix B . D. Barden has given another [Bar] . On the other hand , we can get $\bar{\alpha}$ from immersion theory [Hi_1] [Ha_2] , since T^n (like $S^1 = T^1$) has a trivial tangent bundle . We can arrange $\bar{\alpha}\,\bar{e} \mid B^n = \mathrm{id} \mid B^n$, since two degree +1 CAT imbeddings of B^n into R^n are related by a CAT automorphism of R^n . It is immaterial that $\alpha : B_+ \times (T^n - p) \to B_+ \times R^n$ be CAT .

Construction of $j : B_+ \times R^n \to B_+ \times R^n$. Since $I = [0,1] \subset R$ and $B^k \subset R^k$ we have a natural inclusion

$$B_+ \times R^n = I \times B^k \times R^n \subset R^1 \times R^k \times R^n = R^{m+1} , \qquad m = n + k .$$

And we let $I \times B^k \times R^n$ inherit the metric of the norm $|x| = (x_1^2 + \ldots + x_{m+1}^2)^{½}$ on R^{m+1} . Let $d = 2 \sup\{ |x| ; x \in I \times B^k \times 2B^n\}$. Choose a homeomorphism $\delta : (0, \infty) \to (0, d)$ that fixes $(0, d/2)$ pointwise and satisfies $\delta(t) \leqslant t$ for all $t \in (0, \infty)$. Define

† Since the source and target of $\bar{\alpha}$ are of the same dimension , $\bar{\alpha}$ is necessarily locally an imbedding ; thus $\bar{\alpha}$ is an *immersion* as well as a submersion .

$J : R^{m+1} \to R^{m+1}$ by $J(x) = \frac{\delta(|x|)}{|x|}x$ for $x \in R^{m+1}$. This is a topological imbedding onto the interior of the smooth disc D of radius d. Note that J maps any star-shaped set, e.g. $I \times B^k \times R^n$ into itself and fixes $I \times B^k \times 2B^n$ pointwise. We define $j : I \times B^k \times R^n \to I \times B^k \times R^n$ as the restriction of J. If $f : R^{m+1} \to R^{m+1}$ is a homeomorphism that is *bounded* in the sense that $\sup\{|f(x) - x| \; ; \; x \in R^{m+1}\} < \infty$ then one easily checks that the homeomorphism $J f J^{-1} : \text{int } D \to \text{int} D$ extends by the identity to a homeomorphism $F : R^{m+1} \to R^{m+1}$, simply because J respects rays from $0 \in R^{m+1}$. It follows that, *if* $f_0 : (I, 0) \times B^k \times R^n \to (I, 0) \times B^k \times R^n$ *is a bounded homeomorphism that is the identity on* $\sqsupset$, *then* $J f_0 J^{-1} : J(B_+ \times R^n) \to J(B_+ \times R^n)$ *extends by the identity outside* $J(B_+ \times R^n) \subset D$ *to a homeomorphism* $F_0 : B_+ \times R^n \to B_+ \times R^n$. To see this, note that f_0 extends to a *bounded* homeomorphism $f : R^{m+1} \to R^{m+1}$ by setting $f(-t, x, y) = \omega f(t, x, y)$ for $(-t, x, y) \in [-1, 0] \times B^k \times R^n$, where ω operates merely by changing the sign of the first coordinate, and by making f the identity outside $[-1, 1] \times B^k \times R^n$. Then F_0 is just the restriction of F mentioned above.

Construction of Σ and Σ_1. They are the unique CAT structures making h and α locally CAT imbeddings in

$$[B_+ \times (T^n - p)]_{\Sigma_1} \xrightarrow{\alpha} [B_+ \times R^n]_{\Sigma} \xrightarrow{h} X .$$

Construction of Σ_2. This will be a structure standard near $\sqsupset$ so that $[B_+ \times (T^n - p)]_{\Sigma_1} \subset [B_+ \times T^n]_{\Sigma_2}$ is a CAT imbedding except in a small neighborhood of $B_+ \times p$ chosen disjoint from $i'(B_+ \times 2B^n)$.

The construction will be based on the following very special case of the noncompact s-cobordism theorem (cf. §1).

PROPOSITION 3.1.1. *Let $(Y; V, V')$ be a CAT relative*† *cobordism that is homeomorphic to (or even proper homotopy equivalent to) the product cobordism $(I ; 0, 1) \times V$. Suppose also that $V = M \times R$ where M is a connected compact CAT manifold with free abelian fundamental group. Provide that $\dim Y \geqslant 6$.*

† This means $\partial Y - (\text{int } V \cup \text{int } V') \cong \partial V \times [0,1]$.

Then $(Y; V, V')$ is a CAT product cobordism. Thus, given a CAT open imbedding f_0 of a neighborhood of $\sqsupset$ in $(I; 0, 1) \times V'$ into $(Y; V, V')$, there exists a CAT isomorphism

$$f : (I; 0, 1) \times V' \to (Y; V, V')$$

equal f_0 near $\sqsupset$.

Proof of proposition. The non-compact s-cobordism theorem [Si_7] applies since $\widetilde{K}_0 Z[\pi_1 Y] = 0$ by [BHS] [Bas]. The argument in [Si_7] involves only classical handlebody theory. ■

If we are willing to use some engulfing, the gluing device of [Si_6] reduces 3.1.1 to the compact s-cobordism theorem (for free abelian fundamental group). Indeed, it produces a compact relative CAT s-cobordism $(Y_0 ; V_0, V_0')$ whose ∞-cyclic covering is isomorphic to $(Y ; V, V')$ – by gluing together the two ends of Y, as those of a bandage around your finger. For details see [Si_6, §6].† ■

To begin the construction of Σ_2, note that Σ_1 extends by the standard structure to all of $B_+ \times T^n - (\lambda B_+) \times p$ for some $\lambda < 1$ near 1. We derive a CAT structure $\Sigma_1' = \psi^{-1}\Sigma_1$ on $B_+ \times T^n - 0 \times p$ by choosing an open imbedding ψ of the latter set into the former fixing points outside a small neighborhood of $\lambda B_+ \times p$ in $B_+ \times T^n$. Σ_1' would serve as Σ_2 except that it is not defined at the one point $0 \times p$ of the boundary of $B_+ \times T^n$. The following lemma lets one remedy this to define Σ_2. The R^m in it corresponds to a small open ball about $0 \times p$ in $0 \times B^k \times R^n$.

LEMMA 3.1.2. *Consider a CAT structure σ' on $I \times R^m - 0 \times 0$ standard near $1 \times R^m$. If $m \geq 5$, handlebody theory and the topological Shoenflies theorem can be used to obtain a CAT structure σ on $I \times R^m$ equal σ' near $I \times R^m - [0,1) \times \mathring{B}^m$.*

Proof of lemma. Applying the non-compact h-cobordism theorem 3.1.1, we obtain a CAT isomorphism

† Alternatively, see a variant of this argument in part B of the proof of Theorem 1.1 in Essay II.

$$\phi : I \times (R^m - 0) \to \{I \times (R^m - 0)\}_{\sigma'}$$

equal the identity near $1 \times R^m$. We can extend ϕ over $0 \times R^m$ by mapping 0×0 to itself and we can extend ϕ by the identity to $(\lambda, 1] \times 0$, for λ near 1. Now ϕ is a topological imbedding into $I \times R^m$ defined on $I \times R^m - (0, \lambda) \times \tfrac{1}{2}\mathring{B}^m$ (and more). The m-sphere $\phi(\partial([0, \lambda] \times \tfrac{1}{2}B^m))$ bounds a $(m+1)$-ball in $I \times R^m$ by the Shoenflies theorem [Brn_1]. Thus $\phi|\{I \times R^m - (0, \lambda) \times \tfrac{1}{2}\mathring{B}^m\}$ extends by coning to a homeomorphism $\Phi : I \times R^m \to I \times R^m$. The image of the standard structure under Φ will serve as σ. ∎

Construction of g. Applying *the s-cobordism theorem (see §1)* to $[(I; 0, 1) \times B^k \times T^n]_{\Sigma_2}$, we get a CAT isomorphism $g : [I \times B^k \times T^n]_{\Sigma_2} \to I \times B^k \times T^n$ that is the identity near $\sqsupset$. No torsion obstruction occurs because the Whitehead group $\mathrm{Wh}(\pi_1(B^k \times T^n)) = \mathrm{Wh}(Z^n)$ is zero. ∎

Construction of G. Define G to be the unique self-homeomorphism of $(I, 0) \times B^k \times R^n$ that fixes $1 \times B^k \times R^n$ (pointwise) and covers g, i.e. $eG = ge$. It exists because e is a covering. It is easy to see that G fixes a neighborhood of $\sqsupset$.

LEMMA 3.1.3. *G is bounded in the sense that*

$$\sup\{|G(x) - x|;\ x \in I \times B^k \times R^n \subset R^{m+1}\} < \infty .$$

Proof of lemma. Consider any covering translation T of e, $T \in 8Z^n$. The commutator $TGT^{-1}G^{-1} \equiv [T, G]$ is a covering translation as it covers $g \circ g^{-1}$ = identity. Also $[T, G]$ fixes $1 \times B^k \times R^n$; so $[T, G]$ = identity.

As G commutes with covering translations, the supremum of $|G(x)-x|$ is attained on the compact fundamental domain $I \times B^k \times [0, 8]^n$; hence it is indeed finite. ∎

Construction of Σ_3. It is merely the CAT structure making $G : [B_+ \times R^n]_{\Sigma_3} \to B_+ \times R^n$ a CAT isomorphism. As $eG = ge$ one sees that $e : [B_+ \times R^n]_{\Sigma_3} \to [B_+ \times R^n]_{\Sigma_2}$ is a CAT covering. ∎

Construction of H . We first prove

Assertion : *There is a CAT isomorphism*

$$G' : [B_+ \times R^n]_{\Sigma_3} \to B_+ \times R^n$$

equal G near ∞, *equal the identity near* ⊐, *and such that* $G'(B_+ \times 2\mathring{B}^n) \supset B_+ \times B^n$.

Proof of assertion : G' can clearly be $G\alpha'$ where α' is any CAT automorphism of $(B_+ \times R^n)_{\Sigma_3}$ fixing points outside a compactum in $[B_+ \times (R^n - B^n)]_{\Sigma_3}$ disjoint from ⊐ , such that $\alpha'(B_+ \times 2\mathring{B}^n) \supset$ $\supset G^{-1}(B_+ \times B^n)$. It is easy to produce α' by CAT engulfing in $[B_+ \times (R^n - B^n)]_{\Sigma_3}$. However to accomplish this by handlebody methods we suggest the reader use a CAT isomorphism $\psi : [B_+ \times (R^n - B^n)]_{\Sigma_3} \to B_+ \times (R^n - B^n)$ fixing a neighborhood of ⊐ as provided by Proposition 3.1.1 , in conjunction with the strictly elementary

LEMMA 3.1.4. *Consider a closed subset A of* $B_+ \times (R^n - B^n)$ *with a* $\lambda \in (1,2)$ *such that* $A - B_+ \times \lambda\mathring{B}^n$ *is compact and disjoint from* ⊐ . *Then there exists a CAT automorphism* α *of* $B_+ \times (R^n - B^n)$ *fixing points outside some compactum disjoint from* ⊐ *so that* $\alpha(B_+ \times (\lambda\mathring{B}^n - B^n)) \supset A$.

The proof is trivial when we express $B_+ \times (R^n - B^n)$ as $B_+ \times \partial B^n \times (1,\infty)$ in the standard way. ■

To apply the lemma we must choose λ so near 1 that $\psi\{[B_+ \times (2\mathring{B}^n - B^n)]_{\Sigma_3}\} \supset B_+ \times (\lambda\mathring{B}^n - B^n)$. The assertion is proved. ■

Now define $H : B_+ \times R^n \to B_+ \times R^n$ to be $jG'j^{-1}$ on $j(B_+ \times R^n)$. As G' is bounded , and the identity near ⊐ , the construction of j assures that H extends as the identity outside the bounded set $j(B_+ \times R^n) \subset B_+ \times R^n$ to a homeomorphism H equal the identity near ⊐ .

Construction of Σ_4 . It is the unique CAT structure making $H : [B_+ \times R^n]_{\Sigma_4} \to B_+ \times R^n$ a CAT isomorphism . ■

This concludes the construction of Diagram 3-a . The proof of 3.1 is complete . ■

We close with a corollary that illuminates the Product Structure Theorem (5.1 below). The latter's proof is more complicated but its key is isolated here in a simple form.

COROLLARY 3.2 (Stability). *Let* $h: B^k \times R^n \to V$ *be a* CAT *(= PL or DIFF) handle problem,* $k+n \geqslant 5$*, and suppose that* $h \times id: B^k \times R^n \times R \to V \times R$ *can be straightened [smoothed]. Then* h *can also be straightened [smoothed].*

Proof of 3.2. The solution for $h \times id$ plus the device of translation along R gives an isotopy $F_t: B^k \times R^n \times R \to V \times R$ so that $F_0 = h \times id$, and F_1 is a CAT imbedding near $B^k \times B^n \times r$ for some large $r > 0$, and lastly $F_t = h \times id$ near $B^k \times R^n \times 0 \subset B^k \times R^n \times R$. Alter F_1 near $B^k \times R^n \times r$ by an automorphism of domain so that F_1 is a CAT imbedding near $B^k \times \mathrm{int} 2B^n \times r$.

We apply Theorem 3.1 to the CAT structure Σ on $B^k \times \mathrm{int} 2B^n \times ([0,r], 0)$, making F_1 a CAT imbedding, in order to obtain an isotopy whose restriction to $B^k \times \mathrm{int} 2B^n \times 0$ straightens [smooths] h. ■

D. Webster's article [Webs] may improve on the above exposition and/or provide some alternative arguments.

§4. CONCORDANCE IMPLIES ISOTOPY

A CAT (= DIFF or PL) structure Γ on $M\times I$, where M is a topological m-manifold, is said to give a **concordance** from Σ to Σ' if $\Gamma|(M\times 0) = \Sigma\times 0$ and $\Gamma|(M\times 1) = \Sigma'\times 1$. Then Σ and Σ' are said to be concordant, and one writes $\Sigma \triangleq \Sigma'$.

In this situation, for m large, it would follow already from the s-cobordism theorem that $(M\times I)_\Gamma$ is CAT isomorphic to $M_\Sigma\times I$, by an isomorphism fixing $M\times 0$ and respecting $M\times 1$. † ‡ The Concordance Implies Isotopy Theorem below improves this isomorphism, chiefly by making it majorant small, that is, ϵ - near to the identity where ϵ is a continuous 'majorant' function $M\times I \to (0,\infty)$.

It is always worth emphasizing that a CAT isomorphism $h: M_\Sigma \to M_{\Sigma'}$ does *not* entail a concordance $\Sigma \triangleq \Sigma'$. For example take as M two copies of S^7 consider DIFF structures [Mi_1] . ‡‡ However, as soon as h is sufficiently (majorant) near to $id|M$, there is an isotopy $H: M\times I \to M\times I$ from $H_0 = id|M$ to $H_1 = h$, assured by the local contractibility of the homeomorphism group of M [Če] [EK] ; then $\Gamma = H^{-1}(\Sigma\times I)$ is a concordance from Σ to Σ' .

† If M and ∂M are not simply connected and compact, one would need for this some topological invariance of Whitehead torsion; a result for which a very efficient proof independent of these essays can be found in [Ch_2] . For the noncompact s-cobordism theorem, see [Si_7] .

‡ *Beware that counterexamples exist in low dimensions.* Indeed, by [Si_8 , Theorem 1] , $(M\times I)_\Gamma \cong M_\Sigma\times I$ is false with M equal one or perhaps either of T^3 , T^4 , for suitable Γ . Thus, even $M_\Sigma \cong M_{\Sigma'}$ fails, with M_Σ equal to one or perhaps either of $T^3\times I$, $T^4\times I$, and Σ' equal the Γ of the last sentence, so that $\Sigma \triangleq \Sigma'$ by a the 'corner turning' device in Case 3 of the proof of 4.1 below.

‡‡ Even a single copy of S^7 gives examples. The set of concordance classes of DIFF structures on S^7 maps surjectively (by the DIFF h-cobordism theorem) to the abelian group $\Theta_7 \cong Z_{28}$ of [KeM] . If ρ is a standard reflection, one has $S^7_\Sigma \cong S^7_{\rho\Sigma}$ (by ρ) . But it is easy to see that $[\rho\Sigma] = -[\Sigma]$ in Θ_{28} .

CONCORDANCE IMPLIES ISOTOPY THEOREM 4.1 .

Consider the data: M^m *a topological manifold of dimension* m ; Γ *a* CAT (= DIFF *or* PL) *structure on* $M \times I$, *giving a concordance* $\Sigma \simeq \Sigma'$, *so that* Γ *is equal to* $\Sigma \times I$ *near* $M \times 0$; C *a closed subset of* M ; $U \supset C$ *open subset of* M *such that* $\Gamma = \Sigma \times I$ *on* $U \times I$; D *a closed subset of* M ; V *an open subset of* M *containing* $D-C$ (*but perhaps not* $D-\mathring{C}$); $\epsilon : M \times I \to (0, \infty]$ *a positive continuous function (called a majorant)* .

Suppose $m \geq 6$ (*or* $\dim M = 5$ *and* $\partial M \subset U$).

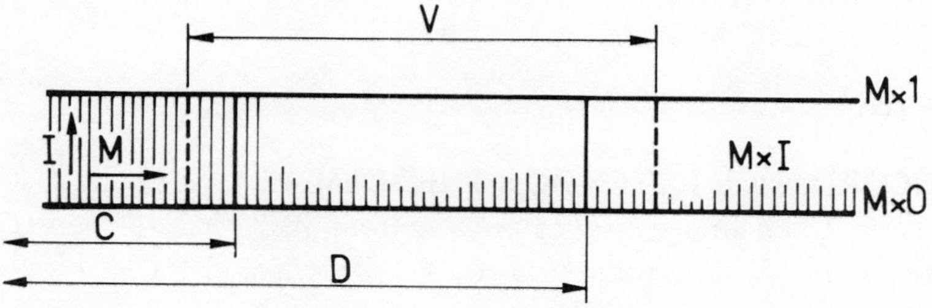

Then there exists an isotopy $h_t : M \times I \to M \times I$, $0 \leq t \leq 1$, *of* $id|(M \times I)$ *such that* :

(i) $h_1 : M_\Sigma \times I \to (M \times I)_\Gamma$ *is a* CAT *imbedding near* $(C \cup D) \times I$.

(ii) *The isotopy* h_t *fixes (pointwise) a neighborhood of* $(M - V) \times I \cup M \times 0 \cup C \times I$, *and it respects* $M \times 1$, *i.e.* h_t *is rel this tripple union and respects* $M \times 1$.

(iii) $d(h_t(x),x) < \epsilon(x)$ *for all* $x \in M \times I$, *and all* t ; *i.e.* h_t *is an* ϵ-*isotopy*.

This result will be deduced quite directly from the Handle Lemma 3.1 , of which it is clearly a generalization. But first we note that it is a little stronger than it first seems, by adding two complements.

Complement 4.1.1 .

Theorem 4.1 remains valid if we add the condition:

(iv) $h_1(N) \supset D \times I$ *for some neighborhood* N *of* $D \times I$ *on which* h_1 *is a* CAT *imbedding, i.e.* h_1 *is a* CAT *isomorphism over a neighborhood of* $D \times I$.

One quickly deduces this complement from 4.1 , as follows. Let C' be any closed neighborhood of C such that $\Gamma = \Sigma \times I$ near $C \times I$, and let D′ be a closed neighborhood of $D-\mathring{C}'$ with $D' \subset V$. Then apply 4.1 with the substitutions $C \leftrightarrow C'$, $D \leftrightarrow D'$ making $\epsilon(x)$, for x in $(M-D')\times I$, less than the distance from x to $(D-\mathring{C}')\times I$. This will establish 4.1.1 with $N = \mathring{C}' \cup \mathring{D}'$. ■

Complement 4.1.2 .

Theorem 4.1 and Complent 4.1.1 remain valid when Γ is not a product along I near M×0 , provided that, in conclusion (ii), one replaces the condition $h_t = id$ near $M\times 0$ by $h_t = id$ on $M\times 0$.

This is an easy corollary of the original version, the CAT collaring existence theorem, and the topological collaring uniqueness theorem (local version in Appendix A). ■

Here is a useful corollary of 4.1 . †

CONCORDANCE EXTENSION THEOREM 4.2 .

Let U be an open set in the manifold M^m , and Γ be a CAT (= DIFF or PL) structure on U×I , for which $\Gamma|(U\times 0)$ extends to a CAT structure Σ on M×0 .

Suppose $m \geqslant 6$ (or $m = 5$ and $\partial M \subset U$).

Then there exists a CAT structure Γ_+ on M×I coinciding with Γ on U×I , and with Σ on M×0 .

Proof of 4.2 . Apply 4.1 to obtain a CAT isomorphism h $h : U_\Sigma\times I \to (U\times I)_\Gamma$, with $d(h(x),x) < \epsilon(x)$, where $\epsilon(x)$ is the distance from x to $(M-U)\times I$ in M×I . Then h extends to a homeomorphism $h : M\times I \to M\times I$, by the identity rule $h(x) = x$ for $x \notin U\times I$; and we can define $\Gamma_+ = h(\Sigma\times I)$. ■

Remarks .

1) This Concordance Extension Theorem holds also without the abiding assumption that M be metrisable. Indeed Zorn's Lemma provides a maximal extension of Γ , say to Γ_μ on $M_\mu\times I$ with M_μ open in M . The metric case then shows that $M_\mu = M$.

2) Exploiting the D and V in 4.1 with $\epsilon = \infty$, one gets a less precise version of the Concordance Extension Theorem in which $\Gamma_+ = \Gamma$ only on a prescribed closed set D×I contained in U×I . This is the best result one can obtain (when M is PL) for CAT = DIFF

† Conversely, one readily shows that 4.2 ⇒ 3.1 (using the s-cobordism theorem) . As 3.1 ⇒ 4.1 (via the proof to follow), one might regard 4.2 and 4.1 as equivalent.

structures Whitehead compatible with the given PL structure on $M \times I$ (cf. §1, §5). †

PROOF THAT CONCORDANCE IMPLIES ISOTOPY 4.1 .

Case 1. *When* $\partial M = \phi$, $\epsilon = \infty$, *and* V *is a chart with compact closure.*

We have an open neighborhood U of C such that $\Gamma = \Sigma \times I$ on $U \times I$. Choose a simplex-wise linear triangulation of the chart V so fine that any closed simplex meeting C lies in U . Let K be the union of the closed simplices contained in U . This is a closed subcomplex of V and a neighborhood of $C \cap V$ in V lying in U . Let L be the union of the closed simplices that meet $D-C$. Then $K \cup L$ contains $(C \cup D) \cap V$. Also $L-K$ is finite for the following reason: the set $D-\overset{\circ}{K} \subset V$ is closed in M ; hence it is compact and only finitely many simplices of V can meet it.

Suppose now that $L-K$ consists of a *single* open simplex $\mathrm{int}\,\sigma$, say of dimension k . We choose some CAT open imbedding of $R^k \times R^n$, $m = k+n$, into V sending $R^k \times 0$ onto $\mathrm{int}\,\sigma$, and we identify $R^k \times R^n$ with its image. Since $\partial\sigma \subset U$ we easily arrange this imbedding so that $(R^k - \mathrm{int}\,B^k) \times R^n$ lies in U .

† Let $H : (-\infty, 0) \times I \to (-\infty, 0) \times I$ be a PL isotopy of $H = \mathrm{id}|(-\infty, 0)$ such that, for $t > 0$, $H_t : (-\infty, 0) \to (-\infty, 0)$ is non-smooth at an unbounded set of points. Let $\{(-\infty, 0) \times I\}$ be the DIFF structure on the target of H making H a diffeomorphism, and let Σ be the standard structure on $(-\infty, \infty) \times 0$. This Γ cannot be extended to any open set containing $(-\infty, 0) \times I$.

To explain this let N be a triangulated PL manifold (= a combinatorial manifold), and let Σ be a DIFF structure on an open subset $V \subset N$, Whitehead compatible with the PL structure of N (not the triangulation). Say that a point $u \in V$ is a *smooth point* if, for some open neighborhood $U \subset V$ of u , the inclusion $U \hookrightarrow V_\Sigma$ is smooth and nonsingular on each set $U \cap \sigma$ for each closed simplex of σ of N (not of some subdivision!) . The complement $Cr(\Sigma, N)$ in V of all the smooth points can be called the *set of crease points* ; it is closed in V . If Σ extends to a Whitehead compatible structure Σ_+ on all of N , observe that $Cr(\Sigma_+, N)$ lies in the $(n-1)$-skeleton of any subdivision N^* of N such that $N^* \hookrightarrow N_{\Sigma_+}$ is smooth and nonsingular on each closed simplex of N^* . Since $Cr(\Sigma_+, N) \cap V = Cr(\Sigma, N)$ we conclude that Σ cannot extend to a compatible Σ_+ on N unless $Cr(\Sigma, N)$ lies in a (closed) codimension $\geqslant 1$ subpolyhedron of N .

Apply Theorem 3.1 to the identity homeomorphism $h : B^k \times R^n \times I \to (B^k \times R^n \times I)_\Gamma$. We can extend the resulting isotopy by the identity to all of $M \times I$; this clearly gives the isotopy of $\mathrm{id} \mid (M \times I)$ asserted by 4.1 in this situation.

In general $L-K$ consists of several open simplices, say s, and we proceed by induction on s. Thus, suppose for induction that we can find an isotopy of $\mathrm{id} \mid M \times I$

$$g_t : M \times I \to (M \times I)_\Gamma \ , \ 0 \leqslant t \leqslant 1 \ ,$$

rel $\{K \cup (M-V)\} \times I$ and rel $M \times 0$ so that g_1 is a CAT imbedding near $(K \cup L_o) \times I$, where L_o is L minus a principal open simplex int σ. Applying to $(M \times I)_{g_1^{-1}(\Gamma)}$ the case $s = 1$ dealt with above, we obtain an isotopy h_t^o, $0 \leqslant t \leqslant 1$ of $\mathrm{id} \mid (M \times I)$ such that the composed isotopy

$$h_t = g_t h_t^o \ , \ 0 \leqslant t \leqslant 1 \ ,$$

is also rel $\{K \cup (M-V)\} \times I$ and rel $M \times 0$ and such that h_1 is a CAT imbedding near $(K \cup L) \times I$. This completes the induction to prove case 1. ■

A Modification. We presently begin to establish the majorant smallness condition (iii) in 4.1. *The proof is facilitated* † *by replacing* h_t *by* h_t^{-1} *in this condition; we shall assume henceforth that this has been done.* The force of the theorem is unchanged, as one sees by applying the following lemma to $M \times I$.

Lemma 4.3. *For any metric space* X, *the group* $H(X)$ *of homeomorphisms is a topological group when endowed with the majorant topology.*

The lemma is proved in Appendix C.

Case 2. $\partial M = \phi$, *and conclusion (iii) is weakened to*

(iii)' $d(p_1 h_t^{-1}(x), p_1(x)) < \epsilon(x)$ *for all* x *in* $M \times I$ *and all* t, *where* $p_1 : M \times I \to M$ *is projection, and* d *is now distance in* M.

† The reason why this switch facilitates the proof is not far to seek. Majorant smallness is measured in the target, and is most easily assured by building h_t as a composition of successive homeomorphisms of the target $h_t = \ldots h_t^3 h_t^2 h_t^1$ each well-behaved with respect to a fine covering of the target. However as the formula $h_t = g_t h_t^o$ in case 1 reveals, our h_t will naturally evolve as a composition $h_t = h_t^1 h_t^2 h_t^3 \ldots$ of homeomorphisms on the source. Hence the switch $h_t \leftrightarrow h_t^{-1}$ in the smallness condition, interchanging source and target.

The simplest proof would perhaps run as follows. Choose a fine triangulation of M . Define K and L as one would for Case 1 with $V = M$. Construct h_t by induction much as in Case 1, but induct over the skeleta of $L-K$ rather than one simplex at a time, and choose the handles $B^k \times R^n$ about the k-simplices pairwise disjoint and small enough that they form a discrete collection in M . This process will yield the required isotopy as soon as the triangulation is sufficiently fine.

We shall leave the details of this particular argument to the reader because it is subsumed in a principle of general topology. This principle, approximately stated, asserts that a result which is proved in a strongly relative *local* (or compact) form (such as Case 1), implies a parallel strongly relative *global* result with a majorant smallness condition (such as Case 2). Appendix C gives a precise formulation of it, plus a proof using nerves if need be in place of triangulations, and gives enough indications to apply it to prove Case 2. It will be reused several times in these essays.

Case 3 . *When* $\partial M = \phi$.

The appropriate device is to apply Case 2 to a concordance

$$\Gamma' \ : \ \Sigma \times I \simeq \Gamma \quad \text{rel} \ \ M \times 0 \cup C \times I \ ,$$

observing that the nearness condition provided by Case 2 is genuine majorant nearness when attention is restricted to the Γ-end of the concordance.

The concordance Γ' is derived from $\Gamma \times I$ by a ‘corner turning’ trick. To define Γ' for CAT = PL , let f_0 be a self-homeomorphism of the square $I \times I = I^2$ fixing $I \times 1$ and mapping $0 \times I \cup I \times 0$ ‘around the corner’ onto $0 \times I$; then set $\Gamma' = f^{-1}(\Gamma \times I)$ where

$$f = (\text{id}\,|M) \times f_0 \ : \ M \times I \times I \ \rightarrow \ M \times I \times I \quad .$$

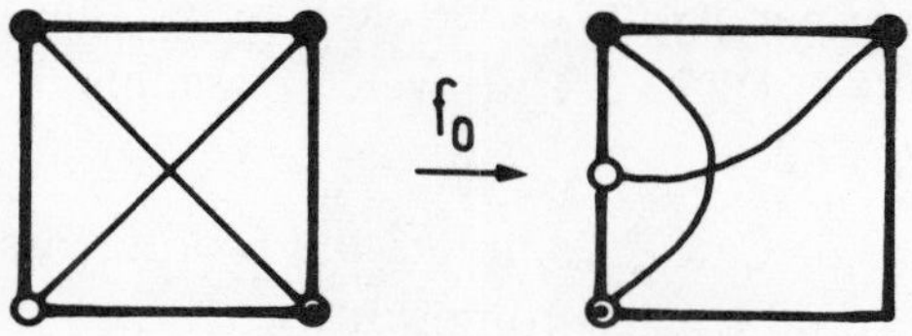

If CAT = DIFF , one proceeds similarly to begin, but one must alter Γ' before applying Step 2, by rounding (unbending) the corner along

$f^{-1}(M\times 1\times 0)$, and introducing one along $M\times 0\times 0$, in the manner of say [III, §4.3] .

It is no loss of generality to assume Γ' is a product along I (first factor) near $D\times 1\times I$. To see this, use the local collaring uniqueness theorem, Appendix A.1 .

We now apply Case 2 twice, once for boundary, once for interior. First apply it to $\Gamma'|M\times 1\times I$ producing an isotopy of f_t of $id|(M\times I^2)$ so that the isotopy f_t is a product along I (first factor) near $M\times 1\times I$. Second, one applies Case 2 to $f_1^{-1}\Gamma'$ on $(M\times \mathring{I})\times I$ using $V\times \mathring{I}$ for V in 4.1 , to get an isotopy g_t of $id|(M\times I^2)$.

The wanted isotopy h_t of $M\times I = M\times I\times 1$ is the restriction of

$$h_t = f_t g_t \quad , 0 \leqslant t \leqslant 1 \ .$$

It is majorant as near as we please to the identity because, on $M\times I\times 1$, both f_t and g_t are as near as we please to the identity (use Lemma 4.3).

For completeness we record the substitutions to be used in applying Case 2 .

First application : $M \leftrightarrow M\times 1$; $\Gamma \leftrightarrow \Gamma'|(M\times 1\times I)$; $C \leftrightarrow C\times 1$; $D \leftrightarrow D\times 1$; $V \leftrightarrow V\times 1$; $\epsilon \leftrightarrow$ a sufficiently small majorant .

Second application : $M \leftrightarrow M\times \mathring{I}$; $\Gamma \leftrightarrow f_1^{-1}(\Gamma')$; $C \leftrightarrow C'$ a sufficiently small closed neighborhood in $M\times \mathring{I}$ of $(M\times 0) \cup (C\times I) \cup (M\times 1)$; $D \leftrightarrow D\times \mathring{I}$; $V \leftrightarrow V\times \mathring{I}$; $\epsilon \leftrightarrow$ a sufficiently small majorant . ■

General Case . (Relaxing the condition that $\partial M = \phi$)

Let $\partial M\times J \subset M$ be a CAT collar neighborhood of ∂M_Σ in M_Σ . Adjusting Γ by a (majorant) small self-homeomorphism of $M\times I$ which is the identity on $M\times 0$ and near $\{C \cup (M-V)\}\times I$ (using the collaring uniqueness theorem Appendix A.1) we can assume that Γ is a product along J near $(\partial M)\times I \cap D\times I$. Thus we can assume Γ has this property from the outset.

Two applications of Case 3 complete the proof. First apply it to $\Gamma|(\partial M \times I)$, and extend the resulting isotopy (f_t say) naturally to one of $id|(M\times I)$ with support in the collaring $(\partial M)\times J\times I$ of $(\partial M)\times I$. Then a second application to $f_1^{-1}\Gamma$ on $M\times I$ (with $V-\partial M$ for V) completes the proof. The details are much as for Case 3 . ■

§5. THE PRODUCT STRUCTURE THEOREM

THEOREM 5.1 *(Product Structure Theorem)* .

Let M^m be a TOP manifold and Σ be a CAT $(=$ DIFF or PL$)$ structure on $M \times R^s$, $s \geqslant 1$. Let U be an open subset of M so that there is a CAT structure ρ on U with $\Sigma \mid U \times R^s = \rho \times R^s$.

Provide that $m \geqslant 6$, or $m = 5$ and $\partial M \subset U$. Then there exists a CAT structure σ on M extending ρ and a conditioned concordance rel $U \times R^s$ from Σ to $\sigma \times R^s$.

Complement 5.1.1. *A structure σ as described , with $\sigma \times R^s$ concordant rel $U \times R^s$ to Σ , is unique up to concordance rel U .*

Remark 1. By Theorem 4.1 there is an ϵ-isotopy from the identity to a CAT isomorphism between $(M \times R^s)_\Sigma$ and $M_\sigma \times R^s$. If C is a prescribed closed subset of U , the isotopy can fix $C \times R^s$.

Remark 2. Together 5.1 and 5.1.1 show that the rule $\sigma \mapsto \sigma \times R^s$ is a one-to-one correspondence between concordance classes of CAT structures on M and on $M \times R^s$, under the above dimensional restrictions . The relative version of this , i.e. the version rel U , is also important .

Remark 3. Consider 5.1 without the proviso that $m \geqslant 6$, or $m = 5$ and $\partial M \subset U$, and *call that statement* (P) . Using (a) the classification theorem $[KS_1][IV]$, which follows from 5.1 , (b) the calculation that $\pi_i(TOP/PL)$ and $\pi_i(TOP/O)$ are 0 for $i \leqslant 2$ and are $Z_2 = Z/2Z$ if $i = 3$ $[KS_1]$, and (c) the uniqueness up to isotopy of CAT structures on metrizable manifolds of dimension $\leqslant 3$ (cf. §3) , one easily finds that *for $m \leqslant 3$ and $m+s \geqslant 6$ (or $m+s = 5$ and $\partial M \subset U$) the statement (P) holds if and only if the cohomology group $H^3(M, U; Z_2)$ is zero* . For example (P) fails if $M^3 = S^3$, $s \geqslant 2$, and $U = \phi$! Statement (P) is undecided if $m = 4$.

Remark 4. Clearly 5.1 is modeled on the famous Cairns-Hirsch theorem stated at the end of this section (5.3). It is the PL-DIFF analogue of our TOP-CAT theorem 5.1. We obtain a new proof of it *in high dimensions*, based on 5.1 and 4.1. We likewise obtain a proof of the PL-DIFF Concordance Extension Theorem; however the PL-DIFF Concordance Implies Isotopy Theorem will elude us until §3.6 of Essay II.

Remark 5. By a happy accident our proof of 5.1 (and 5.1.1) works equally well without the abiding assumption that M be metrizable (or even Hausdorff), when we use Zorn's Lemma in place of a countable induction over charts.

Let us begin with

PROOF OF 5.1.1 FROM 5.1. It is the one most usual when there is an existence theorem in relative form. Let σ and σ' be two CAT structures on M offered by 5.1. The conditioned concordances $\sigma \times R^s \simeq \Sigma \simeq \sigma' \times R^s$ rel $U \times R^s$ provide a CAT structure Γ on $R \times M \times R^s$ giving on $I \times M \times R^s$ a concordance from $\sigma \times R^s$ to $\sigma' \times R^s$ rel $U \times R^s$.

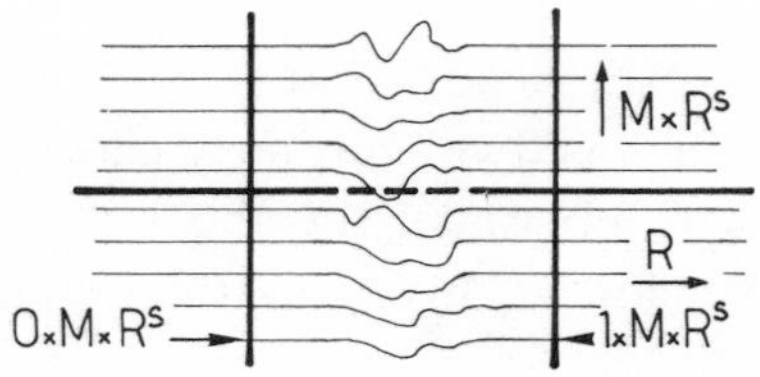

We arrange that Γ is a product along R and along R^s, on all of $U' \times R^s \equiv (R - [1/4, 3/4]) \times M \times R^s \cup R \times U \times R^s$. Then Γ gives $U' \times 0$ a CAT structure $\gamma \times 0$ which 5.1 extends to all $R \times M \times 0$ giving a concordance $\sigma \simeq \sigma'$ rel U. ■

PROOF OF 5.1 *(using 4.1, 4.2 and the Stable Homeomorphism Theorem)*. In our proofs so far, handlebody theory has been the only difficult technique involved; in particular, we have avoided immersion theory and surgery. However, now, the STABLE Homeomorphism Theorem [Ki_1], which involves the nonsimply connected surgery of Wall [Wa], is needed in order to show that

(H) M admits a STABLE structure θ (= a maximal atlas of charts to R^m_+ related by STABLE† *homeomorphisms) compatible with Σ in the sense that Σ is contained in the (maximal) STABLE atlas $\theta \times R^s$.*

The proof as we give it then yields a structure σ automatically satisfying the condition

(C) σ agrees with the STABLE structure θ of hypothesis (H).

For the proof of 5.1 we admit (H) as a hypothesis. (It is redundant by [Ki_1].) This causes all further mention of the STABLE Homeomorphism Theorem of [Ki_1] to vanish. For convenience of reader *any detail of the proof that could be eliminated by free use of it will henceforth be enclosed in square brackets.* ⟦In short, what we now prove is the version of 5.1 for (compatible) CAT structures on STABLE topological manifolds – the STABLE-CAT version rather than the TOP-CAT version.⟧

We may (and shall) assume $s = 1$ in proving 5.1, because a finite induction using $M \times R^s = (M \times R^{s-1}) \times R$ retrieves the general case.

The best part of the proof is

The Case of 5.1 where $M = R^m, m \geqslant 5$, or $R^m_+, m \geqslant 6$, ⟦and where θ agrees with the linear structure⟧.

⟦Since $\theta \times R$ agrees with Σ, our definition of STABLE makes it clear that⟧ there is a concordance (not rel $U \times R$) from the standard structure on $M \times R$ to Σ. Apply 4.1 to this concordance using for examples the substitutions $C \nleftrightarrow \phi$, $D \nleftrightarrow M \times [1,\infty)$, $V \nleftrightarrow M \times (½,\infty)$, in order to replace Σ by a CAT structure Σ_1 that is standard near $M \times [1,\infty)$, and equals Σ near $M \times (-\infty,0]$. Examine Figure 5-a below.

Since $\Sigma | U \times R = \rho \times R$, the set $U \times [0,1]$ inherits from Σ_1 a CAT manifold structure written $(U \times [0,1])_{\Sigma_1}$. Apply 4.1, the Concordance Implies Isotopy Theorem, for the second time to find a

† A homeomorphism $h: U \to V$ of open subsets of R^m_+ is STABLE if and only if, for each point $x \in U$, there exists an open neighborhood U_x in U and topological isotopy $h_t: U_x \to R^m_+$ through open imbeddings from $h_0 = h | U_x$ to a linear imbedding h_1. See [BrnG] [Ki_1], where equivalent definitions are discussed, and also [III, Appendix A].

CAT isomorphism $h : U \times [0,1] \to (U \times [0,1])_{\Sigma_1}$ so small that $d(h(x),x) < \epsilon(x)$ where $\epsilon : M \times [0,1] \to [0,\infty)$ is a continuous map with $\epsilon^{-1}(0,\infty) = U \times [0,1]$.†

FIGURE 5-a

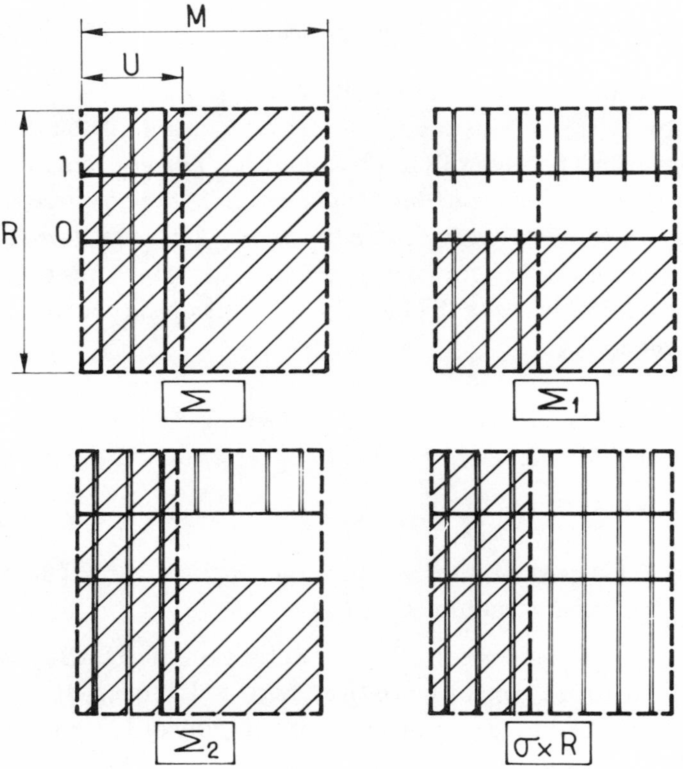

† If at this point one prefers to use the version without ϵ of the Concordance Implies Isotopy Theorem 4.1 , (but with a careful choice of V) , one succeeds easily enough at the cost of weakening the final result of 5.1 to be a concordance rel $C \times R$ to a product structure where $C \subset U$ is a prescribed closed subset of M . The interested reader might pause to investigate two advantages that accrue, compensating this loss of precision . First , the technically delicate parts of §4 are avoided , while one obtains a version of 5.1 adequate for the applications in [III] to topological manifolds . Second , the argument without ϵ applies *tel quel* to prove the PL - DIFF Product Structure Theorem of $[Hi_7]$ from the (more basic?) PL - DIFF Concordance Implies Isotopy Theorem of $[Hi_7]$.

Extend h to a self-homeomorphism h of $M \times R$ so that one has $h(x,r) = (x,r)$ if $r \leq 0$ or $x \notin U$, and $h(x,r) = (p_1 h(x,1),r)$ if $r \geq 1$. With the help of the (local) CAT collaring uniqueness theorem A.1 we arrange that h is a product with R near $U \times 0$ and $U \times 1$. Then define Σ_2 to make $h : (M \times R)_{\Sigma_2} \to (M \times R)_{\Sigma_1}$ a CAT isomorphism. Clearly $\Sigma_2 | M \times (1,\infty)$ is of the form $\sigma \times (1,\infty)$. This defines σ.

The properties of Σ_2 indicated in Figure 5-a let one construct more or less standard conditioned concordances $\Sigma_2 \triangleq \sigma \times R$ rel $U \times R$ and $\Sigma_2 \triangleq \Sigma$ rel $U \times R$ using the following easy

Windowblind Lemma 5.1.2.

Let Σ' and Σ'' be two CAT structures on $M \times R$ so that $\Sigma' = \Sigma''$ on $M \times (a,b)$, $a < b$, so that both Σ' and Σ'' are a product along R on an open subset $U \times R$. Then there is a conditioned concordance

$$\Sigma' \triangleq \Sigma'' \ \textit{rel}\ U \times R .$$

Proof of lemma. Using a conditioned CAT homotopy h_t, $0 \leq t \leq 1$, of $\mathrm{id} | R$ through imbeddings to an imbedding onto (a,b), form an open imbedding

$$H : I \times M \times R \to I \times M \times R$$

with $H_t(x,r) \equiv H(t,x,r) = (t,x,h_t(r))$. The structure $H^{-1}(I \times \Sigma')$ on $I \times M \times R$ making H a CAT imbedding into $I \times (M \times R)_{\Sigma'}$ is a conditioned concordance $\Sigma' \triangleq H_1^{-1}(\Sigma')$. It is rel $U \times R$ since Σ' is a product along R on $U \times R$. Similarly $\Sigma'' \triangleq H_1^{-1}(\Sigma'')$ rel $U \times R$. But $H_1^{-1}(\Sigma') = H_1^{-1}(\Sigma'')$, so we find $\Sigma' \triangleq \Sigma''$ rel $U \times R$ as desired. ∎

The first case $M = R^m$ or R^m_+ of 5.1 is now established. ∎

The general case follows by an easy chart by chart induction. We present the details in a novel fashion since the induction that the reader first thought of, probably does not work immediately if M is non-metrizable.

Let $\mathcal{S}$ be the set of pairs (U', Σ') consisting of an open subset U' of M containing U, and a CAT structure Σ' on $U' \times R$ such that $\Sigma' = \Sigma$ on $U \times R \cup U' \times (-\infty,0)$, and Σ' is a product along R on $U' \times (1,\infty)$. (Compare behavior of Σ_2 in Figure 5-a.)

According to Lemma 5.1.2, the theorem is proved when we have found a pair $(M, \Sigma') \in \mathcal{S}$.

Now $\mathcal{S}$ is partially ordered: $(U', \Sigma') \leqslant (U'', \Sigma'')$ meaning that $U' \subset U''$ and $\Sigma' = \Sigma''$ on $U' \times R$. Every totally ordered subset has an upper bound (the 'union'). Hence Zorn's Lemma (transfinite induction) asserts that $\mathcal{S}$ has a maximal element (M_o, Σ_0).

Assertion: $M_o = M$.

It will suffice to show that for any open subset V of M that is [[STABLE]] homeomorphic to R^m or R^m_+, one can find an element $(M_o \cup V, \Sigma_1) \in \mathcal{S}$ with $\Sigma_1 = \Sigma_0$ on $M_o \times R$. First extend Σ_0 over $M_1 \times (-\infty,0)$, $M_1 \equiv M_o \cup V$, letting it equal Σ there. Next further extend Σ_0 arbitrarily over all of $M_1 \times R$. This is possible using the Concordance Extension Theorem 4.2, because Σ_0 is *concordant* to a structure that extends – concordant via an isotopy h_t, $0 \leqslant t \leqslant 1$, of $id | R$ through imbeddings with $h_t(-\infty,0) \subset (-\infty,0)$ and $h_1 R = (-\infty,0)$.

Σ_0 is now defined on $M_1 \times R$. The proved case of 5.1 offers a concordance of $\Sigma_0 | V \times (1,\infty)$ rel $(M_o \cap V) \times (1,\infty)$ to a product structure. Extend this by the identity concordance of $\Sigma_0 | M_o \times R \cup M_1 \times (-\infty,0)$. Then further extend, by 4.2 again, to a concordance of $\Sigma_0 | M_1 \times R$ to a structure Σ_1 on $M_1 \times R$. Clearly $(M_o, \Sigma_0) \leqslant (M_1, \Sigma_1) \in \mathcal{S}$ completing the proof of 5.1. ∎

Applications of the product structure theorem frequently demand the following "local" version, which the reader will recognize as an easy consequence of the original version and the concordance extension theorem.

THEOREM 5.2. Local version of Product Structure Theorem.

Consider the following data:
M^m a TOP manifold; W an open neighborhood of $M \times 0$ in $M \times R^s$, $s \geqslant 1$; Σ a CAT structure on W; $C \subset M \times 0$ a closed subset such that Σ is a product along R^s near C; D another closed subset of $M \times 0$; $V \subset W$ an open neighborhood of $D - C$.

Suppose that $m \geqslant 6$, or $m \geqslant 5$ and $\partial M \subset C$. Then there exists a concordance rel $(W-V) \cup C$ from Σ to a CAT structure Σ' on W so that Σ' is a product along R^s near D. ∎

Figure 5-b

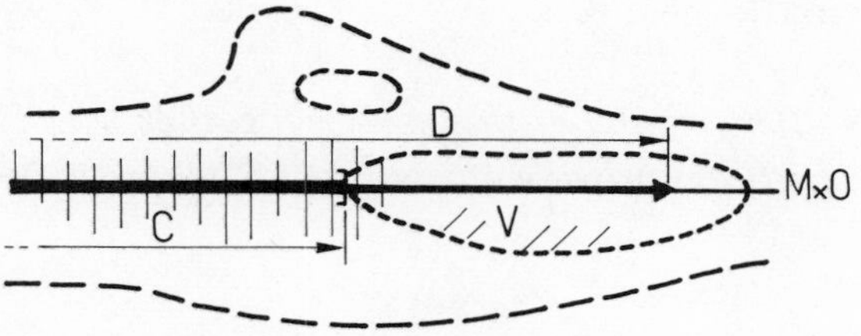

The analogue of the Product Structure Theorem 5.1 for compatible DIFF structures on PL manifolds is the venerable

CAIRNS-HIRSCH THEOREM 5.3 *(see [Ca] [Hi_2] [Hi_7]).*

Let M^m have a PL structure s and $M \times R^s$ a DIFF structure Σ which is (Whitehead) compatible with $s \times R^s$. Suppose ρ is a DIFF structure compatible with s on a neighborhood of C in M such that $\Sigma = \rho \times R^s$ near $C \times R^s$. Then there exists a DIFF structure σ on M compatible with s, and extending ρ near C; furthermore there exists a conditioned DIFF concordance Γ rel $C \times R^s$ from Σ to $\sigma \times R^s$ such that Γ is compatible with $I \times s \times R^s$.

First recall the meaning of compatibility. Let S and Σ be respectively a PL and a DIFF structure on a manifold M. We say that S and Σ, are (Whitehead) *compatible*, or that S is a DIFF triangulation† of Σ, if for each co-ordinate chart (h, U) of S the composition $R^m_+ \supset h(U) \xrightarrow{h^{-1}} M_S \xrightarrow{id} M_\Sigma$ is a C^∞ non-singular imbedding on each closed simplex of some linear triangulation of $h(U) \subset R^m_+$. Note that if $f : M_S \to M'_{S'}$, is a PL isomorphism, then the DIFF structure $f\Sigma$ on M′ is compatible with S'.

A basic theorem of J. H. C. Whitehead [Wh_1] [Mu_1] states that every DIFF manifold M_Σ has a DIFF triangulation S. In fact if C

† This is a convenient abuse of language; S is at best a maximal family of piecewise-linearly related triangulations.

is closed in M , and R is a DIFF triangulation of $\Sigma|U$ where U is a neighborhood of C , then there exists a DIFF triangulation S of M_Σ that agrees with R near C .

ASSERTION: *The results of this essay , together with Whitehead's triangulation theorem provide a proof of 5.3 when* $m \geqslant 6$, *or* $m \geqslant 5$ *and* $\partial M \subset C$.

Proof of assertion .

By 5.1 , (ignoring s) there is a DIFF structure σ' on M and a DIFF conditioned concordance Γ' rel $C \times R^s$ from Σ to $\sigma' \times R^s$. However , σ' may not be compatible with s , nor Γ' with $I \times s \times R^s$. Using Whitehead's result find s' a DIFF triangulation of σ' equal to s near C , and find G' a DIFF triangulation of Γ' , giving a conditioned PL concordance from $s \times R^s$ to $s' \times R^s$ rel $C \times R^s$.

We shall construct a PL isomorphism $F : (I \times M \times R^s)_{G'} \to I \times M_s \times R^s$ which fixes a neighborhood of $0 \times M \times R^s \cup I \times C \times R^s$ and is a product with I and with R^s near $1 \times M \times R^s$. Then $F(\Gamma') = \Gamma$ will clearly be the required concordance compatible with $I \times s \times R^s$, and $F(1 \times \sigma' \times R^s) = 1 \times \sigma \times R^s$ will define σ .

We proceed to define F as a composition

$$(I \times M \times R^s)_{G'} \xrightarrow{H} (I \times M)_g \times R^s \xrightarrow{h \times id} I \times M_s \times R^s .$$

By Theorem 5.1 , Remark 1 , applied to $I \times M$, there is a PL structure g that is a conditioned concordance s to s' rel C , and there is a PL homeomorphism H equal to the identity near $\{0,1\} \times M \times R^s \cup I \times C \times R^s$. Finally by Theorem 4.1 , there is a PL homeomorphism $h : (I \times M)_g \to I \times M_s$ so that $h = id$ near $0 \times M$ and h is a product with I near $1 \times M$. This establishes 5.3 in case $m \geqslant 6$, or $m = 5$ and $\partial M \subset C$. ∎

5.4. OBSERVATION . *In the Concordance Extension Theorem 4.2.1 (rel a closed subset) suppose M is a PL manifold . For "CAT structure" read "compatible DIFF structure". There results a valid concordance extension theorem for compatible DIFF structures .* Its proof is an easy consequence of 4.2 for DIFF reinforced by 4.1 , for PL and we leave it as an exercise . Again some low dimensions are not covered by this proof .

As a corollary one deduces a *local* version of the Cairns-Hirsch theorem analogous to 5.2 .

These results provide , for dimension $\geqslant 5$, the basis for Hirsch's first obstruction theory [Hi_3] for introducing and classifying up to concordance compatible DIFF structures on a given PL manifold . More elaborate theories of Hirsch and Mazur [Hi_4] [HiM] [Mor_1] [Hi_7] [Hi_8] and of Lashof and Rothenberg [LR_1] also use these results .

It is of course a telling fault that we do not deal with compatible DIFF structures on PL manifolds of dimension $\leqslant 4$. But the theory is rather joyless there ; compatible structures exist and are unique up to concordance [Ce_1] .

Appendix A. COLLARING THEOREMS

We define a **local collaring** of a closed subset M of a metric space W to be an open neighborhood U of $M \times 0$ in $M \times [0, \infty)$ and an open topological imbedding $f : U \to W$ satisfying $f(p \times 0) = p$ for all p in M. In case W and M are CAT objects (CAT = TOP, PL or DIFF), and f is in addition a CAT embedding, then f is called a CAT local collaring.

THEOREM A.1. Local collaring uniqueness.

Consider the following data: M a CAT manifold (CAT = TOP, PL or DIFF); W a CAT manifold, with metric d, possibly with corners if CAT = DIFF, and containing M as a closed subset; $f,g : U \to W$ two CAT local collarings of M in W; C a closed subset of M such that $f = g$ near $C \times 0$ in $C \times [0, \infty)$; D another closed subset of M; $\epsilon : W \to [0, \infty)$ a continuous function, positive on D.

With these data there exists an isotopy h_t, $0 \leqslant t \leqslant 1$, of $id|W$ fixing M so that $f = h_1 g$ near $D \times 0$ and $d(h_t(x), x) \leqslant \epsilon(x)$ for all $x \in W$. The isotopy $h_t g$ is constant near $C \times 0$ in $C \times [0, \infty) \cap U$.

We will indicate a proof of this presently. Familiar (relative) collaring existence theorems follow from this, (see [Ar] [Brn] [Cny] for TOP; [Hu_2] for PL; [Mu_1] for DIFF.)

It is not difficult to deduce a global uniqueness result.

DEFINITION. If a CAT local collaring $f : M \times [0, \infty) \to W$ restricts to a closed imbedding of $M \times [0, 1]$, we call the restriction $f | M \times [0, 1]$ a CAT **clean collaring.**

THEOREM A.2. Collaring uniqueness.

Consider CAT clean collarings $f,g : M \times [0, 1] \to W$ of $M \subset W$. Suppose $f = g$ near a closed subset $C \times [0, 1]$. Let $D \subset M$ be closed and let $V \subset W$ be an open neighborhood of $f(D \times [0, 1]) \cup g(D \times [0, 1])$.

Then there exists a CAT isotopy h_t, $0 \leqslant t \leqslant 1$, of $id|W$, fixing all points near $f(C \times [0,1])$ and all points outside V, such that $f = h_1 g$ near $D \times [0,1]$. Further, if $f = g$ near $D \times 0$, then $h_t = id$ near M.

The reader should be able to obtain the local collaring uniqueness theorem from the following key lemma, suggested by [Cny]. (There is an extra step unique to DIFF concerning angle of departure from the boundary; cf. [Mu_1, §6].)

LEMMA A.3 .

Let U be an open neighborhood of $M \times (-\infty, 0]$ in $M \times R$ and let $f: U \to M \times R$ be an open CAT imbedding equal the identity on $M \times (-\infty, 0]$; let $D \subset M$ be a closed set and let $\epsilon : M \times R \to [0, \infty)$ be a continuous function positive on $D \times 0$.

Then there exists a CAT ϵ-isotopy h_t, $0 \leqslant t \leqslant 1$, of $id|M \times R$ fixing $M \times (-\infty, 0]$ so that $h_1 = f$ near $D \times 0$.

If f is the identity on a neighborhood of $C \times (-\infty, 0]$ in $C \times R$ for some closed set $C \subset M$, then the isotopy h_t, $0 \leqslant t \leqslant 1$, can be so chosen that the same is true of it.

Proof of Lemma (in outline). Concerning smallness, it will suffice to show that the isotopy h_t can be made to have support in $\epsilon^{-1}(0,\infty)$ and can simultaneously be made as small as we please for the fine C^0 topology, equivalently for the majorant topology, cf. §4 and App. C .

Using normality, find a closed neighborhood Y of $D \times 0$ in $M \times R$ contained in the open set $\epsilon^{-1}(0,\infty)$, disjoint from the closed set $M \times R - f(U)$, and so small that $f \mid \{Y \cap (C \times R)\} =$ identity . Define $X = f^{-1}(Y)$.

Construct 'sliding' CAT isotopies

$$\sigma_t : M \times R \to M \times R \ , \ 0 \leqslant t \leqslant 1 \ ,$$

of id $| M \times R$ such that:

(a) $\sigma_t =$ identity on $M \times R - (X \cap Y)$;

(b) $\sigma_t(x \times R) = x \times R$ and $\sigma_t(x \times (-\infty,0]) \supset x \times (\infty,0]$ for all t and x ;

(c) $\sigma_1(M \times (-\infty,0)) \supset D \times (-\infty,0]$;

(d) σ_t is as small as we please for the majorant topology.

Consider the rule

$$f_t = \sigma_t f \sigma_t^{-1} \ , \ 0 \leqslant t \leqslant 1 \ .$$

By (a) , f_t is an isotopy of f through open imbeddings $U \to M \times R$, with support in $X \cap Y$ and disjoint from $C \times 0$. By (b) , f_t fixes $M \times (-\infty, 0]$. By (c) , f_1 is the identity near $D \times (-\infty, 0]$. By (d) f_t is as small as we please for the majorant topology.

By (a) again, $f_t U = fU$ and $f_t = f$ outside X . Since $Y = f(X)$ is closed, and f(U) is open we can define h_t , $0 \leqslant t \leqslant 1$, by the rules

$$h_t(y) = f f_t^{-1}(y) \quad \text{for } y \in f(U)$$
$$h_t(y) = y \quad \text{for } y \notin f(X) \ .$$

This is the isotopy required. It is as small as we please for the majorant topology. ■

A.4 GENERALIZATIONS

There are two ways in which it is often useful to generalize the above results . Neither generalization requires much more than notational change in the proofs .

(a) TOP can be enlarged to the category of continuous maps of metric spaces ; PL can be enlarged to the category of piecewise-linear maps of locally compact polyhedra .

(b) *(Respectful versions)*. One supposes W equipped with a family $\{W_\alpha\}$ of closed subsets . Then one insists in hypotheses and conclusions that each isotopy mentioned respect each set W_α or $M_\alpha \times R$ as the case may be . For our purposes $\{W_\alpha\}$ is usually the set of fibers of a CAT submersion $p : W \to \Delta$ to a CAT object Δ .

Appendix B : SUBMERGING A PUNCTURED TORUS

This contains verbatim a letter from J. Milnor of October, 1969, which gives an elementary construction of a submersion of the punctured torus T^n–point into euclidean space R^n. It is used in §3. A different elementary construction was found by D. Barden [Bar] [Ru] earlier in 1969, and another by S. Ferry, [Fe] 1973†. Milnor produces a smooth C^∞ (= DIFF) submersion. A secant approximation to it in the sense of J. H. C. Whitehead [Mu_1, §9] provides a piecewise-linear (= PL) submersion.

"Let M be a smooth compact manifold.

HYPOTHESIS. M has a codimension 1 embedding in euclidean space so that, for some smooth disk $D \subset M$ and some hyperplane P in euclidean space, the orthogonal projection from $M - D$ to P is a submersion.

THEOREM. *If M satisfies this hypothesis, so does $M \times S^1$.*

It follows inductively that every torus satisfies the hypothesis.

PROOF. Suppose that $M = M^{k-1}$ embeds in R^k so that $M - D$ projects submersively to the hyperplane $x_1 = 0$. We will assume that the subset $M \subset R^k$ lies in the half-space $x_k > 0$. Hence, rotating R^k about R^{k-1} in R^{k+1}, we obtain an embedding $(x, \theta) \mapsto (x_1, ..., x_{k-1}, x_k \cos\theta, x_k \sin\theta)$ of $M \times S^1$ in R^{k+1}. This embedding needs only a mild deformation in order to satisfy the required property.

Let $e_1, ..., e_{k+1}$ be the standard basis for R^{k+1}. Let r_θ be the rotation

$$e_i \mapsto e_i \text{ for } i < k, \quad \begin{aligned} e_k &\mapsto e_k \cos\theta + e_{k+1} \sin\theta, \\ e_{k+1} &\mapsto -e_k \sin\theta + e_{k+1} \cos\theta. \end{aligned}$$

Let $n(x) = n_1(x)e_1 + \cdots + n_k(x)e_k$ be the unit normal vector to M in R^k.

† And still another by A. Gramain [Gra] 1973.

For $x \in M-D$ we can assume that n_1 is bounded away from zero . Say $n_1 \geqslant 2\alpha > 0$.

Suppose that M lies in the open slab $0 < x_k < \beta$ of R^k . Choose $\epsilon > 0$ so that the correspondence $(x, t) \mapsto x + tn(x)$ embeds $M \times (-\epsilon, \epsilon)$ diffeomorphically in this slab .

Choose a smooth map $t : S^1 \to (-\epsilon, \epsilon)$ so that

$$\frac{dt}{d\theta} \geqslant 2\beta/\alpha \text{ when } \theta = 0 ; \quad \cos\theta \frac{dt}{d\theta} \geqslant 0 \text{ always .}$$

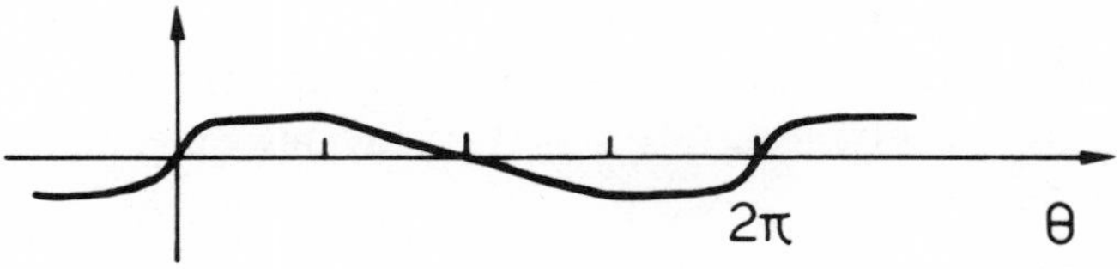

The required embedding $M \times S^1 \to R^{k+1}$ is now given by

$$(x, \theta) \mapsto r_\theta(x + t(\theta)n(x)) .$$

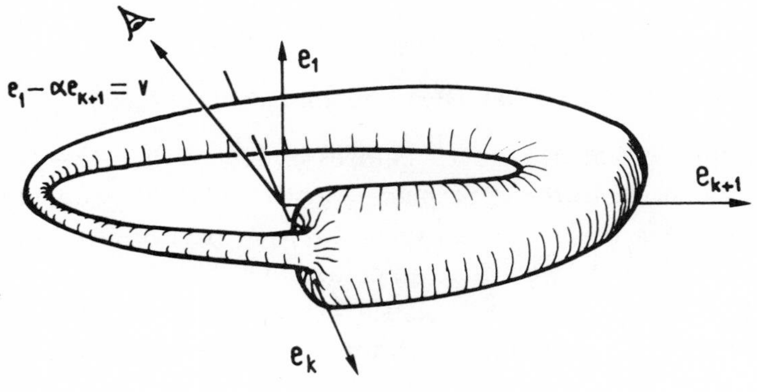

Computation shows that the normal vector to this embedding is $p/\|p\|$ where

$$p(x, \theta) = (x_k + tn_k)r_\theta(n) - \frac{dt}{d\theta}$$

Let $v = e_1 - \alpha e_{k+1}$. Then $p \cdot v = A + B$ where

$$A = (x_k + tn_k)(n_1 - \alpha \sin\theta\, n_k) \quad \text{and} \quad B = \alpha \cos\theta \frac{dt}{d\theta} \geqslant 0 .$$

Thus if $x \in M - D$ we have

$$A \geqslant (x_k + tn_k)(2\alpha - \alpha) > 0$$

hence $p \cdot v > 0$. On the other hand, for any $x \in M$, if $\theta = 0$, we have

$$A \geqslant -\beta , \qquad B \geqslant \alpha(2\beta/\alpha) .$$

Hence $p \cdot v > 0$ for $\theta = 0$, and therefore $p \cdot v > 0$ for all sufficiently small θ; say for $|\theta| \leqslant \eta$.

It now follows that the complement $(M \times S^1) - (D \times [\eta, 2\pi - \eta])$ projects submersively to the hyperplane $v^\perp$. This completes the proof.
■

Appendix C. MAJORANT APPROXIMATION

This appendix collects some basic observations about majorant approximation of functions, and it proves a theorem that provides a local criterion for majorant approximating a 'good' function by a 'better' function.

Let X and Y be topological spaces. If $\mathfrak{U}$ is an open cover of Y and f , g are two continuous functions (= maps), we say that f and g are $\mathfrak{U}$ *-near* if for each point x in X , both f(x) and g(x) belong to some one set of the cover $\mathfrak{U}$.

We topologize the set C(X,Y) of continuous maps $X \to Y$ in terms of the sets

$$N(f,\mathfrak{U}) = \{g \in C(X,Y) \mid f \text{ and } g \text{ are } \mathfrak{U}\text{-near}\}$$

by defining a set $N \subset C(X,Y)$ to be open if for each $f \in N$ there exists $\mathfrak{U}$ so that $N(f,\mathfrak{U}) \subset N$. This is called the *(target) majorant topology*.

Fact 1 . *If Y is a fully normal topological space, each set $N(f,\mathfrak{U})$ is a neighborhood of f ; then for fixed f the sets $N(f,\mathfrak{U})$ form a basis of neighborhoods of f* . Proof: Full normality means precisely that, for any open cover $\mathfrak{U}$, there exists an open cover $\mathfrak{W} = \{W_i\}$ such that the starred cover $\mathfrak{W}^*$ refines $\mathfrak{U}$. ‡ (By definition $\mathfrak{W}^* = \{W_i^*\}$, where $W_i^* = \cup_j \{W_j \mid W_i \cap W_j \neq \phi\}$.) Then observe that $h \in N(g,\mathfrak{W}) \Rightarrow N(h,\mathfrak{W}) \subset N(g,\mathfrak{U})$. It follows that, for any $N \subset C(X,Y)$, the subset N^{Δ} of all g in N , such that there exists a $N(g,\mathfrak{U}) \subset N$, is an open set. This in turn implies that each $N(f,\mathfrak{U})$ is a neighborhood of f . ■

Fact 2 . *For $f \in C(X,Y)$ and $g \in C(Y,Z)$ the rule of composition $(f,g) \mapsto gf \in C(X,Z)$ is a continuous mapping $C(X,Y) \times C(Y,Z) \to C(X,Z)$ provided Z is fully normal.*

‡ Full normality (as defined above) is easy to establish from metrizability. It also follows from the existence of fine partitions of unity (by regarding them as maps to nerves as in the proof of C.1 below). Two equivalences are worth noting. 1) full normality & paracompactness $\Leftrightarrow$ existence of fine partitions of unity . 2) For spaces that are regular (all points have small closed neighborhoods) , full normality is equivalent to paracompactness (= existence of fine locally finite open covers). See [Sto_1] or any treatise, shunning superfluous separation axioms.

Proof: Given $N(gf,\mathfrak{V})$, find an open cover $\mathfrak{W}$ such that $\mathfrak{W}^* < \mathfrak{V}$. Then $f' \in N(f, g^{-1}\mathfrak{W})$ and $g' \in N(g,\mathfrak{W})$ imply $gf' \in N(gf,\mathfrak{W})$ and $f'g' \in N(gf',\mathfrak{W})$, whence $g'f' \in N(gf, \mathfrak{W}^*)$. ■

Fact 3. *The group of homeomorphisms $H(X) \subset C(X,X)$ is a topological group, for the inherited majorant topology, provided again that X is assumed to be fully normal.* Proof: Given fact 2 we have only to verify the continuity of the rule $f \mapsto f^{-1}$. But f^{-1}, g^{-1} are $\mathfrak{V}$-near $\Leftrightarrow$ $f^{-1}g$, id are $\mathfrak{V}$-near $\Leftrightarrow$ g, f are $f(\mathfrak{V})$-near. ■

In case Y is a metric space there is an alternative definition of our topology on $C(X,Y)$ as follows. Call positive continuous functions $\delta : Y \to (0,\infty]$ *majorants*, and for f in $C(X,Y)$ define

$$N_\delta(f) = \{ g \in C(X,Y) \mid d(f(x),g(x)) < \delta(f(x)) \text{ for all } x \text{ in } X \} .$$

Fact 4. *As δ varies, the sets $N_\delta(f)$ give a basis of neighborhoods of f in $C(X,Y)$.* Proof: (i) Given δ, one has $N(f,\mathfrak{V}) \subset N_\delta(f)$ for any cover $\mathfrak{V}$ of Y such that, for each $V \in \mathfrak{V}$, diameter $(V) < \inf \{\delta(y) \mid y \in V\}$. To obtain such a $\mathfrak{V}$, we simply cover each open set $\delta^{-1}(1/n,\infty)$ with balls of radius $< 1/2n$.
(ii) Given $\mathfrak{V}$, one has $N_\delta(f) \subset N(f,\mathfrak{V})$ for any majorant δ such that, for all $y \in Y$, the ball $B_{\delta(y)}(y)$ of radius $\delta(y)$ about y lies in some $V \in \mathfrak{V}$. To obtain such a δ first find a locally finite refinement $\mathfrak{W} = \{W_i\}$ of $\mathfrak{V}$ (recalling that metric $\Rightarrow$ paracompact); then define $\delta_i : Y \to [0,\infty)$ to be the distance to $Y - W_i$; and finally let $\delta(y) = \max_i \delta_i(y)$. Since $\mathfrak{W}$ is locally finite the maximum $\delta(y)$ exists and is continuous. ■

The following lemma shows that, in case $f : X \to Y$ is *proper* (i.e. K compact $\Rightarrow$ $f^{-1}K$ compact), one could as well replace majorants δ on Y by majorants δ' on X in the definition of $N_\delta(f)$, substituting $< \delta'(x)$ for $< \delta(f(x))$.

Lemma. *Let $f : X \to Y$ be a proper continuous map to a metric space Y, and let $\delta' : X \to (0,\infty)$ be continuous. Then there exists a continuous map $\delta : Y \to (0,\infty)$ such that $\delta(f(x)) < \delta'(x)$ for all $x \in X$.*

Proof. Since Y admits fine partitions of unity to help us build δ, it is enough to check that each $y \in Y$ has a neighborhood V_y such that $\inf \{\delta'(x) \mid x \in f^{-1}(V_y)\} > 0$. Since $f^{-1}(y)$ is compact, it has an open neighborhood U_y such that $\inf\{\delta'(x) \mid x \in U_y\} > 0$. But, as f is proper, f maps closed sets to closed sets, so we can take $V_y = Y - f(X - U_y)$. ■

Putting these introductory observations aside, we proceed to formulate the promised approximation theorem.

Let X and Y be topological spaces. We shall be working with the collection $\mathcal{S}$ of continuous maps $f: U \to V$ where U is open in X and V is open in Y . Letting $\mathcal{S}(V)$ denote the set of all such f having as target the set V , We observe that the rule $V \mapsto \mathcal{S}(V)$, together with the restriction maps $\mathcal{S}(V) \to \mathcal{S}(V')$ for $V' \subset V$, constitute a *sheaf of sets* † on Y^*, that is also denoted by $\mathcal{S}$.

Let $\mathcal{E}_+ \subset \mathcal{E} \subset \mathcal{S}$ be two subsheaves of $\mathcal{S}$. (Thus, for example, $\mathcal{E}(V)$ is a subset of $\mathcal{S}(V)$ for all open $V \subset Y$.)

Also let $\pi: Y \to Y^*$ be a continuous map to a space Y^* . It serves to give a second, coarser topology on $C(X,Y)$; the simplest case $Y = Y^*$ with π = identity is perhaps the most usual, and we choose our language to conform to it.

We pose the following

Approximation Problem . *Given $f: X \to Y$ in $\mathcal{E}(Y)$ and given an open cover $\mathcal{V}$ of Y^*, under what conditions can we find a map $g: X \to Y$ in $\mathcal{E}_+(Y)$ such that f and g are $\mathcal{V}$-near in the sense that for all x in X , the two points $f(x)$ and $g(x)$ lie in some one set $\pi^{-1} V$ with $V \in \mathcal{V}$?*

Example . Given a homeomorphism $f: X \to Y$ of smooth manifolds and open cover $\mathcal{V}$ of Y^*, when can we find a diffeomorphism $g: X \to Y$ that is $\mathcal{V}$-near to f ? This question fits into the above framework when one defines $\mathcal{E}$ [respectively $\mathcal{E}_+$] to be the subsheaf of $\mathcal{S}$ consisting of all maps that are homeomorphisms, [respectively diffeomorphisms].

The answer we can give in general terms, is merely that it usually suffices to verify the strongly relative local criterion below. Let $\mathcal{U}$ be

† A sheaf of sets on (or over) Y is a contravariant functor $\mathcal{S}$ from the inclusions of open subsets of Y to the category of sets, such that the following 'unique pasting' axiom is verified. For any open $V \subset Y$ and any open cover $\{V_i\}$ of V the elements of $\mathcal{S}(V)$ correspond bijectively via restriction to the indexed collections $\{x_i \mid x_i \in \mathcal{S}(V_i)\}$ such that for any pair of indices i,j the elements x_i, x_j restrict to the same element of $\mathcal{S}(V_i \cap V_j)$.

a fixed open cover of Y^* . (When Y^* is Hausdorff ($= T_2$) and locally compact a popular choice for $\mathfrak{U}$ is the open sets in Y^* that have compact closure.) The 'strongly relative' local criterion is stated as follows in terms of the data X , $\pi: Y \to Y^*$, $\mathcal{E}$, $\mathcal{E}_+$, $\mathfrak{U}$:

(♀) *Let C and D be closed subsets of Y^* with D contained in some $U \in \mathfrak{U}$. Let V be an open neighborhood of $D - C$. Consider any $f: X \to Y$ in $\mathcal{E}(Y)$ that is in $\mathcal{E}_+$ near C* † . *Then there exists $g: X \to Y$ in $\mathcal{E}$ equal to f near $C \cup (Y^* - V)$ such that g is in $\mathcal{E}_+$ near $C \cup D$.*

APPROXIMATION THEOREM C.1 .

Given the data $X, \pi: Y \to Y^*$, $\mathcal{E}$, $\mathcal{E}_+$, $\mathfrak{U}$, as presented above, suppose that (♀) is verified.

Provide that Y^* is fully normal and paracompact.

Then, given $f: X \to Y$ in $\mathcal{E}$, together with any open cover $\mathfrak{V}$ of Y^* , there exists $f_+: X \to Y$ in $\mathcal{E}_+$, such that f and f_+ are $\mathfrak{V}$-near .

Furthermore suppose that f is in $\mathcal{E}_+$ near a closed set $C \subset Y^*$. Let D be any closed set in Y^* , and V any open neighborhood of $D-C$. Then there exists $g: X \to Y$ in $\mathcal{E}$, $\mathfrak{V}$-near to f , such that g is in $\mathcal{E}_+$ near $C \cup D$ and equals f near $C \cup (Y^* - V)$.

Example . In §4 , this theorem can provide part of the proof of the Concordance Implies Isotopy Theorem. With the notation used there, Y^* will be the manifold M , X and Y will be $I \times M \times I$ and π will be the projection to $M = Y^*$. For V open in Y , an element $f: U \to V$ of $\mathcal{E}(V)$ will be a homeomorphism respecting projection to the first interval factor, and equal to the identity both on $U \cap (0 \times M \times I)$ and near $U \cap (I \times M \times 0)$; this f lies in $\mathcal{E}_+(V)$ if it gives a CAT isomorphism $U \cap (1 \times M \times I)_\Gamma$ to $V \cap (1 \times M_\Sigma \times I)$. Thus a $f: X \to Y$ in $\mathcal{E}_+$ is a topological isotopy of $\mathrm{id} \,|\, (M \times I)$ rel $M \times 0$ to a CAT isomorphism $(M \times I)_\Gamma \to (M_\Sigma \times I)$.

Remarks. We have made the approximation theorem just general enough to cover the handful of applications in these essays. One can generalize it by allowing $\mathcal{E}$, $\mathcal{E}_+$ to be arbitrary sheaves of sets, given together

† This means that for some open neighborhood N_C of C , the map $f^{-1}\pi^{-1}N_C \to \pi^{-1}N_C$, obtained by restricting f , lies in $\mathcal{E}_+(\pi^{-1}N_C)$. Similar 'abuse' of language will persist.

with sheaf morphisms $\mathcal{E}_+ \to \mathcal{E} \to \mathcal{S}$ that are not necessarily subsheaf inclusions. This generalization is somewhat more awkward to discuss †, but it permits one simplification: the (generalised) case where $Y = Y^*$ and π = identity immediately implies the most general case.

PROOF OF THE APPROXIMATION THEOREM .

We need only prove the first assertion of C.1 ; the stronger assertion involving given subsets C , D and V of Y^* can then be deduced by applying the first assertion to suitable subsheaves $\mathcal{E}'$ and $\mathcal{E}'_+$ of $\mathcal{E}$ and $\mathcal{E}_+$, namely:

$\mathcal{E}'$ = the maps in $\mathcal{E}$ equal to f near $C \cup (Y^* - V)$ (wherever defined).

$\mathcal{E}'_+$ = the maps in $\mathcal{E}'$ that lie in $\mathcal{E}_+$ near D (wherever defined).

This works because the local criterion (♀) is verified for $\mathcal{E}'$ and $\mathcal{E}'_+$ in place of $\mathcal{E}$ and $\mathcal{E}_+$ above.

To prove the first assertion we begin by using paracompactness of Y^* to find an open cover $\mathfrak{W} = \{W_i\}$ of Y^* that refines the two covers $\mathfrak{U}$ and $\mathfrak{V}$.

Since Y is fully normal as well as paracompact, there exists a partition of unity $\{\alpha_i\}$ on Y^* with $\alpha_i^{-1}(0,1] \subset W_i$ for each i . This can be interpreted as a continuous map

$$\alpha : Y^* \to |\mathfrak{W}|$$

to the nerve of the cover $\mathfrak{W}$; namely the one such that, for each y in Y^* and each index i , the point $\alpha(y)$ has $\alpha_i(y)$ for its barycentric coordinate with respect to the vertex w_i corresponding to W_i .

The appropriate topology for $|\mathfrak{W}|$ is the 'weak' topology ‡ , for which a set is closed if and only if it meets each closed simplex of $|\mathfrak{W}|$ in a closed set. The continuity of the functions α_i implies that $\alpha^{-1}(\sigma)$ is closed for each closed simplex σ and that α is continuous over each such σ . Local finiteness of the partition $\{\alpha_i\}$ implies local finiteness 'value-wise' of the collection $\{\alpha^{-1}(\sigma)\}$ in Y^* —i.e. , given x in Y^*

† For example, $f : X \to Y$ in $\mathcal{E}$ comes to mean an element f of $\mathcal{E}(Y)$ whose image in $\mathcal{S}(Y)$ is a continuous map $X \to Y$.

‡ The three standard topologies on $|\mathfrak{W}|$ (cartesian, metric, weak) coincide precisely if $\mathfrak{W}$ is *star-finite* . The covering $\mathfrak{W}$ can always be chosen to be star-finite precisely if Y^* is *strongly paracompact*. Every locally separable metric space, and also every locally compact Hausdorff space is strongly paracompact, see [Na_2, p.172, p.182] .

there is an open set U containing x so that $\{ \alpha^{-1}(\sigma) \cap U \}$ (σ varying) is a finite collection. Thus, a set in Y^* is closed precisely if it meets each set $\alpha^{-1}(\sigma)$ in a closed set. It follows that $\alpha^{-1}(A)$ is closed if A is closed; so the continuity of α is now explained.

Writing $|\mathfrak{W}| = K$ for convenience, suppose for an induction over the skelleta $K^{(r)}$, $r \geqslant -1$, that we have built $f_r : X \to Y$ in $\mathcal{E}$ that is in $\mathcal{E}_+$ near $K^{(r)}$. The induction begins with $f_{-1} = f$.

For any closed simplex σ of $|\mathfrak{W}|$ we have a standard open 'spindle neighborhood' $N(\sigma)$ of $\operatorname{int}\sigma = (\sigma - \partial\sigma)$, namely the open star of $\operatorname{int}\sigma$ in the first barycentric subdivision of $|\mathfrak{W}|$ †.

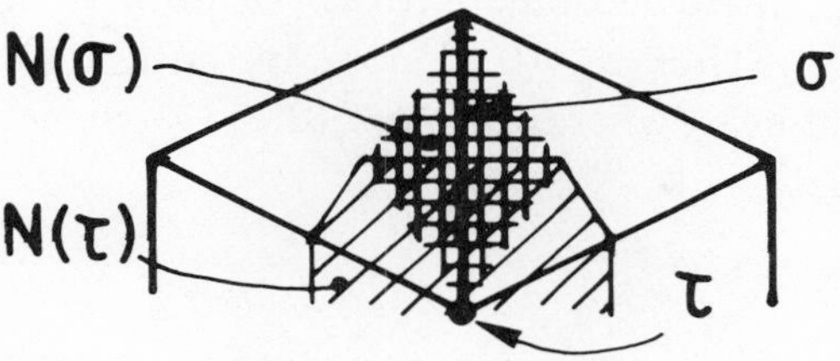

Observe that, if $N(\sigma)$ meets $N(\tau)$, then either $\sigma < \tau$ or $\tau < \sigma$.

For each closed (r+1)-simplex σ of $|\mathfrak{W}| = K$, apply the criterion (♀) with the substitutions $C \leftrightarrow \alpha^{-1}K^{(r)}$, $D \leftrightarrow \alpha^{-1}(\sigma)$, $V \leftrightarrow \alpha^{-1}N(\sigma)$, to obtain a map $g_\sigma : X \to Y$ in $\mathcal{E}$ equal to f_r near $Y^* - \alpha^{-1}(N(\sigma))$ and lying in $\mathcal{E}_+$ near $\alpha^{-1}(K^{(r)} \cup \sigma)$. Since the open sets $\alpha^{-1}N(\sigma)$ are pairwise disjoint, we can ‡ piece together a unique map $f_{r+1} : X \to Y$ such that f_{r+1} coincides with f_r near $Y^* - N_{r+1}$ where

$$N_k = \cup \{ N(\sigma) \mid \dim\sigma = k , \sigma < K\}$$

and f_{r+1} coincides with g_σ over $\alpha^{-1}N(\sigma)$. Clearly f_{r+1} is in $\mathcal{E}_+$ over $\alpha^{-1}(K^{(r+1)})$. This completes the induction to construct a sequence $f_0, f_1, f_2, \ldots$ of maps $X \to Y$.

† If σ has vertices $w_0, \ldots, w_s$ then $N(\sigma)$ is the set of all points $z \in |\mathfrak{W}|$ such that if w_i is not a vertex of σ then $w_i(z) < \min\{w_0(x), \ldots, w_s(x)\}$. This $N(\sigma)$ is clearly open in the *weak* topology; it is often not open for the other two.

‡ The collection of all $\alpha^{-1}N(\sigma)$, σ a simplex of $|\mathfrak{W}|$, is easily seen to be a locally finite open cover of Y^*, because $\mathfrak{W}$ is locally finite. But beware that, in general, the open covers $\{N(\sigma)\}$ and $\{N_r\}$ of $|\mathfrak{W}|$ are not locally finite.

The collection $\{\alpha^{-1}N_r \mid r = 0, 1, \ldots\}$ is locally finite in Y^* because $\mathfrak{W}$ is locally finite . Thus any point $y \in Y^*$ has a neighborhood V_y meeting only finitely many sets $\alpha^{-1}N_r$. It follows that the sequence $f_r \mid V_y$, $r = 1, 2, \ldots$, is eventually constant. This shows that $g = \lim f_r$ is a well defined map $X \to Y$ in $\mathcal{E}$. It clearly lies in $\mathcal{E}_+$.

Finally, we show that g is $\mathfrak{V}$-near to f . For any point $x \in X$, suppose the eventually constant sequence $f(x), f_0(x), f_1(x), \ldots$ moves just before $f_{r(n)}(x)$, $n = 0, 1, \ldots, s$. Then $f_{r(n-1)}(x)$ and $f_{r(n)}(x)$ lie in some $\alpha^{-1}N(\sigma_n)$. Thus $N(\sigma_n)$ meets $N(\sigma_{n+1})$, $n = 0, \ldots, s-1$, and clearly this implies that $\sigma_0 < \sigma_1 < \ldots < \sigma_s$. If w_i is a vertex of σ_0 , then the barycentric coordinate for w_i is positive on $N(\sigma_0) \cup \ldots \cup N(\sigma_s)$. Thus $\alpha_i^{-1}(0,1]$ contains both f(x) and g(x) , i.e. f and g are $\mathfrak{A}$-near, where $\mathfrak{A} = \{\alpha_i^{-1}(0,1]\}$. Since $\mathfrak{A} < \mathfrak{W} < \mathfrak{V}$, the proof is complete. ■

Source majorant approximation *(supplementary remarks)* .

There is a companion to the majorant approximation problem discussed above; it arises when one wishes to measure progress in the source space X rather than the target Y .

To simplify we assume henceforth that X *and* Y *are metric and that* X *is locally compact.*

One gives to C(X,Y) the *source majorant topology* , letting a basis of neighborhoods of $f : X \to Y$ be the sets

$N(f;\epsilon) = \{ g \in C(X,Y) \mid d(f(x),g(y)) < \epsilon(x)$ for all x in X $\}$

where $\epsilon : X \to (0,\infty)$ ranges over all positive continuous functions (majorants). This is also known as the 'fine Whitney topology' or the 'fine C^0 topology', cf. [Mu_1, p.29] . It is in general finer than the (target) majorant topology first considered, as the example X = R , Y = [0,1] amply illustrates. Of course, when X is compact, both majorant topologies coincide with the compact-open topology.

Fixing attention on the source space X , let U be any open subset of X , and denote by $\mathcal{S}(U)$ the set of continuous functions $U \to Y$. With obvious restriction maps, the sets $\mathcal{S}(U)$ form a sheaf of sets on X as U varies; it is denoted $\mathcal{S}$. We consider subsheaves

$\mathcal{E}_+ \subset \mathcal{E} \subset \mathcal{S}$ of 'good' functions $\mathcal{E}$ and 'better' functions $\mathcal{E}_+$. Pause here to note that $\mathcal{E}$ could for example be the open *immersions* to Y , but it could not be the open *embeddings*. (Just inspect the basic axiom for sheaves on X !)

Approximation Problem: *Is* $\mathcal{E}_+(X)$ *dense in* $\mathcal{E}(X)$ *for the source majorant topology?*

To be sure, one can pose this problem in a strongly relative form, but, as noted in the proof of C.1, such generality is illusory.

We indicate how to solve this problem in two steps.

Step I). $\mathcal{E}_+(X)$ *is indeed dense in* $\mathcal{E}(X)$ *provided the following strongly relative local criterion is verified:*

(♂) *Given* $f : X \to Y$ *in* $\mathcal{E}$ *that is in* $\mathcal{E}_+$ *near a closed set* $C \subset X$, *and given a compactum* $D \subset X$ *and an open neighborhood* V *of* D , *there always exist approximations* g *to* f *that are in* $\mathcal{E}_+$ *near* $C \cup D$ *and are equal to* f *near* $C \cup (X - V)$.

Note that this property (♂) remains just as strong if we suppose that V has compact closure, or if we suppose that V is contained in some set of a fixed open covering of X .

It is still unsatisfactory that in (♂) we should have to hypothesize approximations; nevertheless it is encouraging to note that it does not matter which topology we use on $\mathcal{E}(X)$, since V may as well have compact closure.

The proof of **I)** is left as an exercise, cf. [Si_{12} , Proof of 6.3] . *Hint:* It can exploit a well known 'concentric annuli' trick to express X as $X_0 \cup X_1$ where X_0 and X_1 are closed sets each a *discrete* sum of compacta. (For $X = R^2$ these compacta could be concentric annuli.)

Step II). Verification of (♂) .

Frequently, when one is faced with a specific approximation problem, the criterion (♂) is directly verifiable . This will be the case for approximation of maps by ones transversal to a TOP microbundle

[III, §1] ; of for approximation of continuous functions by TOP Morse functions [III, §2] .

However, it is reassuring to observe that C.1 reduces (♂) to a criterion not involving approximation.

Assertion: (♂) *is implied by the weaker version* (♂)* *where* g *is no longer required to be an approximation to* f *but instead one insists that* g(V) *lie in a prescribed neighborhood of* f(V) .

Proof: Regard (♂) as a target majorant approximation problem as follows. With the notation of (♂) let $\mathcal{E}^!$ be the sheaf on Y consisting of continuous maps $h : U' \to V'$ with U' open X and V' open in Y , so that $U' \xrightarrow{h} V \hookrightarrow Y$ is in $\mathcal{E}$ and h equals f near $U' \cap \{C \cup (X-V)\}$. Also let $\mathcal{E}^!_+ \subset \mathcal{E}^!$ be the subsheaf of maps h such that $U' \xrightarrow{h} V' \hookrightarrow Y$ is in $\mathcal{E}_+$ near $U' \cap (C \cup D)$. Then in view of C.1 , the criterion (♂) for $\mathcal{E}_+ \subset \mathcal{E}$ is implied by (♀) for $\mathcal{E}^!_+ \subset \mathcal{E}^!$. This (♀) is in turn implied by (♂)* . ∎

Essay II

DEFORMATION OF SLICED FAMILIES OF MANIFOLD STRUCTURES

by

L. Siebenmann

Section headings in Essay II

§0. INTRODUCTION

A **sliced family** of CAT (= DIFF or PL) structures on a manifold M with parameters in Δ (a simplex or any CAT manifold) is a CAT manifold structure Γ on the product $\Delta \times M$ that is **sliced over** † Δ in the sense that the projection $(\Delta \times M)_\Gamma \to \Delta$ is a CAT submersion. Then, for each point u in Δ , Γ gives to $u \times M$ a CAT structure Γ_u . Beware that a family Γ is decidely more than the collection $\{\Gamma_u \mid u \in \Delta\}$ of its 'members', whenever $\dim M \geqslant 1$. ‡

A key result of §1 is that the projection $p_1 : (\Delta \times M)_\Gamma \to \Delta$ is a CAT bundle projection provided $\dim M \neq 4 \neq \dim \partial M$. When Δ is contractible, this means that there is a CAT isomorphism $h : (\Delta \times M)_\Gamma \to \Delta \times (M_\gamma)$ that is **sliced** over Δ (in the sense that $p_1 h = p_1$). This result requires no dimension restriction if M is compact, since every proper CAT submersion is easily seen to be a CAT bundle projection. In the noncompact case we shall use an engulfing procedure in order to revert to the compact case; a simple covering trick is involved. As first sketched in [KS$_4$] , this argument enjoyed considerable excess generality that was then intended to let one pursue the study of limits of homeomorphisms, begun in [Si$_{11}$] (see [Si$_{11}$,end] for an example). We retain that generality here, but we have tampered with the original proof to make it self-contained, and in addition we have found a short-cut for the case at hand. R. Lashof and D. Burghelea have in the meantime devised an alternative argument [BuL,I] .

The above bundle theorem of §1 is extensively used in classifying sliced families in Essay V .

† Beware of confusion with the notion of a "slice" in the theory of group actions.

‡ There is for example a DIFF structure Γ on $T^1 \times S^6$ so that Γ_u is standard for all $u \in \Delta = T^1 = R/Z$, but $(T^1 \times S^6)_\Gamma \not\cong T^1 \times S^6$. To build Γ , let it be standard near the complement of a disc $(-\frac{1}{4},\frac{1}{4}) \times R^6$ in $T^1 \times S^6$, with $S^6 = R^6 \cup \infty$, on which disc we make it the image of the standard structure under a TOP Alexander isotopy H with compact support, to $\mathrm{id} \mid R^6$ from $h \mid R^6$, where h is an exotic DIFF automorphism [Mi$_6$] of S^6 rel ∞ ; more specifically H can map $(t,x) \mapsto (t, 8th(x/8t))$ for $0 < t < 1/4$ and $(t,x) \mapsto (t,x)$ for $-1/4 < t \leqslant 0$.

In §2 we refine the bundle theorem in case Δ is a simplex or cube, showing that there is a majorant-small **TOP** isotopy h_t, $0 \leqslant t \leqslant 1$, of $\mathrm{id}\,|(\Delta \times M)$ *sliced* over Δ, so that $h_1 \Gamma$ is of the form $\Delta \times \Sigma$, with Σ a CAT structure on M. If Γ was of the form $\Delta \times \Sigma$ near $\Lambda \times \Sigma$, where Λ is a retract of Δ, this isotopy can be the identity near $\Lambda \times M$. There are further refinements, giving, in all, a complete analogue for "sliced" concordances of the Concordance Implies Isotopy Theorem of [I, §4].

This result of §2 is useful in converting information presented in terms of such sliced concordances Γ into geometrically useful isotopies. For example it lets one approximate topological isotopies of the identity by CAT isotopies of the identity.

The 1972 version of this essay contained two further sections† that studied the space of CAT concordances rel ∂ of the structure of a CAT manifold M, by generalizing directly the proof of the Concordance Implies Isotopy Theorem of Essay I (cf. the announcement [KS_4]). In particular, provided $\dim M \geqslant 5$, this space (semi-simplicially or semi-cubically defined using sliced families) is contractible for CAT = PL and 1-connected for CAT = DIFF. The reader will find that such results can equally well be obtained by combining §2 with the classifications of Essay V. Certainly the approach via Essay V is more sophisticated; but on the other hand it is probably more enlightening.

† To which, no doubt, stray references persist. Perhaps these sections will be published in Manuscripta Mathematica.

§1. BUNDLE THEOREMS

A CAT map $p:E \to B$ (for CAT = DIFF, PL, TOP, T_2 †) is called a CAT **submersion** if for each point x in E there is a CAT object U, an open neighborhood B_o of p(x) in B, and a CAT open imbedding $f: B_o \times U \to E$ onto a neighborhood of x such that pf is projection $B_o \times U \to B_o \subset B$. Such an imbedding f is called a **submersion chart**. When f is **normalized** so that U is an open subset of $p^{-1}p(x)$ and $f(p(x),u) = u$ for u in U, then we sometimes call f a **product chart** about U for p. Note that every fiber of a CAT submersion is a CAT object. Also, for any open set $E_o \subset E$, the restriction $p|E_o$ is a CAT submersion, and its image $p(E_o)$ is open in B.

Using the CAT isotopy extension principle (CAT $\neq T_2$) one can show that any compactum in a fiber $p^{-1}(y)$ of a CAT submersion p is contained in the image of some submersion chart.‡‡ Hence, if the submersion p has compact fibers and is closed, i.e., maps closed sets to closed sets, it follows that p is a locally trivial CAT bundle projection.

The aim of this section is to give conditions assuring that a CAT submersion with *noncompact* fibers is a locally trivial CAT bundle projection.

Our main result is a technical one, formulated so as to avoid restrictions of dimension or category.

† T_2 is the category of all Hausdorff topological spaces and continuous maps. PL could be enlarged in this section to contain all locally compact polyhedra and piecewise linear maps.

‡ For CAT = T_2 this applies nevertheless, if for example $p^{-1}(y)$ is locally triangulable; see [Si_{12}] for the relevant isotopy extension principle.

‡ See [Si_{12}; 6.14, 6.15] for this reduction to the CAT isotopy extension theorem. But note that, for CAT = DIFF or PL, very direct proofs exist. For CAT = PL, [III, Lemma 1.7] provides one. For CAT = DIFF, any smooth neighborhood retraction $r: E_0 \to p^{-1}(y)$ provides one since r is a DIFF submersion near $p^{-1}(y)$.

TECHNICAL BUNDLE THEOREM 1.1 .

Consider a CAT object E (CAT = DIFF , PL , TOP or T_2) , equipped with
(a) a CAT submersion $p : E \to \Delta$ onto a CAT object Δ . (For $u \in \Delta$ write $F_u = p^{-1}(u)$ for the fiber of p over u .)
(b) a merely continuous map $\pi : E \to R$. (For a,b in R write $F_u(a,b) = F_u \cap \pi^{-1}(a,b)$; adopt similar notation with any subset of R in place of the open interval (a,b)) ,

We suppose verified an 'engulfing' condition somewhat weaker than the statement that each $F_u(a,b)$ is a CAT product with R :

(*) For any $u \in \Delta$ and any pair of integers $a \leqslant b$, there exists a CAT isotopy h_t , $0 \leqslant t \leqslant 1$, of F_u with compact support in $F_u(a-1, b+1)$ such that

$$h_1 F(-\infty,a) \supset F(-\infty,b]$$

For CAT = T_2 we must assume that each fiber F_u is locally triangulable † , and that, for each pair of integers $a < b$, the set $F_u[a,b]$ is connected and compact. ‡ Also we must assume that Δ is paracompact and ǂ of covering dimension $d < \infty$.

Under these hypotheses $p : E \to \Delta$ is a locally trivial CAT bundle.

Remarks on 1.1 .

1) If CAT = DIFF , PL , or TOP the special case where Δ is a simplex (or cube) implies the general case.

2) The TOP version is clearly subsumed in the T_2 version.

3) The T_2 version could probably be proved differently in case E and Δ are complete metric spaces, by using a selection theorem of E. Michael [Mic] .

4) Whether the hypothesis $\dim\Delta < \infty$ is necessary is unknown. For the applications of the T_2 version to the closure of the homeomorphism group of a manifold envisaged in [Si_{11}] , it is a most undesirable restriction.

† Or more generally verifies the general isotopy (local) extension principle of [Si_{12}; §0,§6] for parameters in Δ .

‡ This compactness, in fact, follows easily from (*) .

ǂ i.e., each covering of Δ admits a refinement whose nerve is a simplicial complex of dimension $\leqslant d$, and d is the least such number. One could equivalently use finite coverings only, see [Nag, p.22] [Dow] .

5) The requirements on some new category CAT, in order that 1.1 hold, could be axiomatized without difficulty.

6) The fibers of p may indeed not be a product with R. For examples in dimension $\geqslant 5$ detected by class group obstructions, see [Si_6]. A suitable $\pi : F_u \to R$ can be gotten by identifying F_u with the natural infinite cyclic covering of a band formed by gluing together the ends of F_u, see [Si_6, §2].

7) In stating this theorem in [KS_4], we forgot to mention the isotopy h_t to h_1 in (*); our proof definitely needs it.

PROOF OF THEOREM 1.1.

⟦ Certain parts of the proof not needed to establish the important DIFF and PL Bundle Theorem 1.8 below are marked off by double brackets. ⟧

We shall show that p factorizes as

$$p = p' \circ q : E \xrightarrow{q} B \xrightarrow{p'} \Delta \quad ,$$

where q is a CAT infinite cyclic covering, classified by a map $f : B \to S^1$, while p' is a closed CAT submersion with compact fibers, and hence a CAT bundle projection.

It follows that p is also a CAT bundle projection. *Proof:* As the condition is local in Δ, we can assume p' is trivial, i.e., $B \cong F' \times \Delta$ so that p' becomes projection to Δ. To verify local CAT triviality of p at $u \in \Delta$, consider $F_u = q^{-1}(F' \times u)$ and the infinite cyclic covering

$$q_o : F_u \times \Delta \xrightarrow{(q|F_u)\times id} F' \times \Delta = B \quad .$$

This covering q_o is clearly classified by the map $f_o : F' \times \Delta \to S^1$ sending $(x,v) \mapsto f(x,u)$, and f_o agrees with f on $F' \times u$. Since close maps $F' \to S^1$ are canonically homotopic, these classifying maps f and f_o are homotopic when we cut Δ down to a small neighborhood of u. Then, by the bundle homotopy theorem [Hus], q and q_o give isomorphic coverings of $F' \times \Delta$, and so $E \cong F_u \times \Delta$ making q correspond to q_o, and hence $p'q = p$ correspond to the projection $p'q_o = p_2 : F_u \times \Delta \to \Delta$. This proves the wanted CAT local triviality.

The proposed factorization of p will be obtained by patiently constructing an infinite cyclic covering translation on E.

PART A). *The following global sliced version* $\#(\gamma)$ *of the engulfing property* (*) *holds true, with* $\gamma = \dim\Delta + 1 = d+1$.

$\#(\gamma)$: *For any pair of integers* $a \leqslant b$, *there exists a CAT isotopy* h_t , $0 \leqslant t \leqslant 1$, *of* $id|E$ *sliced over* Δ , *such that*

$$h_1 \pi^{-1}(-\infty,a) \supset \pi^{-1}(-\infty,b] ,$$

and the support of the isotopy h_t *is contained in* $\pi^{-1}[a-\gamma, b+\gamma]$.

To prove this we must consider also a weaker condition

$$\#_r(\gamma, C)$$

for C a closed subset of Δ and $r \in (0,\infty]$. This differs only through weakening the inclusion condition to be

$$h_1 \pi^{-1}(-\infty,a) \supset \pi^{-1}(-\infty,b] \cap p^{-1}C ,$$

and suppressing even this condition except when

$$[a-\gamma, b+\gamma] \subset [-r+2, r-2] .$$

Clearly $\#_\infty(\gamma,\Delta) = \#(\gamma)$; also $\#_r(\gamma,C) \Rightarrow \#_s(\delta,D)$ if $r \geqslant s$, $\gamma \leqslant \delta$, and $C \supset D$.

Addition Lemma 1.2 . *The conditions* $\#_r(\gamma,C)$ *and* $\#_r(\delta,D)$ *together imply the condition* $\#_r(\gamma+\delta, C \cup D)$.

Proof of 1.2 : Given a pair of integers $a < b$; the isotopy asserted by $\#_r(\gamma+\delta, C \cup D)$ can be defined as

$$h_t = h_t^C \circ h_t^D , \quad 0 \leqslant t \leqslant 1 ,$$

Figure 1-A .

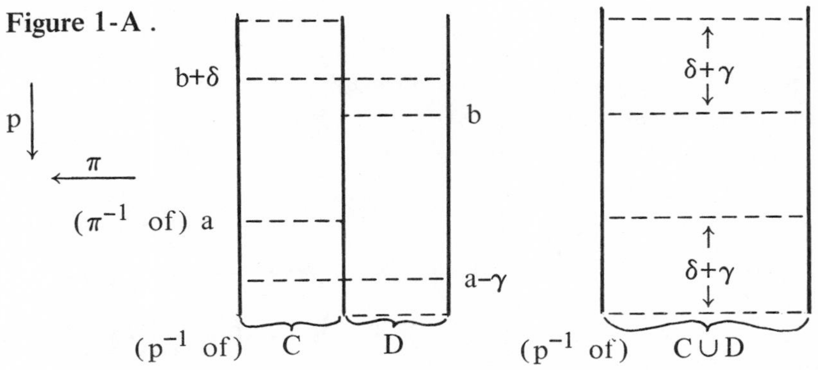

where h_t^C is an isotopy provided by $\#_r(\gamma,C)$ for the pair $(a,b+\delta)$, and h_t^D is an isotopy provided by $\#_r(\delta,D)$ for the pair $(a-\gamma,b)$. Figure 1-A immediately suggests why. ■

Fixing now an integer $r<\infty$, we seek closed sets $C_0, C_1, \ldots, C_d$ with union Δ so that $\#_r(1,C_i)$ holds for each C_i. Then this addition lemma (applied d times) establishes $\#_r(d+1,\Delta)$. Since r is as large as we please we shall have $\#(d+1)$ as required for Part A.

Consider a (normalized) CAT product chart

$$\varphi : F_u(-r,r)\times\Delta_u \to E$$

for the CAT submersion p, the set Δ_u being an open neighborhood of a point $u\in\Delta$.

Consider also a CAT isotopy g_t of $id|F_u(-r,r)$, with compact support.

For any CAT function

$$\alpha:\Delta\to[0,1] \quad ,$$

with support contained in Δ_u, we then define a map h_t on Image(φ)

$$h_t : \varphi(x,v) \mapsto \varphi(\, g_{\alpha(v)t}(x)\, , v\,)$$

and extend as the identity outside Image(φ) to an isotopy h_t of $id|E$, sliced over Δ.

For the moment we fix φ and let α and g_t vary.

Fact 1. *There exists an open neighborhood U of u in Δ_u so small that*

$$support(\alpha)\subset U$$

will imply that, as g_t ranges over the finitely many isotopies promised by (*), one for each integral interval $[a,b]\subset[-r+2,r-2]$, the corresponding isotopies h_t establish $\#_r(1,C)$, with $C=\alpha^{-1}(1)$.

(This is easily verified given the proviso that E is topologically $M\times R\times\Delta$ with π and p projections to R and Δ; for then h_t can be regarded as an isotopy of $M\times R$ with parameters in $[0,1]\times\Delta$. Since this case is adequate for the important Bundle Theorem 1.8 below, we postpone the general argument to the end; see Lemma 1.3 below.)

Next, letting u and φ vary for the first time (r still fixed), we form an open covering $\{U_k\}$ of Δ by sets U as above.

Fact 2 . *We can decompose* Δ *into* $d+1$ *closed sets* $C_0, \ldots, C_d$, *each set* C_i *being a disjoint discrete sum of closed sets* C_{ij} *each contained in some* $U_k = U_{k(ij)}$.

(In case CAT = DIFF or PL we can assume Δ is the d-cube; then the C_{ij} can be the i-handles of a fine standard handle decomposition of the cube . Again, the general argument is postponed.)

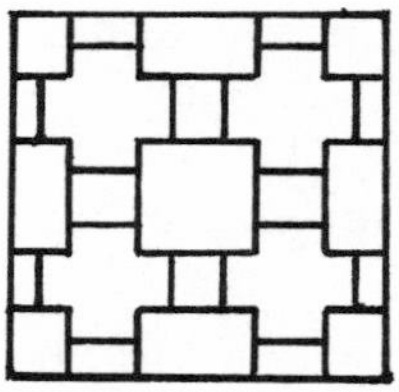

For each C_i we now form CAT functions $\alpha_{ij} : \Delta \to [0,1]$ with $\alpha_{ij}(C_{ij}) = 1$, each having as support a set A_{ij} contained in $U_{k(ij)}$, so that $A_i = \cup_j A_{ij}$ is again a disjoint discrete sum. Then the isotopies corresponding to the α_{ij} , j varying, compose disjointly to establish $\#_r(1,C_i)$.

Finally, as promised, d applications of the addition lemma 1.2 deduce $\#_r(d+1,\Delta)$ and hence $\#(d+1)$. ■

⟦ It remains only to prove facts 1 and 2 in full generality.

To prove Fact 1 in full generality, one needs:

Lemma 1.3 . *Let* $\varphi : F_u(-r,r) \times \Delta_u \to E$ *be a normalized CAT product chart for our CAT submersion* $p : E \to \Delta$. *Consider integers* $a < b$ *with* $[a,b] \subset (-r,r)$, *together with a compactum* $X \subset F_u(a,b)$, *and an open set* W *of* $F_u(-r,r)$ *containing* $F_u[a,b]$. *Then for all* $v \in \Delta$ *sufficiently near* u *we have*

(i) $\varphi(X \times v) \subset F_v(a,b)$, *and*

(ii) $\varphi(W \times v) \supset F_v[a,b]$.

Proof of Lemma 1.3 .

Condition (i) is equivalent to $X \times v \subset \varphi^{-1}\pi^{-1}(a,b)$. As this last set is open and contains $X \times u$, this inclusion holds for all v near u in virtue of the compactness of X .

To prove (ii) recall that all $F_u[a',b']$ are *compact*, and choose a compact neighborhood $Y \subset W$ of $F_u[a,b]$. Choose also a point

$x \in F_u(a,b)$. Arguing as for (i), we have

(a) $\varphi(\delta Y \times v) \cap F_v[a,b] = \phi$, δ indicating frontier in F_u ,

(b) $\varphi(x,v) \in F_v(a,b)$,

for all v near u . Since $\varphi(\delta Y \times v)$ is the frontier $\delta\varphi(Y \times v)$ taken in F_v , the sets $\varphi(\mathring{Y} \times v)$ and $F_v - \varphi(Y \times v)$ then give an open partition of $F_v[a,b]$, which is *connected* by hypothesis. But $\varphi(\mathring{Y} \times v)$ does meet $F_v[a,b]$ since (b) says $\varphi(x,v)$ is in both. Hence $F_v - \varphi(Y \times v)$ does not meet $F_v[a,b]$, and so

$$F_v[a,b] \subset \varphi(\mathring{Y} \times v) \subset \varphi(W \times v) ,$$

for v near u , as required. ■

The derivation of Fact 1 using Lemma 1.3 is left as an exercise for the reader; it is straightforward but does require some patience.]]

[[To prove Fact 2 in general, choose a refinement $\mathcal{R}$ of the covering $\{U_k\}$ that has nerve $|\mathcal{R}|$ of *dimension* $\leqslant d$, and then choose a locally finite partition of unity on Δ subordinated to $\mathcal{R}$ (using *paracompactness* of Δ). This partition can be regarded as a continuous map

$$\rho : \Delta \to |\mathcal{R}| ,$$

when $|\mathcal{R}|$ is given the least topology making all the barycentric coordinate functions continuous. Let H_i be the disjoint discrete sum in $|\mathcal{R}|$ of all standard i-handles of $|\mathcal{R}|$. (A standard i-handle is a closed star in the second barycentric subdivision $|\mathcal{R}|''$ of a barycenter of an i-simplex of $|\mathcal{R}|$.) Clearly $H_0, \dots, H_d$ are closed with union $|\mathcal{R}|$ so we can define $C_i = \rho^{-1}(H_i)$, $i = 0, \dots, d$. ■]]

This completes Part A.

The following shortcut should be taken by readers interested in the Bundle Theorem 1.8 alone.

Shortcut 1.4. For CAT = DIFF or PL with Δ a simplex or cube the property #(d+1) established by Part A implies (almost immediately) : *For any pair of integers* $a < b$ *there exists an open set* E_{ab} *in* E *containing* $\pi^{-1}[a,b]$ *such that*

$$p : E_{ab} \to \Delta$$

is a locally trivial CAT bundle.

Proof: Any automorphism h_1 of E sliced over Δ with $h_1\, \pi^{-1}(-\infty,a) \supset \pi^{-1}(-\infty,b]$, as provided by #(d+1), already is infinite cyclic covering translation when restricted to the following open set $E_{ab} \supset \pi^{-1}[a,b]$:

$$E_{ab} = \cup \{h_1^n Z_{ab} \mid n = 0\,,\ \pm 1\,,\ \pm 2\,, \ldots\}\ ,\ \text{where}$$
$$Z_{ab} = h_1\, \pi^{-1}(-\infty,a] - \pi^{-1}(-\infty,a)\ .$$

Indeed the compactum Z_{ab} is a fundamental domain. Local triviality of $p|E_{ab}$ as asserted now follows via the argument given before Part A. ■

In this situation one can go on to prove that $p : E \to \Delta$ itself is a locally trivial bundle projection, by using an infinite collection of such E_{ab} together with the *bundle homotopy theorem* † ; the argument is given for the Bundle Theorem 1.8 below. It is really just this last step that breaks down in general, necessitating Part B.

⟦ PART B . The covering translation . ⟧

⟦ We use a classical meshing process to build it in general.

Adjoin two ideal points ϵ_- and ϵ_+ to E, forming a Hausdorff topological space $\hat{E} = E \cup \{\epsilon_-,\epsilon_+\}$, by specifying the sets $\epsilon_+ \cup \pi^{-1}(a,\infty)$, $a \in Z$, to be a basis of neighborhoods for ϵ_+ in $\hat{E}$, and similarly specifying $\epsilon_- \cup \pi^{-1}(-\infty,a)$, $a \in Z$, to be a basis for ϵ_- .

Complementation in E will be indicated by the superscript c ; thus $S^c = E - S$.

$\mathcal{E}$ will denote the set of CAT automorphisms of E sliced over Δ that each fix some neighborhood of $\{\epsilon_-,\epsilon_+\}$.

Given $U \supset U'$, two neighborhoods of ϵ_+ in E, we say that U' is *compressible* toward ϵ_+ in U and write

$$U' \searrow \epsilon_+ \ (U)$$

if, for any neighborhood U'' of ϵ_+, there exists $h \in \mathcal{E}$ fixing U^c so that $h(U') \subset U''$.

Define $U_k = \pi^{-1}(k(d+1),\infty)$, $k \in Z$, and observe that #(d+1) of Part A implies

† This theorem shows that bundles over a cube or a simplex are trivial.

$$U_k \searrow \epsilon_+ \ (U_{k-1}) \quad \text{and} \quad U_k^c \searrow \epsilon_- \ (U_{k-1}^c) \quad .$$

Insertion Lemma 1.5 . *Given four neighborhoods U, U', V, V' of ϵ_+ in E so that $U' \searrow \epsilon_+ \ (U)$ and $V' \searrow \epsilon_+ \ (V)$, there exists $h \in \mathcal{E}$ fixing $(U \cup V)^c$, such that*

$$U \supset h(V') \supset U' \quad .$$

Similarly with ϵ_- in place of ϵ_+ .

Proof of Lemma : Find $h_1 \in \mathcal{E}$ fixing V^c such that $h_1(V) \subset U$. Find $h_2 \in \mathcal{E}$ fixing U^c such that $h_2(U') \subset h_1(V)$. Then $h = h_2^{-1} \circ h_1$ will serve. ■

Assertion . *There exists a CAT automorphism h of E sliced over Δ such that for all odd k*

$$U_k \supset hU_{k-3} \supset U_{k+2} \quad .$$

This h will be the covering translation.

Proof of Assertion.

We will say that $g \in \mathcal{E}_r$ (with $r \leqslant 1$ odd) if $g \in \mathcal{E}$ and $U_{-r} \supset V_{-r+1} \supset U_{-r+2} \supset \cdots \supset U_{r-2} \supset V_{r-1} \supset U_r$ where V_k denotes $g(U_{k-4})$.

We begin an induction by using the insertion lemma to find $h_1 \in \mathcal{E}_1$. Then suppose we have $h_r \in \mathcal{E}_r$ (r odd) . Since $U_{r-2} \supset V_{r-1} \supset U_r$ while

$$V_{r+1} \searrow \epsilon_+ \ (V_{r-1}) \quad \text{and} \quad U_{r+2} \searrow \epsilon_+ \ (U_r)$$

the insertion lemma provides $f_+ \in \mathcal{E}$ fixing V_{r-1}^c so that

$$U_r \supset f_+ V_{r+1} \supset U_{r+2} \quad .$$

Similarly, passing to complements and changing the sign of subscripts, we obtain $f_- \in \mathcal{E}$ fixing V_{-r+1} so that $U_{-r}^c \supset f_- V_{-r-1}^c \supset U_{-r-2}^c$, i.e.

$$U_{-r-2} \supset f_- V_{-r-1} \supset U_r \quad .$$

Setting $f_r = f_- f_+$ we have

$$h_{r+2} = f_r h_r \in \mathcal{E}_{r+2} \quad .$$

This completes an induction to construct $h_1, h_3, h_5, \cdots$.

Noting that f_s fixes $\pi^{-1}[-r+2, r-2]$ for all $s \geqslant r$, we can define a unique $h = \lim h_r : E \to E$ such that $h = h_r$ over $\pi^{-1}[-r+2, r-2]$ for all odd r . This h establishes the assertion. ■

Let h be the automorphism of E provided by the assertion, and form the quotient

$$q : E \to B = E / \{ h(x) = x \mid x \in E \} \ .$$

Since h is sliced over Δ , there is a unique map $p' : B \to \Delta$ so that $p'q = p$.

The inclusions $U_k \supset hU_{k-3} \supset U_{k+2}$ (k odd) show that q is an *infinite cyclic covering map*, when we recall that U_k
$U_k = \pi^{-1}(k(d+1), \infty)$. Further †, as principal Z-bundle , $q : E \to B$ is induced from $R \to R/Z$ ($\cong S^1$) by a *'classifying' map* $f: B \to R/Z$. Indeed, when B is regarded as a quotient of $h\pi^{-1}(-\infty, 0] - \pi^{-1}(-\infty, 0)$, the desired f can be π itself on $\pi^{-1}[0,1]$ and be zero (in R/Z) elsewhere!

Since h is a CAT automorphism, B inherits a CAT structure making q a *CAT covering* . The map p′ is a CAT *submersion* , because p is one, while q is a CAT covering.

Each fiber of p′ is *compact* since it is a quotient of a compactum $F_u[0,r]$, r large ($r \geqslant 5(d+1)$ will do) . Also the map p′ is *closed*, in the sense that the image of every closed set is closed. *Proof* : It is enough to check that p is closed when restricted some set that maps onto B , for example $\pi^{-1}[0, 5(d+1)]$. In fact, we can show that $p \mid \pi^{-1}[a,b]$ is closed for any integers $a < b$. Being a property that is local in the target Δ , this does follow immediately from Lemma 1.3 and the compactness of the sets $F_u[a-1, b+1]$.

From this much information, we have deduced at the very outset that $p : E \to \Delta$ is a locally trivial bundle. Theorem 1.1 is now established in full generality.]] ■

Remarks on the proof of 1.1 .

The proof in the 1972-3 preprint version of this essay (cf. the 1970 sketch [KS_4]) was slightly different: The exposition of the first half was perhaps harder to understand although still briefer and more explicit. The second half gave a genuinely different construction for the covering translation, by gluing together the ends of E making E , as it were, a bandage about your finger, then identifying the natural

† We cannot expect B to be paracompact, since in general $\Delta \times [0,1]$ may not even be normal [Rud] .

infinite cyclic covering of this band with E . The shortcut was overlooked.

Next we recall a classical engulfing result that helps us apply 1.1 . Consider a connected CAT(=DIFF or PL) manifold F with a proper map $\pi : F \to R$. We are interested in the property:

(**) *Given any neighborhoods U_- and U_+ of $-\infty$ and $+\infty$ in F , one can find a CAT isotopy h_t , $0 \leqslant t \leqslant 1$, of $id \mid F$ with compact support, such that $(h_1 U_-) \cup U_+ = F$.*

Proposition 1.6 . *(CAT = DIFF or PL).*

Suppose this F is homeomorphic to a product P×R for some compact manifold P , and suppose that $dim F \neq 4 \neq dim \partial F$ (or that $dim \partial F = 4$ and ∂F has property (**)).

Then F has the property (**) , π being projection $P \times R \to R$.

Proof of 1.6 for $dim F = 3$: The 2-manifold P has a CAT structure, and so F is CAT isomorphic to P×R by a 3-dimensional Hauptvermutung (Moise), which is provable by the method of [KS_1] , cf. [Si_8, §5] [Essay V, §5.3] . ■

Proof of 1.6 for $dim F \geqslant 5$.

Stallings gave the required argument [St_2] [Hu_2] if CAT = PL . If $\partial F \neq 0$, first do the engulfing in a collar of the boundary (or use (**) there for $dim \partial F = 4$) , then engulf with compact support in intM . This proof is adapted to CAT = DIFF either by using a Whitehead triangulation of F or by doing the engulfing in just one smooth coordinate chart at each stage in the way discovered by Newman [Ne] for topological engulfing. (See the exposition in [Lu] .) The engulfing isotopy is, in either case, a finite composition of standard smooth isotopies that each engulf a linear simplex in a coordinate chart. We recall that around any simplex smoothly imbedded in a DIFF manifold one can build a smooth chart making this simplex linear. ■

Proposition 1.7 . *(CAT = DIFF) .*

Suppose F is homeomorphic to a product P×R for some compact PL manifold P , by a homeomorphism $h : P \times R \to F$ that smoothly imbeds each simplex of some PL triangulation of P×R . (Make no dimension restrictions.)

Then F has the property (**) .

Proof of 1.7 .

The Cairns-Hirsch PL-DIFF product Structure Theorem of [HiM$_{1,2}$] certainly implies this when we regard h as giving a Whitehead compatible DIFF structure Σ on the PL product $P \times R$ (see [Essay I, §5]). But there is a simpler argument as follows that we have from C. Morlet and C. Rourke (separately 1970), see [Ro$_2$] .

For given $a < b$ we find a PL triangulation of $P \times (-\infty, b]$ such that

(i) on each simplex the inclusion into $(P \times R)_\Sigma$ is a DIFF imbedding, and

(ii) there are simplicial expansions

$$\partial P \times (-\infty, a] \nearrow \partial P \times (-\infty, b]$$
$$P \times (-\infty, a] \cup \partial P \times (-\infty, b] \nearrow P \times (-\infty, b] \quad .$$

One first assures (i), then subdivides to assure (ii), see [Hu$_2$] . Given open $U \supset P \times (-\infty, a]$, it is now an elementary matter to convert the given expansions of (ii) , simplex by simplex, into a DIFF isotopy h_t , $0 \leqslant t \leqslant 1$, of $\mathrm{id} \mid (N \times R)_\Sigma$ with compact support so that $h_1(U) \supset P \times (-\infty, b]$. This clearly establishes (**) . ■

The following result concerning CAT (= DIFF or PL) structures on a product $\Delta \times M$ of a simplex with a m-manifold M will be our most basic geometric tool in deriving the classification theorems of Essay V .

BUNDLE THEOREM 1.8 . *(CAT = DIFF or PL) .*

Let Γ be a CAT manifold structure on $\Delta \times M$ sliced over the simplex (or cube) Δ . Suppose that $\dim M \neq 4 \neq \dim \partial M$ (or that $\dim \partial M = 4$ and $\Gamma \mid \Delta \times \partial M$ is of the form $\Delta \times \gamma$).

Then the projection $p_1 : (\Delta \times M)_\Gamma \to \Delta$ is a CAT bundle projection (not just a CAT submersion). Thus $(\Delta \times M)_\Gamma \cong \Delta \times (M_\gamma)$ for some γ , by a CAT isomorphism sliced over Δ .

COMPLEMENT 1.8.1 . *Let Γ be a DIFF structure on $\Delta \times M$ sliced over the simplex Δ . Suppose M is a PL manifold and Γ is*

Whitehead compatible (see [I, § 5.3]*) with the product PL structure on* $\Delta \times M$ *. Then the conclusion of 1.8 holds true (with no dimension restrictions).*

Proof of 1.8 and 1.8.1 .

We shall use 1.6 (or 1.7) in combination with the weak version of 1.1 provided by Shortcut 1.4 .

Fix a point $* \in \Delta$ and set $* \times \gamma = \Gamma | (* \times M)$. Identify $M = M_\gamma$, to make M CAT . We can assume M is sigma compact.

Filter M by compact clean CAT m-submanifolds $M_0 \subset M_1 \subset \ldots$, with $M_i \subset \mathring{M}_{i+1}$ and $\cup_i M_i = M$. For the frontiers δM_i choose disjoint open CAT bicollarings $U_i \cong \delta M_i \times R$ in M , the collar of δM_i in M_i being $\delta M_i \times (-\infty, 0]$.

From this point on, all imbeddings will be normalised to be the identity over $* \in \Delta$. This saves words.

Proposition 1.6 (or 1.7) applies to any set $u \times \delta M_i \times (a,b)$ with the structure inherited from Γ . Hence we have the hypotheses of 1.1 with $E = (\Delta \times \delta M_i \times R)_\Gamma$, the projections p and π of 1.1 being cartesian factor projections. Thus Shortcut 1.4 provides an open subset E_i' of this E containing $\Delta \times \delta M_i \times 0$ so that E_i' is a locally trivial CAT bundle over Δ . The CAT bundle-homotopy theorem † then provides bundle trivializations

$$\varphi_i : \Delta \times F_i' \xrightarrow{\cong} E_i' \subset \Delta \times U_i$$

sliced over Δ . Write M_i' for $F_i' \cap M_i = F_i' \cap \delta M_i \times (-\infty, 0]$.

Define the compact CAT submanifold W_i of $(\Delta \times M)_\Gamma$ by

$$W_i = \{ (\Delta \times M_i - E_i') \cup \varphi_i(\Delta \times M_i') \}_\Gamma$$

noting that $W_i \subset \mathring{W}_{i+1}$ and $\cup_i W_i = (\Delta \times M)_\Gamma$.

Projection $p_1 : W_i \to \Delta$ is a CAT submersion; since W_i is compact it is also a locally trivial (even trivial) CAT bundle projection. Thus we get isomorphisms $\psi_i : \Delta \times M_i \to W_i$ sliced over Δ . Given ψ_i and ψ_{i+1} we can always adjust ψ_{i+1} by the CAT Δ-parametered isotopy extension theorem [Ce_1] [P_1] [$Hu_{1,2}$] so that

† This shows that any CAT bundle over a simplex is trivial. For CAT = DIFF , see the footnote on corners in Essay V after §1.5 .

$\psi_{i+1} | \Delta \times M_i = \psi_i$. This we can arrange inductively to hold simultaneously for all i . Then we have $\psi = \lim \psi_i$ a CAT isomorphism $\psi : \Delta \times M \to (\Delta \times M)_\Gamma$ establishing the theorem. ■

§2. SLICED CONCORDANCE IMPLIES SLICED ISOTOPY

In this section the DIFF or PL Bundle Theorem 1.8 is combined with the TOP isotopy extension principle to derive a result, reinforcing 1.8, that is a sliced analogue of the Concordance Implies Isotopy Theorem of Essay I.

Data. Δ a standard simplex or cube; M^m a TOP manifold; $\epsilon : \Delta \times M \to (0,\infty)$ a continuous function; Σ a CAT (= DIFF or PL) structure on M ; Γ a CAT structure on $\Delta \times M$ *sliced* over Δ, i.e. such that the projection $(\Delta \times M)_\Gamma \to \Delta$ is a CAT submersion ; C a closed subset of M such that $\Gamma = \Delta \times \Sigma$ near $\Delta \times C$. This Γ can be described as a *sliced* Δ *- parametered concordance* constant near C .

SLICED CONCORDANCE THEOREM 2.1 . *(for above data)*

If CAT = PL , let $\Lambda \subset \Delta$ *be any contractible subpolyhedron of* Δ *(= PL retract of* Δ *).*

If CAT = DIFF , assume more conservatively that Λ *is some face of* Δ *(not* ϕ *nor* Δ *), or else is* $\partial\Delta$ *minus the interior of one principal face.*

In either case, suppose that $\Gamma | (\Lambda \times M) = \Lambda \times \Sigma$. *(If CAT = DIFF this means that* $\Gamma | (\Delta_0 \times M) = \Delta_0 \times \Sigma$ *for each face* Δ_0 *of* Δ *lying in* Λ *.)*

Suppose $m \neq 4 \neq \dim(\partial M - C)$.

With these hypotheses, there exists an ϵ*-isotopy* h_t , $0 \leqslant t \leqslant 1$, *of* $id | (\Delta \times M)$ *, sliced over* Δ *(i.e. respecting each slice* $u \times M$ *,* $u \in \Delta$ *) , to a CAT isomorphism* $h_1 : \Delta \times M_\Sigma \to (\Delta \times M)_\Gamma$ *, so that the isotopy* h_t *fixes* $\Lambda \times M$ *and a neighborhood of* $\Delta \times C$.

Remark (explaining the restrictions on Λ). The proof of 2.1 will use the following fact: *Any sliced CAT automorphism*† *h of* $\Lambda \times M_\Sigma$

† For CAT = DIFF we mean by this a self-map that restricts to a DIFF automorphism of $\Delta_0 \times M_\Sigma$ sliced over Δ_0 , for each face Δ_0 of Δ in Λ .

extends to a sliced CAT automorphism H of $\Delta \times M_\Sigma$. Furthermore this automorphism H can be the identity near $\Delta \times C$ if h is .
When Λ is a face, or CAT = PL , there is a CAT retraction $\Delta \to \Lambda$, and this serves to extend h as required. For the remaining DIFF cases recall that CAT automorphisms of $\Delta^k \times M_\Sigma$, $k \geq 0$, sliced over Δ^k and equal the identity near $\Delta^k \times C$, form a semi-simplicial group (css group). Any css group enjoys the Kan extension condition [May, p. p. 67] , which is just the statement italicized for Δ = simplex . When Δ = cube , a similar, less formal argument works; it is left as an exercise.

The following alternative version of 2.1 is sometimes more convenient.

ALTERNATE SLICED CONCORDANCE THEOREM 2.2 . *(same data)*

Let $\Lambda \subset \Delta$ be any continuous deformation retract of Δ . †
Suppose $\Gamma = \Delta \times \Sigma$ near $\Lambda \times M$, and suppose $\dim M \neq 4 \neq \dim(\partial M - C)$.

Then there exists a sliced TOP ϵ-isotopy h_t , $0 \leq t \leq 1$, of $id|(\Delta \times M)$, rel $\Lambda \times M \cup \Delta \times C$, to a CAT isomorphism $h_1 : \Delta \times M_\Sigma \to (\Delta \times M)_\Gamma$.

Remark 2.3 . Along with 2.1 and 2.2 , (or from them) one can obtain more complicated versions of each, parallel to the most general Concordance Implies Isotopy Theorem of [I, §4] , involving a closed set D in M and an open set $V \supset D$. One insists that h_t be rel $\Delta \times (M-V)$, while merely requiring h_1 to be a CAT isomorphism near $\Delta \times (C \cup D)$.

The arguments proving the Concordance Implies Isotopy Theorem in §4 of Essay I , amply show that to establish 2.1 and 2.2 (and also 2.3) , it will suffice to prove the following

HANDLE LEMMA 2.4 .

Make the hypotheses of 2.1 ⟦ or 2.2 ⟧ , but forget about the map ϵ .

Suppose further that $C = \partial M$, and let $D \subset M$ be a compact set.

† Readers familiar with shape theory will find that it is enough to suppose Λ is a compactum in Δ having the shape of a point.

Then there exists a compact support sliced isotopy h_t , $0 \leqslant t \leqslant 1$, *of* $id\,|(\Delta \times M)$, *rel* $\Delta \times \partial M$ *and fixing* $\Lambda \times M$, *such that* $h_1 : \Delta \times M \to (\Delta \times M)_\Gamma$ *is a CAT imbedding on and over a neighborhood of* $\Delta \times D$. ⟦ *If we have the hypotheses of* 2.2 , *the isotopy* h_t *can be the identity on a neighborhood of* $\Lambda \times M$ *(independent of* t *).* ⟧

Remark: The case $(M, D) = B^k \times (R^n, B^n)$, $k+n = m$, of a true handle is really no easier to prove once 1.8 is available.

PROOF OF THE HANDLE LEMMA .

To save notation we identify $M = M_\Sigma$ making M CAT .

The Bundle Theorem 1.8 provides a sliced CAT isomorphism

$$\varphi : \Delta \times M \longrightarrow (\Delta \times M)_\Gamma \quad .$$

If we have the hypotheses of 1.1 , every sliced CAT automorphism of $\Lambda \times M$ extends to a sliced CAT automorphism of $\Delta \times M$. Thus we can arrange that

(1) $$\varphi\,|\,(\Lambda \times M) = \text{identity}$$

⟦ Let us deal in brackets with the variant for the hypotheses of 2.2 .

⟦ By the device of cutting down M we can then arrange that Γ is standard (equals $\Delta \times \Sigma$) on an open neighborhood of $\Lambda \times M$ that is *uniform* , i.e. of the form $\Lambda_+ \times M$.

⟦ From $\varphi\,|\,(\Lambda_+ \times M)$ we deduce a sliced CAT automorphism Φ of $\Delta \times M$ equal to φ near $\Lambda \times M$ – as follows. Form a homotopy of $id\,|\,\Delta$ fixing a neighborhood of Λ to a continuous map $\rho : \Delta \to \Delta_+$. This uses the homotopy extension principle and our hypothesis that Λ is a deformation retract of Δ . We easily make ρ CAT . Then define

$$\Phi(u,x) = (\,u\,,\,p_2\varphi(\rho(u),x)\,) \quad , \; u \in \Delta \; , \; x \in M \; ,$$

p_2 being projection to M .

⟦ Using Φ one adjusts φ so that

(1*) φ *is the identity on a small neighborhood* $\Lambda^* \times M$ *of* $\Lambda \times M$. ⟧

Next (for the hypotheses of 2.1 *or* 2.2) we choose a CAT collaring $\partial M \times I \subset M$ of ∂M so that $\partial M \times 0 = \partial M$. Applying a uniqueness lemma for CAT collarings sliced over Δ (see [I; Appendix A.1, A.4]) we correct φ so that

(2_) *φ is a product along the collaring interval factor I, near $\Delta \times \partial M$.*

⟦ Condition (1*) can be preserved in this process. ⟧ Condition (1) can at least be retrieved after, cf. the remark below 2.1 .

Finally we shall replace (2_) by the stronger condition:

(2) *φ is the identity near $\Delta \times \partial M$.*

To accomplish this by a correction of φ in the collar $\Delta \times \partial M \times I$ of $\Delta \times \partial M$, it will clearly suffice to prove the

Lemma. *Any CAT isotopy f_t, $t \in \Delta$, of $id \mid \partial M$, equal to $id \mid \partial M$ for $t \in \Lambda$ ⟦ alternatively for all t near Λ ⟧, extends to an isotopy F_t, $t \in \Delta$, of $id \mid M$, with support in the collar of ∂M, and equal to $id \mid M$ for $t \in \Lambda$ ⟦ alternatively for all t near Λ ⟧.*

Proof (first for 2.1).

To define F_t on the collar $\partial M \times [0,1]$, choose $r_u : \Delta \to \Delta$, $0 \leqslant u \leqslant 1$, a *conditioned* † CAT deformation of $id \mid \Delta$ respecting Λ to a CAT map *into* Λ. (*Onto* is impossible if CAT = DIFF, unless Λ is merely a face.) Then for $(x,u) \in \partial M \times I$ and $t \in \Delta$, set

$$F_t(x,u) = (f(r_u(t),x), u) \in \partial M \times I \subset M ,$$

where $f_t(x)$ is written $f(t,x)$. ∎

⟦ Given the hypotheses for 2.2 , one proves the lemma rather similarly using in place of r_u a conditioned CAT homotopy ρ_u, $0 \leqslant u \leqslant 1$, of $id \mid \Delta$ rel Λ, to a CAT map ρ_1 with image $\rho_1(\Delta)$ in the neighborhood Λ^* mentioned in (1*). ⟧ ∎

At this point $\varphi : \Delta \times M \to (\Delta \times M)_\Gamma$ is a sliced CAT isomorphism equal to the identity on ⟦ or near ⟧ $\Lambda \times M$, and also equal to the identity near $\Delta \times \partial M$.

Letting the TOP isotopy extension theorem intervene (see [EK] [Si$_{12}$, §6.5]) , we deduce a sliced TOP automorphism φ of $\Delta \times M$ with compact support, that coincides with φ on and over $\Delta \times D$, and that equals the identity near $\Delta \times \partial M$, and on $\Lambda \times M$ ⟦ and even near $\Lambda \times M$ ⟧. This φ will be h_1.

† i.e., constant for u near 0 and for u near 1 .

Writing φ_2 for the component of φ on M, we define the desired isotopy h_t, $0 \leq t \leq 1$, of id | (Δ×M) to be

$$h_t(u,x) = (u, \varphi_2(r_{1-t}(u),x)) \quad , \text{ for } (x,u) \in \Delta \times M ,$$

r_t being any continuous deformation of id | Δ, fixing Λ, to a map onto Λ. This completes the proofs of the Handle Lemma 2.4 to establish the sliced concordance theorems. ■

We conclude this section with some direct applications of the Sliced Concordance Theorem(s).

We shall need a variant of 2.1 to prove a sliced concordance extension theorem; it is simple enough to be of interest in itself. (The reader can check that 2.2 has a similar variant.)

2.5 VARIANT OF 2.1 .

Let Γ and Γ′ be two sliced CAT structures on Δ×M where M is a TOP manifold, $\dim M \neq 4 \neq \dim \partial M$. Let $C \subset M$ be closed and let $\epsilon : M \to (0,\infty)$ be continuous.

Suppose that Γ = Γ′ near Δ×C, and on Λ×M, where Λ is as set out for 2.1.

Then there exists a sliced TOP ε-isotopy h_t, $0 \leq t \leq 1$, of id | M, rel Δ×C and fixing Λ×M, to a sliced CAT isomorphism $h_1 : (\Delta\times M)_{\Gamma'} \to (\Delta\times M)_{\Gamma}$.

Proof of 2.5 . The case where Γ′ is of the form Δ×Σ is covered by 2.1. But we can clearly reduce the proof to this case by finding a sliced CAT isomorphism $(\Delta\times M)_{\Gamma'} \cong \Delta\times(M_\Sigma)$ that is sufficiently near id | M. Such an isomorphism is provided by applying 2.1 with Λ = point. ■

SLICED CONCORDANCE EXTENSION THEOREM 2.6 .

Let Γ be a sliced CAT (= DIFF or PL) structure on Δ×M, where M is a TOP manifold, $\dim M \neq 4 \neq \dim \partial M$.

Let Γ^* be a sliced CAT structure on $\Delta\times M^*$ where M^* an open subset of M. Suppose that Γ equals Γ^* on $\Lambda\times M^*$ where $\Lambda \subset \Delta$ is as set out for 2.1.

Then Γ^* extends to a sliced CAT structure Γ^+ on Δ×M equal to Γ on Λ×M.

Proof of 2.6 . The variant 2.5 provides a sliced CAT isomorphism $h : (\Delta \times M^*)_\Gamma \to (\Delta \times M^*)_{\Gamma^*}$ that is ϵ-near the identity for a continuous map $\epsilon : M \to [0,\infty)$ chosen to be positive on M^* and zero on $M - M^*$. Then h extends to a sliced homeomorphism $H : \Delta \times M \to \Delta \times M$ equal to the identity outside $\Delta \times M^*$, and we can set $\Gamma^+ = H(\Gamma)$. ■

Remarks.
1) This theorem establishes a fibration property for spaces of structures, that has a key role in Essay V .
2) Since this result has the appropriate sharply relative form, Zorn's lemma extends it immediately to manifolds that are non-metrizable (and even non-Hausdorff) .

The last application is a CAT approximation theorem for isotopies that should prove useful in geometric problems.

ISOTOPY APPROXIMATION THEOREM 2.7 .

Let M^m and W^m be CAT (= DIFF or PL) manifolds with $m \neq 4 \neq \dim \partial M$. Let $H : \Delta \times M^m \to \Delta \times W^m$ be a TOP open imbedding, sliced over Δ , a simplex or cube . (This H is to be regarded as a Δ-parametered TOP isotopy through open embeddings). Suppose that H is CAT near $\Delta \times C \cup \Lambda \times M$ where C is closed in M and Λ is a (continuous) retract of Δ .

Then, for any continuous function $\epsilon : M \to (0,\infty)$ we can find a sliced TOP ϵ-isotopy H_t , $0 \leq t \leq 1$, of $H = H_0$ rel $\Delta \times C \cup \Lambda \times M$ running through open imbeddings onto Image H to a CAT open imbedding $H_1 : \Delta \times M \to \Delta \times W$.

The ϵ-smallness condition on H_t asserts that $d(H_t(x), H(x)) < \epsilon(p_2(x))$ for all points x in $\Delta \times M$ and t in $[0,1]$, the metric d being a standard product metric on $\Delta \times W$.

Proof of 2.7 . Simply apply 2.2 to the sliced CAT structure Γ making $H : (\Delta \times M)_\Gamma \to \Delta \times W$ a CAT imbedding. ■

Remarks.
1) This theorem clearly has a slightly different version derived from 2.1 instead of 2.2 .
2) E.H. Connell [Cnl_1] invented radial engulfing to prove the first theorem of this sort .

Essay III

SOME BASIC THEOREMS ABOUT TOPOLOGICAL MANIFOLDS

by

L. Siebenmann and R. Kirby

Section headings in Essay III

§0. INTRODUCTION

The impressive array of existing classification theorems concerning differentiable C^∞(= DIFF) manifolds of finite dimension ≥ 5 are for the most part derived, by methods involving algebraic topology, from a handful of geometric 'tool theorems'. For example, the geometric basis of cobordism classifications is Thom's transversality theorem.

To permit a parallel development of classification theorems for metrizable topological (= TOP) manifolds of finite dimension ≥ 5 we here present proofs of topological analogues of some of the most important tool theorems of differential topology. (Our choice is indicated by the list of section headings opposite.) They are the basic tools required for TOP versions of what are known as cobordism theory, handlebody theory, and surgery. For a sampling of the TOP classification theorems that can be developed with these tools see [BruM] [Si_{10}] [Wa].

Each of our proofs relies heavily on the Product Structure Theorem and/or the Concordance Implies Isotopy Theorem (as proved in Essay I). Typically they serve to lead back a given proof to (relative) applications of the corresponding DIFF theorem. For a quick introduction to this procedure, read from § 6 to § 9 in [Si_{10}].

The results of Essay I just mentioned, were proved using only DIFF handlebody theory and the Stable Homeomorphism Theorem of Kirby [Ki_1]. The dependance of [Ki_1] on the full machinery of non-simply connected surgery [Wa] thus mars our otherwise relatively elementary line of argument. Hence we indicate in Appendix A to what extent this dependence on DIFF † surgery can be eliminated.

The maternal role played by differential topology in this development could be taken over by piecewise linear topology (a willing step-

† That DIFF surgery can replace the traditional PL surgery in proving the Stable Homeomorphism Theorem of [Ki_1], is shown by [V, Appendix B].

mother, so to speak) . † Indeed we often take the trouble to give the required piecewise linear (= PL) arguments in parallel.

Occasionally we do yield to considerations of greater technical convenience and work only from PL or only from DIFF principles. For example we have discussed handle decompositions using PL principles only and TOP Morse functions using DIFF principles only. ‡

Since this essay was first written, the use of Hilbert cube manifolds by T. Chapman, R. Miller, and J. West has brought about a triumphant extension of simple homotopy theory to arbitrary locally compact metrisable ANR's (= absolute neighborhood retracts). See [Ch_2] and other references given in footnotes added to §4 and §5 . Our §4 merely extends simple homotopy theory to topological manifolds. If the reader therefore bypasses §4 , he should probably still read §5 , generalizing the results proved there by replacing topological manifolds where possible ǂ by Hilbert cube manifolds.

This essay will by no means offer all the tools that the study of topological manifolds will demand, and we can only exhort our fellow geometers to fashion new tools as fresh needs arise.

† Beware that the PL surgery used in [Ki_1] requires $\pi_5(G/PL) = 0$, which in turn requires the 5-dimensional PL Poincaré theorem. Now recall that this last depends on very difficult differential topology, indeed J. Cerf's intricate proof that $\Gamma_4 = 0$ [Ce_4] is required to compatibly smooth any PL homotopy 5-sphere, cf. [Mu_3] [Mi_8] . This complication involving [Ce_4] does not arise in applying DIFF surgery .

‡ We must confess that the more refined transversality theorems of § 1 , relying on the PL Appendix B , still seem inaccessible from the DIFF viewpoint.

ǂ Where Poincaré duality is concerned no generalization is possible.

§1. TRANSVERSALITY

If M and N are two affine linear subspaces of an euclidean space R^n, the subspace M can always be perturbed slightly so that M and N meet transversally in the sense that $M \cap N$ is an affine subspace of dimension $\dim M + \dim N - n$. Statements of a similar flavor concerning manifolds are of vital importance both for classification of manifolds and for applications of manifolds in other realms. This long section treats several transversality problems for TOP manifolds, in order of roughly increasing generality and increasing difficulty.

First (Theorem 1.1) we discuss transversality of a *map* to a subspace of its target when the subspace is equipped with a normal *microbundle*. This is a TOP version of Thom's DIFF transversality theorem; indeed it will follow from a simple case of Thom's by an application of the Product Structure Theorem.

We then treat the problem of making submanifolds of a given manifold meet transversally.

Initially one of the two submanifolds is equipped with a normal microbundle of Milnor. To obtain a TOP transversality result (Theorem 1.5) here, we find it convenient to use a codimension $\geqslant 3$ straightening theorem (stated in Appendix B) to help reduce to the corresponding PL result 1.6, which we then easily prove 'by hand'.

Such a TOP transversality result would seem at first sight to be of limited interest – although it is pleasant enough to state and not unreasonably difficult to prove – simply because normal bundles for TOP submanifolds often fail to exist or fail to be (isotopy) unique (see [RS_4], [Stern]). On the contrary, the viewpoint that evolved in discussion with A. Marin maintains that it should (or at least can) play the central role. This is well-illustrated already in the PL category where we observe (1.8) that the Rourke-Sanderson imbedded block bundle transversality theorem follows from the PL imbedded microbundle transversality theorem – which, in turn, we have deduced from the PL analogue of the Sard-Brown theorem. In the TOP category A. Marin has somewhat similarly obtained a general transversality theorem for submanifolds through replacing block bundle transversality by a stable microbundle transversality . (Recall that *stably* normal microbundles do exist and are isotopy unique , cf. [IV , Appendix A] .) We conclude by introducing Marin's work , which is still too difficult to present here in full .

In Appendix C we whall present an ad hoc transversality lemma for immersions designed to adapt directly smooth surgery methods to TOP manifolds . Where it applies many readers will prefer it to the more ambitious approach of this section .

We draw the reader's attention to two **general position** theories , perhaps as important as the transversality results we present here . One for polyhedra in codimension $\geqslant 3$ is mentioned in Appendix B (conclusion) ; another for arbitrary (!) closed subsets is mentioned in Appendix C (see C.3 and [Ed_4]) .

MICROBUNDLE TRANSVERSALITY FOR MAPS

Recall that a TOP n-microbundle ξ^n over a space X can be defined as a total space $E(\xi) \supset X$ together with a retraction $p : E(\xi) \to X$ that , near X , is a submersion whose fibers $p^{-1}(x)$, $x \in X$, are (open) n-manifolds .

A DIFF n-microbundle over a manifold is defined similarly working within the category DIFF of smooth maps of smooth manifolds . Similarly a PL microbundle over a polyhedron .

Consider a pair (Y,X) of topological spaces where X is closed in Y and equipped with a normal microbundle ξ^n , i.e. $E(\xi^n)$ is an open neighborhood of X in Y .

Consider also a continuous map $f : M^m \to Y$ from a TOP m-manifold to Y . Suppose $f^{-1}(X)$ is a TOP submanifold $L \subset M^m$ and ν^n is a normal n-microbundle to L in M such that $f \mid E(\nu)$ is a TOP microbundle map to $E(\xi)$ (i.e. f gives an open TOP imbedding of each fiber of ν into some fiber of ξ) . Then we say that f is (TOP) **transverse to** ξ **(at** ν **)** . We say f is transverse to ξ **on** U open in M , if $f \mid U$ is transverse to ξ . Transverse **near** C means transverse on an open neighborhood of C .

If M^m and ξ^n are DIFF we define DIFF transversality of f at a DIFF normal microbundle ν for $L = f^{-1}(X)$. This means that $f \mid E(\nu)$ is by assumption a DIFF microbundle map to $E(\xi)$. We do not however assume $f : M \to Y$ is DIFF ; it is convenient to allow it to be merely continuous outside $E(\nu)$. Similarly PL transversality .

Let C and D be closed subsets of a (metrizable) TOP m-manifold M^m, and let U and V be open neighborhoods of C and D respectively. Let ξ^n be a normal n-microbundle to a closed subset X of a space Y.

FIRST TRANSVERSALITY THEOREM 1.1.

Suppose $f : M^m \to Y$ *is a continuous map* TOP *transverse to* ξ † *on* U *at* ν_0. *Suppose* $m \neq 4 \neq m-n$, *and either* $\partial M \subset C$ *or* $m-1 \neq 4 \neq m-1-n$.‡

Then there exists a homotopy $f_t : M \to Y$, $0 \leqslant t \leqslant 1$, *of* $f_0 = f$ *fixing a neighborhood of* $C \cup (M-V)$ *so that* f_1 *is transverse to* ξ *on an open neighborhood of* $C \cup D$ *at a microbundle* ν *equal* ν_0 *near* C. *Furthermore, if* Y *is a metric space with metric* d, *and* $\epsilon : M \to (0, \infty)$ *is continuous, then we can require that* $d(f_t(x), f(x)) < \epsilon(x)$ *for all* $x \in M$ *and all* $t \in (0,1]$.

There is (in 1974) no reason to believe that the above dimension restrictions are necessary. ‡ This question is in tight connection with the question whether Rohlin's theorem (on index of almost parallelizable closed 4-manifolds) holds for TOP, see [Si_8, §5] [Mat] [Sch].

We will say no more about the ϵ-smallness condition. It can be carried through the proof, beginning with an application of Sard's theorem to get a similar DIFF result involving ϵ. On the other hand one can always deduce the ϵ-condition from the version without ϵ, by exploiting the strongly relative nature of the theorem. Cf. [I, §4] and [I, Appendix C].

The idea of our proof is to use the Product Structure Theorem of [I] to reduce the proof to chart-by-chart applications of the following easy

Theorem 1.2 (DIFF Transversality).

Consider $f : M^m \to R^n$ *a continuous map of a* DIFF *manifold* M; C,D *closed subsets of* M; U,V *open neighborhoods of* C, D *respectively.*

If f *is* DIFF *transverse on* U *to* 0 (i.e. *to the trivial microbundle* $0 \hookrightarrow R^n \to 0$) *at* ν_0, *then there exists a homotopy* $f_t : M \to R^n$, $0 \leqslant t \leqslant 1$, *of* $f = f_0$ *fixing a neighborhood of* $C \cup (M-V)$ *so that* f_1 *is* DIFF *transverse near* $C \cup D$ *to* 0 *at a* DIFF *microbundle* ν *equal to* ν_0 *near* C.

† Beware of saying transverse to X (ommiting ξ), see discussion above 1.4.
‡ The case $m-n \leqslant 0$ is trivial to prove, even if $m = 4$, (D. Epstein).

Proof of 1.2. This theorem would follow easily from the Sard-Brown theorem [Mi_8] , if we had
(1) assumed f everywhere DIFF , and
(2) required for transversality near $S \subset M$ only differentiability near $f^{-1}(0) \cap S$ and surjectivity of the differential df_1 of f_1 near $f^{-1}(0) \cap S$.

To see the difficulty related to (1) note that f is not necessarily DIFF near C . So we cannot simply make f DIFF everywhere on V without changing it near C . The difficulty is not serious . Let $C' \subset \mathring{U}$ be a closed neighborhood of C in M . Alter f on $E(\nu_0) \cup (V-C') \equiv V'$ by a homotopy f_t , $0 \leqslant t \leqslant 1$, to make it first DIFF on V' [Mu_1 , §4] , then transverse to 0 on V' , cf. [Mi_8] , in the sense of (2) . Neither change need alter f near C (because the transversality condition (2) is stable for the fine C^1 topology) , nor outside V' (since we can perturb less than ϵ where ϵ is a continuous function $M \to [0,\infty)$ with $\epsilon^{-1}(0,\infty) = V'$) .

Now one sees that f_1 is transverse to 0 in the sense of (2) on $\mathring{C}' \cup V \supset C \cup D$. Thus to attain DIFF transversality as we defined it , we need only to equip the DIFF manifold $L = f^{-1}(0) \cap (\mathring{C}' \cup V)$ with a DIFF normal microbundle ν equal to ν_0 near C . This amounts to simply finding a DIFF neighborhood retraction $L \subset E \xrightarrow{r} L$, equal to the projection r_0 of ν_0 near C , since any such r is a DIFF submersion near L by the implicit function theorem . ∎

Another tool we need is the

Pinching Lemma 1.3.

Let ν^n be a microbundle over a paracompact space L . *Consider* D *closed in* L *and a neighborhood* Z *of* D *in* $E(\nu)$. *There exists a homotopy* ρ_t , $0 \leqslant t \leqslant 1$, *of* $\mathrm{id} \mid E(\nu)$ *respecting fibers of* ν *and fixing all points outside* Z *and all points in* L *so that* $\rho_1^{-1} L$ *is a neighborhood of* D .

Proof of 1.3. First suppose $E(\nu) = R^n$, $Z = B^n$ and $C = L = 0 \in R^n$. Then the construction of ρ_t is trivial .

Second suppose $E(\nu)$ is an open sub-microbundle of $L \times R^n$ with $L = L \times 0$. It is this case we shall use for 1.1 . Let $\zeta : L \to [0, \infty)$ be a continuous function positive on D so that $\{ (x,y) \in L \times R^n ; |y| \leqslant \zeta(x) \} \subset Z$. Let $\eta : L \to [0, 1]$ be 1 near

D and zero near $\zeta^{-1}(0)$. Writing σ_t for the homotopy obtained for the first case we can define ρ_t on all $L \times R^n$ by

$$\rho_t(x,y) = (x, \zeta(x)\sigma_{\eta(x)t}(\zeta(x)^{-1}y))$$

if $\zeta(x) > 0$ and by $\rho_t(x,y) = (x,y)$ otherwise. (The continuity is obvious because of η).

In the general case ρ_t can be an infinite but locally finite composition of homotopies obtained by the second case. We leave this unused generality to the reader. ∎

PROOF OF 1.1

Step 1. *The case* M *open in* R^m, $Y = E(\xi) = R^n$ *and* $X = 0$.

Proof: Step 1 follows immediately from the DIFF transversality theorem 1.2 as soon as we find a DIFF structure Σ' on M such that for some open neighborhood N of $f^{-1}(0) \cap C$, the microbundle $\nu_o \cap N : E(\nu_o) \cap N \to L_o \cap N$ is DIFF and $f : M_{\Sigma'} \to R^n$ is DIFF transverse to 0 near C at $\nu_o \cap N$. Here L_o is the base space of ν_o.

To find Σ' we apply the Local Product Structure Theorem [I, §5.2], to $E(\nu_o)$ with the structure Σ inherited from $M \subset R^m$. For this we regard $E(\nu_o)$ canonically as an open sub-microbundle of $L_o \times R^n$ by the rule $E(\nu_o) \ni x \leftrightarrow (p(x), f(x)) \in L_o \times R^n$.

The appropriate substitutions into [I, §5.2] are

$$(M \times R^s, W, \Sigma, D, V) \leftrightarrow (L_o \times R^n, E(\nu_o), \Sigma, C \cap L_o, V')$$

where V' is an open neighborhood of C in $E(\nu_o)$, whose closure in $E(\nu_o)$ is closed in M. Thus [I, §5.2] provides Σ' on $E(\nu_o)$ which extends (by Σ) outside $V' \subset M$ to Σ' on $M \subset R^m$ as required.

If $m - n = 3$ and L_o has closed compact components, [I, §5.2] does not immediately apply. We can quickly recast the argument however by first deleting from M the closed set F consisting of all compact components of L_o meeting C. Then enlarge C to C′ by adding a small closed neighborhood G of F and cut back U to an open neighborhood U′ of C′ such that all compact components of $f^{-1}(0) \cap U'$ lie in F. Then we can repeat the *whole* argument with M−F, C′−F, and U′−F in place of M, C, and U, getting the wanted homotopy without encountering invalid cases of [I, §5.2]. ∎

Remark for $m-n \leqslant 3$: Even the nonstandard cases of the Product Structure Theorem [I , §5] used in the above proof , viz. for structures on $L^k \times R^n$, $k+n \geqslant 5$, $k \leqslant 3$, with $\partial L = \phi$ and L lacking compact components if $k = 3$, can be proved without recourse to a microbundle classification of structures , by solving a sequence of handle problems in dimension $n + k$ corresponding to the critical points of a proper Morse function on L lacking critical points of index 3 . See [KS_1] and [Si_{11}] .

Step 2 . *The case* M *open in* R^m , $Y = E(\xi)$, *and* ξ^n *a standard trivial bundle (over any space* X *)*.

Proof : ξ^n is $X \overset{\times 0}{\rightarrow} X \times R^n \overset{p_1}{\rightarrow} X$. So we can write $f : M \rightarrow E(\xi) = X \times R^n$ componentwise $f = (f_1 , f_2)$. By step 1 there is a homotopy rel $C \cup (M-V)$ from $f_2 : M \rightarrow R^n$ to a map f_2' which is transverse to 0 on a neighborhood W of $C \cup D$ at a normal microbundle ν to $L = f_2'^{-1}(0) \cap W$ so that $\nu = \nu_o$ on a neighborhood $U_1 \subset U$ of C . We can assume $W \subset U_1 \cup V$.

The resulting homotopy from $f = (f_1 , f_2)$ to $f' = (f_1 , f_2')$ does not solve our problem since f_1 is perhaps not constant on the fibers of ν near L . To remedy this , find a neighborhood Z of $L \cap (C \cup D)$ in $E(\nu)$ which is closed in M , and , noting that $E(\nu) \subset L \times R^n$ by $x \mapsto (p(x) , f_2'(x))$, apply the Pinching Lemma 1.3 to obtain a fiber preserving pinching homotopy ρ_t , $0 \leqslant t \leqslant 1$, of $id \mid E(\nu)$ which extends by the identity outside Z to all M . Then $(f_1 \rho_t , f_2')$, $0 \leqslant t \leqslant 1$, is a homotopy of f' to $f'' = (f_1 \rho_1 , f_2')$. If $N \subset Z$ is a neighborhood of $L \cap (C \cup D)$ so that $\rho_1(N) \subset L$, the map $f'' = (f_1 \rho_1 , f_2')$ is transverse to ξ at $\nu \cap N$ since $f_1 \rho_1$ is constant on fibers of $\nu \cap N$. The homotopy f' to f'' is constant near C since f' is already constant on fibers of $\nu \cap U_1$ (where $f' = f$ and $\nu = \nu_o$) . Hence the homotopy f' to f'' has support in $Z - U_1 \subset W - U_1 \subset V$. (The *support* is the closure of the set of points moved) . Recall that $W \subset U_1 \cup V$ by choice .

The composed homotopy f to f' to f'' establishes Step 2 . ■

Now we let Y grow larger than $E(\xi)$.

Step 3 . *The case* M^m *open in* R^m , *and* ξ^n *a trivialized bundle* , i.e. $E(\xi^n)$ *contains* $X \times R^n$ *as an open sub-microbundle* .

Proof: Apply Step 2 with the substitutions

Proof: Apply Step 2 with the substitutions

$$M \leftrightarrow f^{-1}(X \times R^n),$$

$$C \leftrightarrow \{C \cup f^{-1}(X \times (R^n - \mathring{B}^n))\} \cap f^{-1}(X \times R^n),$$

$$D \leftrightarrow D \cap f^{-1}(X \times R^n),$$

etc. to obtain a homotopy which extends as the constant homotopy outside $f^{-1}(X \times R^n)$ to the required homotopy to transversality. The normal microbundle ν obtained may at first be smaller than ν_o near C if $X \times R^n \neq E(\xi)$, but this is trivial to remedy by adding on to this ν the restriction of ν_o to a small neighborhood of C. ∎

Next we allow ξ to be arbitrary and M to be an open manifold.

Step 4. *The case* $\partial M = \phi$.

Proof: Let X be covered by open sets X_α, α in some index set, where ξ^n is trivialized over each X_α, i.e. a microbundle map $X_\alpha \times R^n \hookrightarrow E(\xi)$ is given extending the inclusion $X_\alpha \times 0 = X_\alpha \hookrightarrow E(\xi)$.

There is a locally finite collection of co-ordinate charts in M, R^m_j, $j = 1, 2, \cdots$ such that $D \subset \cup_j B^m_j$, each $R^m_j \subset V$, and each set $p\{f(R^m_j) \cap E(\xi)\} \subset X$ lies in some set X_α, which will be denoted X_i.

Suppose now for a construction by induction on $i \geqslant 0$ that we have constructed a continuous map $f'_i : M \to Y$ transverse to ξ on an open subset $U_i \subset M$, at ν_i, where $U_i \supset C_i \equiv C \cup \{B^m_1 \cup \cdots \cup B^m_i\}$. At $i = 0$, we begin with $U_0 = U$ and $f'_0 = f$. Apply Step 3 with the substitutions $M \leftrightarrow R^m_{i+1}$; $C \leftrightarrow C_i \cap R^m_{i+1}$; $D \leftrightarrow B^m_{i+1}$; $V \leftrightarrow 2B^m_{i+1}$; $Y \leftrightarrow (Y - E(\xi)) \cup E(\xi | X_{i+1})$; $\xi \leftrightarrow \xi | X_{i+1}$; $f \leftrightarrow f'_i | R^m_{i+1}$. There results a homotopy f'_t, $i \leqslant t \leqslant i+1$, constant outside R^m_{i+1}, an open neighborhood U_{i+1} of C_{i+1}, and a microbundle ν_{i+1}, such that f'_{i+1} is transverse to ξ on U_{i+1} at ν_{i+1}, while U_{i+1} and ν_{i+1} coincide outside R^m_{i+1}. This completes the induction to construct f'_t, $0 \leqslant t < \infty$, U_i and ν_i for $0 \leqslant i \in Z$.

To complete the proof of Step 4, we have only to let f_t, $0 \leqslant t \leqslant 1$, be the unique conditioned homotopy so that $f_t = f'_{\alpha(t)}$, $0 \leqslant t < 1$, where $\alpha(t) = t/(1-t)$, and near any compact set let ν equal ν_j for j large enough. ∎

Step 5 . *The general case* .

Proof : We are now allowing M a non-empty boundary . One simply applies Step 4 twice , once to ∂M , trivially if $\partial M \subset C$, (hence the condition $m - 1 \neq 4 \neq m - 1 - n$, or $\partial M \subset C$) ; then once again to intM . To prepare the application to intM one uses a collaring $\partial M \times [0, 1)$ of $\partial M \times 0 = \partial M$ in M so chosen that ν_0 coincides near $\partial L_o \cap C$ with $(\nu_o \mid \partial L_o) \times [0, 1)$ under this collaring . In providing such a collaring one should apply the bundle homotopy theorem to the restriction of ν_o to a collar of ∂L_o in L_o , and then apply relative collaring theorems . ■

MICROBUNDLE TRANSVERSALITY OF SUBMANIFOLDS

Two CAT submanifolds U^u and V^v (all manifolds without boundary) in a CAT manifold W^w are **locally** CAT **transverse** if the triad (W ; V , V′) is locally CAT isomorphic to a triad $(R^w; L^u, L'^v)$ given by transversally intersecting affine linear subspaces with $\dim(L^u \cap L'^v) = u + v - w$.‡ J.F.P. Hudson [Hu_3] (cf. [RS_5]) established the disturbing fact that , for CAT = PL or TOP , there exists no relative transversality theorem involving this notion : If U and V are locally transverse near a closed set $C \subset W$ it is in general *impossible* to move U and/or V relative to C (or even alter them rel C) to make U and V everywhere geometrically transverse . In his examples U and V are certain topologically unknotted euclidean spaces of codimension $\geqslant 3$ and closed in $R^w = W$, with $(u + v) - w = 4k + 1$, $k \geqslant 1$, while C is a neighborhood of ∞ .

In the PL category there is nevertheless a perfectly satisfactory relative transversality theorem due to Rourke and Sanderson [RS_0,II] ,

‡ For manifolds with boundary we use a model of the form $(R^w ; L^u , L'^v) \times [0,\infty)$.

which we will reprove as 1.8 . It involves a more strigent (non-local) notion of transversality called block-transversality .

In the topological category , the theorem which we are led to prove first is analogous to that of Rourke and Sanderson ; but (TOP) microbundles replace (PL) block bundles .

DEFINITION 1.4.

Consider a CAT map $f : M^m \to Y$ and a CAT normal n-microbundle ξ^n to X in Y as given for our definition of transversality of f to ξ above 1.1 . We suppose that (Y,X) is a CAT manifold pair and that f is a proper inclusion $f : M \hookrightarrow Y$ onto a (clean) CAT submanifold M = fM of Y .

We shall call M CAT **imbedded-transverse** to ξ in Y if $f : M \hookrightarrow Y$ is CAT transverse to ξ^n (see 1.0), and $M \cap X$ is a CAT submanifold of X (and hence also of Y) . Note that M is then CAT *locally* transverse to X in Y .

We say M is CAT imbedded-transverse to ξ in Y *near* $C \subset Y$ if , for some open neighborhood Y_o of C in Y , one finds $Y_o \cap M$ imbedded-transverse to $\xi \cap Y_o$ in Y_o .

TOP IMBEDDED MICROBUNDLE TRANSVERSALITY THEOREM 1.5.

In this situation , let M^m be imbedded-transverse to ξ^n near a closed subset $C \subset Y$, and let $D \subset Y$ be a closed set . Suppose M , X and Y without boundary . Suppose also $m \neq 4 \neq m-n$ (as for Theorem 1.1) , and either $\dim Y - \dim M \geqslant 3$ or $\dim Y \neq 4 \neq \dim X$. †

Then there exists a homotopy f_t, $0 \leqslant t \leqslant 1$, of $f : M \to Y$ rel C so that :

(a) $f_1(M)$ is imbedded-transverse to ξ near $C \cup D$,

(b) the homotopy f_t, $0 \leqslant t \leqslant 1$, is realized by an ambient isotopy F_t of $id \mid Y$, i.e. $f_t = F_t f$, and

(c) the ambient isotopy F_t is as small as we please with support in a prescribed neighborhood of D − C in Y .

Results for manifolds with boundary can be deduced with the help of collaring theorems . (See last step of proof of 1.1 .)

†The case $m - n = 0$, $\dim Y \neq 4$, partially excluded in 1.5 , can be proved easily with the help of the Stable Homeomorphism Theorem of [Ki_1] , see [Si_{10}, § 7.2] . The case $m - n < 0$ is covered by Hommas's method [Hom] , or by general position (see Appendix C) .

Just as for 1.1 , the dimension conditions may prove unnecessary .

The parallel DIFF and PL theorems hold true (without the dimension conditions) ; the PL version is proved below as 1.6 , and the DIFF one similarly .

The pattern of proof for the given (TOP) version is as for 1.1 , but it is the PL version we fall back on ; the DIFF version would appear to leave us with a metastability condition $(m-n) \leqslant 2q-3$ on the co-dimension $q = \dim Y - \dim M$, instead of $q \geqslant 3$. The condition $q \geqslant 3$ is used to straighten a PL piece of M in a PL piece of Y ; this is nontrivial (cf. Appendix B) and contributes to make the proof more 'expensive' than for 1.1 .

PROOF OF THEOREM 1.5 .

We perform a cumulative sequence of normalizations until we reach a case that is clearly implied by the PL version . Each successive normalization (α) , (β) , . . . is added *without loss (of generality)* – in the sense that the general case remains a consequence .

(α) *Without loss , D is compact .*

(β) *Without loss , we can leave aside the smallness conditions (c) on the isotopy F_t if instead we make it have compact support in Y .*

(γ) *Without loss , M and X are PL (indeed have a single chart) .*

(δ) *Without loss , Y is open in a PL product $X_+ \times R^n$ and $X = Y \cap (X_+ \times 0)$ while ξ^n is the inherited trivial normal microbundle to X .*

The normalizations (α) – (δ) are clearly possible because of the strongly relative nature of the theorem . (We could in fact have $Y = X \times R^n$ without loss , but that would soon become a nuisance .)

(ϵ) *Without loss , M is a PL product along R^n near $C \cap X$, where M is already transverse to ξ .*

Condition (ϵ) is realized through altering the PL structure on M using the (local) Product Structure Theorem [I , §5.2] , which requires $m \neq 4 \neq m-n$ (and for $m-n = 3$ the device of neglecting compact transverse dim 3 intersection components , as for 1.1 , Step 1) .

At this point one has an open subset L_o of $M \cap X$ containing $(M \cap X) \cap C$ with PL manifold structure so that , near L_o , M coincides as PL manifold with $L_o \times R^n$.

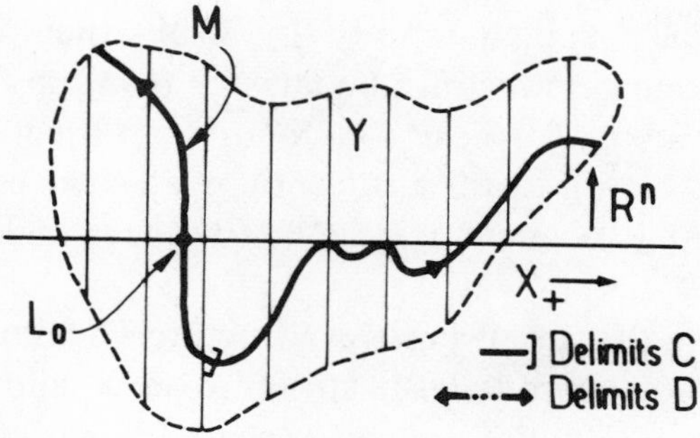

Figure 1-a

(ζ) *Without loss, $L_o \hookrightarrow X$ is PL and locally flat (although not closed).*

(η) *Without loss, $M \hookrightarrow Y$ is PL and locally flat.*

With the normalizations (α) – (η) the theorem obviously follows from its PL version (1.6 below). It remains to verify (ζ), (η), and for this we distinguish two cases.

Codimension M $\geqslant$ 3.

In case dim Y – dim M $\geqslant$ 3, we realise (ζ) and (η) without loss, using the codimension $\geqslant$ 3 straightening theorem first proved by R.T. Miller (see Appendix B for the precise statement). A first application alters the PL structure on X to realize (ζ). Then $M \hookrightarrow Y$ is PL near $C \cap X$. A second application alters $M \hookrightarrow Y$ by a compact support ambient isotopy of Y rel C so that $M \hookrightarrow Y$ is PL locally flat near $D \cap X$. By then cutting back Y (and hence X etc.) to a sufficiently small open neighborhood of $D \cap X$, we realize (η) without loss, as well. This proves 1.5 for dimY – dimM $\geqslant$ 3. ■

Codimension M $\leqslant$ 2.

In case q = (dimY – dimM) $\leqslant$ 2 and dimY $\neq 4 \neq$ dimX, we shall realise (ζ) and (η) by building a normal bundle to M and exploiting the Product Structure Theorem, as follows. The PL structure on M promised by (γ) should be discarded now.

In codimension q $\leqslant$ 2, there is a relative existence theorem for normal R^q-bundles; for q = 2 it requires ambient dimension $\neq 4$ as the proof in [KS_5] is based on handlebody theory and torus geometry ; for $q = 1$ it follows trivially from collaring theorems ; and for $q \leqslant 0$ it is vacuous . Applying this to L_o in X , then to M in Y rel $C \cap X$ we obtain a normal R^q-bundle ν^q to M in Y such that ν^q is a product along R^n near $C \cap X$. As our theorem normalized by (α) , . . . ,(ϵ) is still sufficiently relative we can assume *without loss* that $M \approx R^m$; then ν^q can be trivialized making $E(\nu) \approx M \times R^q$.

This lets us apply the (local) Product Structure Theorem twice .

First we use it to concord the PL structure on X and on X_+ so that $L_o \hookrightarrow X$ is PL locally flat, for some (new) PL structure on L_0 .

Next we use it to find a concordance rel $C \cap X$ of the structure just obtained on $Y \subset X_+ \times R^n$ so that $M \hookrightarrow Y$ becomes PL locally flat for some (new) PL structure on M while M coincides PL with $L_o \times R^n$ near $C \cap X$.

To this second concordance we apply the Concordance Implies Isotopy Theorem [I, §4.1] to produce a compact support ambient isotopy of Y rel C making $M \hookrightarrow Y$ locally flat PL near $D \cap X$. (We don't alter the structure on $Y \subset X_+ \times R^q$ at this point) .

Finally we cut back Y to a sufficiently small open neighborhood of $D \cap X$ in $X_+ \times R^q$ to realise (ζ) and (η) without loss of generality . This closes the proof of 1.5 for $\dim Y - \dim M \leqslant 2$. ■

For completeness we present the

PL IMBEDDED MICROBUNDLE TRANSVERSALITY THEOREM 1.6.

The transversality theorem 1.5 holds true in the PL category without any restrictions concerning dimension .

The proof is based on the following simplicial lemma , which will play the role the Sard-Brown theorem had in 1.1 .

Lemma 1.7. (well known) *Every simplicial map* $f : K \to \Delta^d$ *of a finite complex onto the standard d-simplex is a PL (trivial) bundle map over* $\operatorname{int}\Delta^d$. *There are standard bundle charts which respect every subcomplex of* K .

Proof of 1.7. Topologically this is clear since K is canonically a subcomplex of the join

$$K_0 * K_1 * \dots * K_d , \qquad K_i = f^{-1}(v_i) ,$$

where $v_0, \dots, v_d$ are the vertices of Δ^d . For the PL version we can clearly assume that K is all of this join. Then there is a canonical homeomorphism

$$\psi : (\mathrm{int}\,\Delta) \times K_0 \times \ldots \times K_d \to f^{-1}(\mathrm{int}\,\Delta)$$

such that $f\psi$ = projection, but it is not PL, by the standard mistake (even if f is $\Delta^2 \to \Delta^1$). If Δ_0 is any closed linear d-simplex in Δ we form the standard simplicial subdivision of the regular convex linear cell complex $\Delta_0 \times K_0 \times \ldots \times K_d$ and replace ψ on this by the simplex-wise linear map φ equal ψ on the vertices. Then

$$\varphi : \Delta_0 \times K_0 \times \ldots \times K_d \to f^{-1}(\Delta)$$

is a typical standard PL bundle chart respecting each subcomplex of K. ∎

PROOF OF 1.6.

Without loss (of generality) we normalize imposing the conditions (α), (β) and (δ) used in the TOP proof. Further normalization is required.

(i) *Without loss, we can omit the condition that* f_t, F_t *be rel* C. (Recall that $F_1 M$ is required to be transverse to ξ^n near $C \cup D$).

To see this choose a small closed neighborhood C′ of C in M near which M is transverse to ξ^n and apply the new version after replacing C by C′ and D by $D - \mathring{C}'$, and then delete C from Y (and from everything else).

We can now choose a compact polyhedral neighborhood K of D in Y and endow it with a finite PL triangulation such that $M \cap K$ is a subcomplex and the projection $p_2 | K$ to R^n is simplicial for some triangulation of R^n. Then we can replace Y by the interior $\mathring{K}$ of K in Y establishing via the simplicial lemma 1.7 that:

(ii)′ *Without loss, over the interiors of the open n-simplices of a triangulation of* R^n, *the PL projection* $Y \xrightarrow{p_2} R^n$ *is a PL bundle with charts that respect* M.

We would like such a chart over a neighborhood of the origin 0 in R^n. So we choose an open relatively compact neighborhood X_1 of $D \cap X$ in X, then a small neighborhood R_1 of 0 in R^n so that $D \cap p_2^{-1} R_1 \subset X_1 \times R_1 \subset Y \subset X_+ \times R^n$.

There clearly exists a PL isotopy G_t, $0 \leq t \leq 1$, of id | Y such that G_t has compact support in $X_1 \times R_1$, G_t maps each fiber $(x \times R^n) \cap Y$ of ξ^n into itself and (most important) for all x in a neighborhood of $D \cap X$ in X one has

(*) $$G_1(x,v) = (x,g^{-1}(v))$$

where $g:R^n \to R^n$ is a PL automorphism of R^n with $g(0)$ in the interior of an n-simplex of the mentioned triangulation of R^n . See Figure 1-b .

Note that, provided R_1 is small, since G_1 respects each fiber of ξ , the image $G_1 M$ will be transverse to ξ^n near C , as is M itself. Thus, in view of (*), we can have in place of (ii)′ the condition:

(ii) *Without loss, for some small open neighborhood $X_o \times R_o$ of $(D \cap X) \times 0$ in Y , there exists a PL submersion chart*† *$\varphi: X_o \times R_o \to Y$ about X_o for $p_2|: Y \to R^n$ such that $\varphi\,|(X_o \cap M) \times R_o$ is a submersion chart for $p_2|: M \to R^n$ about $X_o \cap M$. See Figure 1-c .*

At this point M is necessarily *locally* PL transverse near $C \cup D$ to X in Y . Indeed the existence of φ in (ii) assures this; the well-known PL lemma needed is $\Sigma^n(A,B) \cong (S^y,S^x) \Rightarrow (A,B) \cong (S^{y-n},S^{x-n})$ and can be proved by induction on n using 'pseudo-radial projection' [Hu_2, pg. 21].

Although M is not yet transverse to ξ^n near $C \cup D$, we are not far from this goal. Let $\delta_t : R_o \to R^n$, $0 \leqslant t \leqslant 1$, be a PL deformation of δ_1 = inclusion , fixing the origin, to the retraction

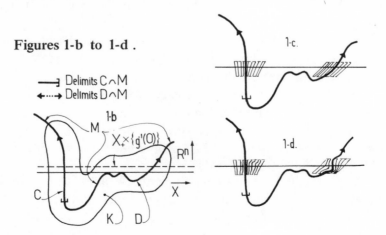

Figures 1-b to 1-d .

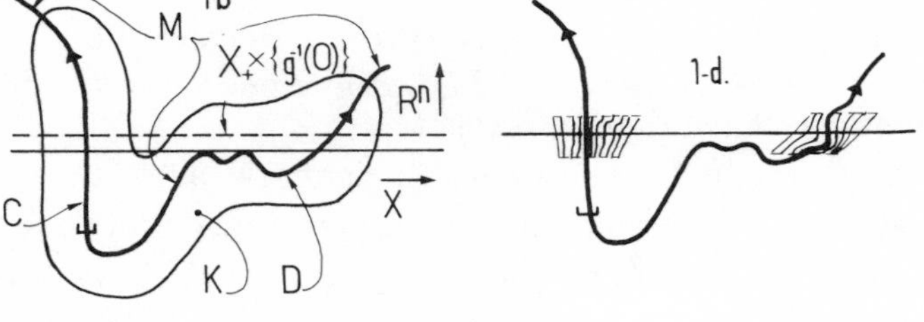

†This means that φ is a PL imbedding $\varphi(x,0) = x$ and $p_2\varphi(x,v) = v$ for all $(x,v) \in X_o \times R_o$.

ρ_0 onto the origin. This determines an isotopy
$\varphi_t : X_o \times R_o \to Y \subset X_+ \times R^n$ by the rule

$$\varphi_t(x,y) = (p_1\varphi(x,\rho_t(y)), y) \in X_+ \times R^n$$

Now φ_t is a PL isotopy of open imbeddings fixing $X_o \times 0 = X_o$, and respecting M near $C \cap X$, that runs from φ_0 = inclusion to $\varphi_1 = \varphi$. Applying the PL (1-parameter) isotopy extension theorem [RS_3, pp.56-59] we obtain PL isotopy Φ_t of id|Y with compact support in Image(φ) such that Φ_t equals φ_t near $D \cap X$, while † Φ_t fixes X pointwise and near C respects M . The isotopy Φ_t^{-1} , $0 \leqslant t \leqslant 1$, moves M imbedded-transverse to ξ near $C \cup D$ in the required fashion, cf. Figure 1-d. This completes the proof of the PL transversality theorem 1.6 . ■

Remark: The above argument rehabilitates (by simplification) R . Williamson's approach to transversality [Wi] .

† The method of proof in [RS_3] readily permits this (not the statements).

GENERAL TRANSVERSALITY OF SUBMANIFOLDS

First we discuss PL block transversality, the general PL transversality notion of Rourke and Sanderson, illustrating our contention that microbundle transversality can play the central role.

We alter the PL data of 1.6 by making the normal microbundle ξ^n to X in Y rather a PL normal block bundle. Then $E(\xi^n)$ is a closed PL regular neighborhood of X in Y ; X has a given PL regular cell-decomposition; for each k-cell σ of X there is a given PL (n+k)-cell B_σ in $E(\xi)$ called the block over σ , so that $B_\sigma \cap X = \sigma$; $(B_\sigma, \sigma) \cong (B^{n+k}, B^k)$; $E(\xi) = \cup_\sigma B_\sigma$; $B_\sigma \cup B_{\sigma'} = \cup\{B_\tau \mid \tau$ a cell in $\sigma \cap \sigma'\}$; and $\mathrm{int}B_\sigma \cap \mathrm{int}B_{\sigma'} = \phi$ for $\sigma \neq \sigma'$. See [RS_0,I] .

One says that M is **block transverse** to ξ near C in Y if for some subdivision ξ' of ξ , making $M \cap X$ a subcomplex of X (subdivided), M coincides near $C \cap X$ in Y with the restriction $\xi' | (M \cap X)$, which is just the union of the blocks over $M \cap X$.

The following result, due to Rourke and Sanderson [RS_0,II] , was first proved using an argument involving induction on dimensions.

PL IMBEDDED BLOCK BUNDLE TRANSVERSALITY THEOREM 1.8. *Theorem 1.5 holds true in the PL category without any dimension restrictions, even when ξ^n is a PL block bundle and imbedded-transverse means block transverse as formulated above.*

A block bundle transversality theorem for maps is deduced in [RS_0,II].

Proof of 1.8 from the microbundle analogue 1.6 : As the result is strongly relative we can assume ξ is trivial. Then it can be trivialized respecting the blocks of any subdivision ξ' , in particular one so that $M \cap X$ is a subcomplex of X and M coincides with $\xi' | (M \cap X)$ near $C \cap X$, cf. [RS_0,I] . This makes ξ' simultaneously a microbundle, to which 1.7 applies. Its conclusion gives 1.8 as soon as we further subdivide X to make $F_1(M) \cap X$ a subcomplex of X . ■

Still another proof of 1.8 (a very pleasing global one) has recently been given in [BuRS, II Mockbundles] .

Note that, in the PL definitions of transversality and in 1.6 & 1.8, all mention of PL local flatness can be suppressed. Then 1.6 and 1.8 hold good (both statements and proofs) if M and X are arbitrary polyhedra. Further generality still is possible in the PL category! See [Ar_2] [Sto] [McR] [BuRS] [Milt]. Our grasp of TOP transversality is still modest by comparison.

We return now to the topological category. As usual M and X denote TOP locally flat submanifolds of Y, both with boundary. But now no normal microbundle ξ to X in Y is given.

In case X has codimension $\leqslant 2$ in Y, a normal microbundle ξ for X in Y can always be introduced and is isotopy unique, except perhaps if $\dim Y = 4$ and $\dim X \leqslant 2$, see [KS_5]. Thus the microbundle transversality theorem 1.5 is quite adequate when X has codimension $\leqslant 2$ (granting the other dimension restrictions). Again, if M is of codimension $\leqslant 2$ in Y, we get a theorem by exchanging the roles of X and M. When both M and X are of codimension $\leqslant 2$, the two notions of transversality that result are equivalent by the following easy lemma.

Lemma 1.9. *Let M and N be submanifolds of W such that M has a normal microbundle μ in W to which N is transverse, and $M \cap N$ has a normal microbundle ν' in M. Then ν' extends to a normal microbundle ν_o normal in Y, to a neighborhood N_o of $M \cap N$ in N, such that the projections of μ and ν_o commute near $M \cap N$.*

Proof of Lemma 1.9: Let r_t be a homotopy of $\mathrm{id}|M$ to a map $r_1 : M \to M$ which coincides with the projection of ν' near $M \cap N$. The microbundle homotopy theorem, cf. [IV, §1] provides a microbundle homotopy $\rho_t : \mu \to \mu$ over r_t. If W_o is now a sufficiently small open neighborhood of $M \cap N$ in W, the required microbundle ν_o over $N_o = W_o \cap N$ is

$$\nu_o : N_o \hookrightarrow W_o \xrightarrow{\rho_1} N_o$$

as the reader will easily verify. ∎

Remark: This lemma admits a relative form; it also admits a slice preserving form showing that the ν_o is unique near $M \cap N$ up to isotopy fixing $M \cup N$.

Let us therefore assume henceforth that both M and X are of codimension $\geqslant 3$ in Y. Then an adequate substitute for a normal bundle to X in Y turns out to be a stable normal microbundle to $X \times 0$ in $Y \times R^s$, s large, (which *does* exist and *is* isotopy unique).

DEFINITION 1.10. M is **stably microbundle transverse** to X in Y if M is locally transverse to X in Y and if for some integer $s \geqslant 0$, there exists a normal microbundle ξ to $X \times 0$ in $Y \times R^s$ so that $M \times R^s$ is imbedded microbundle transverse to ξ in $Y \times R^s$. We write $M \pitchfork X\ (\xi)$ to express this, or simply $M \pitchfork X$ when ξ is not specified.

GENERAL TRANSVERSALITY THEOREM 1.11. (A. Marin [Mar]).

Suppose $M \pitchfork X$ near a closed subset C of Y, and let $D \subset Y$ be another closed subset.

Under the same dimension restrictions as for the microbundle analogue 1.5, there exists a homotopy f_t, $0 \leqslant t \leqslant 1$, of $f: M \to Y$ rel C so that:

(a) $f_1 M \pitchfork Y$ near $C \cup D$,

(b) the homotopy f_t is realized by an ambient isotopy F_t of $id \mid Y$, i.e. $f_t = F_t f$,

(c) the ambient isotopy F_t is as small as we please, with support in a prescribed neighborhood of $D - C$ in Y.

Furthermore, given (from the outset) a normal microbundle ξ_0 to $X \times 0$ in $Y \times R^s$, s large, such that $M \times R^s$ is transverse to ξ_0 near C, one can construct along with F_t another normal microbundle ξ, such that $F_1(M) \times R^s$ is transverse to ξ near $C \cup D$, and $\xi = \xi_0$ near C.

The possibility of a general TOP transversality theorem has been surmised since 1969 by several mathematicians. However, even the formulation of a suitable definition has been a major stumbling block; the only one seriously proposed hitherto is due to Rourke and Sanderson [RS_2]. Stable microbundle transversality seems simpler and more maniable. It shares with PL block transversality one feature vital for any chart by chart approach to proving a transversality theorem – namely the local character of the (ternary) relationship $M \pitchfork X\ (\xi)$: this means that if $\{U_\alpha \mid \alpha \in A\}$ is an indexed open covering of Y, then the relationship $M \pitchfork X\ (\xi)$ implies and is implied by the collection of relationships $M_\alpha \pitchfork X_\alpha\ (\xi_\alpha)$, $\alpha \in A$, where subscripts α are attached to indicate intersection with U_α , e.g. $M_\alpha = M \cap U_\alpha$. Beware that the (binary) relationship $M \pitchfork X$ (omitting ξ) is not of local character, as Hudson's examples showed.

It would be possible to give a proof of Marin's general transversality theorem in parallel with that of the PL result 1.8 above, by making strenuous use of the Rourke-Sanderson block bundle technique $[RS_1]$ $[RS_2]$† in local charts, where enough triangulations are available. Marin's proof (the first proof) is quite remarkable for not using block bundle theory in earnest – although it does still appeal to some PL topology. It too reduces the result finally to the microbundle transversality theorem 1.5 , although along the way there is a surprising interchange in the roles of M and X . We will not attempt to prove Marin's theorem here; by either approach it is nontrivial, and unfortunately involves at present the surgically proved Casson-Sullivan imbedding theorem, cf. $[RS_5]$. Instead we shall conclude by indicating the scope of Marin's theorem and by describing phenomena that seem essential to an understanding of transversality.

Marin shows (nontrivially!) that the relationship $M \pitchfork X$ is symmetric. Note that the nonstable relationship is already symmetric by the elementary Lemma 1.9 as soon as normal microbundles to $M \cap X$ in M and in X are known to exist.

The conclusion of 1.11 being thus in effect symmetric in X and M (via the device of replacing F_t by F_t^{-1}), we can symmetrize the hypotheses on dimension in 1.11 ‡. The reader will thus verify that 1.11 plus 1.5 fail to give together an adequate general transversality theorem only for the following combinations of dimension involving 4 . We write $y = \dim Y$, $m = \dim M$, $x = \dim X$.

Excluded values of $(y, m, x, m+x-y)$:

$(y; m, x; 4)$, $(4; m, x; m+x-4)$, $(y; 4; y-1, 3)$, $(y; 4, y-2; 2)$

– plus, of course, those obtained by interchanging the values given to m and x .

†To avoid an annoying restriction to fiber dimension $\geqslant 5$ for TOP (micro-) block bundles, that appears in $[RS_2$, § 4] , one should assume any TOP block bundles having fiber dimension 3 or 4 to be reduced to PL block bundles over the 2-skeleton. This applies to all TOP block bundles mentioned in this essay.

‡Where the last statement of 1.11 (involving ξ) is concerned, a sharper version of the symmetry above is needed, see [Mar].

The case (7;4,4;1) should ostensibly be excluded too; but in this case Marin gives a special argument to prove 1.11 , and notes that the parallel *nonstable* result 1.5 is in this case equivalent to the unproved 4-dimensional Annulus Conjecture.

Finally, leaving even the precise formulation aside, we mention that, from 1.11 , one can deduce a *general transversality theorem for maps* parallel to our first transversality theorem for maps 1.1 . In the statement, the normal microbundle ξ to X in Y of 1.1 becomes stable; for the proof one uses a device of thickening X to let one apply 1.11 .

To help the reader acquire a global understanding of transversality, here are two cardinal facts from [RS_5] and [Mar] (or derivable with the help of techniques given there and in [RS,II] [RS_5]). With the codimension $\geqslant 3$ data of Definition 1.10 we consider M *locally* transverse to X in Y . Then there is an obstruction in the group $[M \cap X, G/TOP]$, to M being *stably microbundle transverse* to X in Y , i.e. to having $M \pitchfork X$. Furthermore, when X and $M \cap X$ are prescribed, these obstructions in $[M \cap X; G/TOP]$ classify the locally transverse choices of M near X , up to isotopy fixing X . Secondly, fixing M as well as X , suppose ξ given so that $M \pitchfork X\ (\xi)$. Then all alternative choices ξ' of ξ with $M \pitchfork X\ (\xi')$ are classified up to isotopy in $Y \times R^s$ fixing $X \times 0$ and respecting $M \times R^s$, by the elements of the group $[M \cap X; \Omega(G/TOP)]$.

TOP (micro-) *block bundle* † transversality (compare 1.8) makes good sense in the situation of 1.10 as soon as the pair $(X, M \cap X)$ is triangulated. A block bundle version of the two cardinal facts then holds true – involving the same stable classifying space G/TOP . This in turn reveals a useful stability:

block transversality	$\Leftrightarrow$	*stable* block transversality	$\Leftrightarrow$	stable microbundle transversality

the second equivalence resulting trivally from blockwise application of existence and uniqueness of stable normal microbundles. What is more, the isotopy classes of germs of the three types of bundles, expressing transversality of a fixed M , correspond naturally and bijectively. This last fact is exactly what one needs to establish 1.11 imitating the proof of 1.8 . (The patient reader can pause to verify this for himself.) To

†Recall the footnote to the discussion following 1.9 .

sum up, TOP block bundles are adequate to discuss TOP transversality provided enough triangulations ‡ exist. At worst, TOP block bundles offer a useful tool to be used locally.

We close on an amusing note by comparing PL and TOP transversality. Consider once more the codimension $\geqslant 3$ situation above, with M topologically locally transverse to X in Y, and *suppose all the manifolds and inclusions happen to be PL*. Then M is also PL locally transverse to X in Y, since Hudson has a criterion for this of local homotopy theoretic character [Hu_3, Lemma 1]. There is an obstruction $\theta \in [M \cap X ; G/PL]$ to M being PL block transverse to N in Y (by the PL version of the first cardinal fact, cf. [RS_5]). The map

$$[M \cap X ;\ G/PL] \to [M \cap X ; G/TOP] ,$$

whose kernel is $H^3(M \cap X; Z_2)/ \operatorname{Im} H^3(M \cap X; Z)$, see [$Si_{10}$, §15], carries this θ to the obstruction to TOP block transversality. *Thus it may occur that M is topologically but not piecewise linearly block transverse to X in Y.*

This curiosity is mitigated somewhat by the observation that in this case an ambient TOP isotopy fixing X can move M to coincide *near* X with a piece of PL submanifold that *is* PL block transverse to X (just apply the first cardinal fact!).

It is mitigated further by the fact that, provided $\dim(M \cap X) \neq 4$, new *global* PL manifold structures can be assigned (changing the concordance class of just one PL structure, that on $M \cap X$) so as to make M PL block transverse to X in Y.

‡Whether they do *always* exist is of course the residual triangulation problem for manifolds, cf. [Si_9].

§2. HANDLE DECOMPOSITIONS

Let W be a CAT (= PL or TOP) m-manifold and $M \subset W$ a possibly empty clean m-submanifold (clean merely emphasizes that M is a closed subset and the frontier is a manifold CAT bi-collared in W). A **handlebody decomposition** of W on M is a filtration of W (possibly infinite), $M = M_0 \subset M_1 \subset M_2 \subset \ldots$, with $\cup M_i = W$, by clean submanifolds such that, for each i, H_i = closure $(M_i - M_{i-1})$ is a clean compact submanifold of M_i and $(H_i, H_i \cap M_{i-1}) \cong (B^k, \partial B^k) \times B^{m-k}$ for some k, $0 \leqslant k \leqslant m$. H_i is called a **handle**. It is supposed that the collection $\{H_i\}$ is locally finite, so that each compactum in W meets only finitely many handles. When such a filtration is present we say W is **a handlebody on** M.

EXISTENCE THEOREM 2.1. *(CAT = PL or TOP).*

Consider any CAT m-manifold W and clean CAT m-submanifold M. If CAT = TOP suppose $m \geqslant 6$. Then W is a CAT handlebody on M.

The PL case is easily proved using a triangulation of (W, M) as a combinatorial manifold pair, see [Hu_2, pp.223–227]. We will deduce the TOP case using the Product Structure Theorem.

PROOF OF 2.1 FOR CAT = TOP.

For greater simplicity we suppose at first that $\partial W = \phi$. Using a co-ordinate covering of W of nerve dimension $\leqslant m$ we find a closed covering $A_0, A_1, \ldots, A_m$ and open sets $U_i \supset A_i$ with PL manifold structure such that each closure $\bar{U}_i$ is expressible as a disjoint locally finite union of compacta.

REMARK 2.1.1. $\bar{U}_i$ can lie in the preimage in W, under a standard map $\varphi : W \to |\mathfrak{w}|$ to the nerve of a suitable covering $\mathfrak{w}$, of the open stars of barycenters of i-simplices of $|\mathfrak{w}|$ in the 1st barycentric subdivision $|\mathfrak{w}|'$ of $|\mathfrak{w}|$, see [I, §4]. More precisely, φ can come from a partition of unity subordinate to $\mathfrak{w}$, and $\mathfrak{w}$ can be any dim $\leqslant m$ covering by relatively compact charts. Then the reader will verify that the handle decomposition to be constructed enjoys the property :

Each handle H lies in some chart of $\mathfrak{w}$, and the union of all charts of $\mathfrak{w}$ containing H contains all handles of some finite sub-handlebody on M including H .

Thus the handle decomposition will be 'fine' if $\mathfrak{w}$ is fine .

We propose to define a filtration $M = M_0 \subset M_1 \subset M_2 \subset \ldots \subset M_m = W$ so that M_i is a *handlebody* on M_{i-1} , and so that closure$(M_i - M_{i-1})$, although probably not a single handle , is at least compact or a locally finite disjoint union of compacta . Once M_{i-1} is known to be a handlebody , it is easy to show (by adding the handles in a suitable order) that M_i is one too . Then W is a handlebody on M as required.

Suppose for an inductive construction that we have obtained a clean submanifold M_{i-1} , $i \geqslant 1$, that contains M and is a neighborhood of $A_0 \cup \ldots \cup A_{i-1}$. *Apply the Product Structure Theorem to the boundary of $U_i \cap M_{i-1}$ in U_i to alter the PL structure on U_i making $U_i \cap M_{i-1}$ a PL m-submanifold of U_i* . Recall now that $U_i = U_{i,1} \cup U_{i,2} \cup \ldots$ is a union of open subsets U_{ij} , $j \geqslant 1$, whose closures are compact disjoint and form a locally finite collection . In each U_{ij} we find a clean PL m-submanifold M_{ij} of the form $M_{i-1} \cap U_{ij}$ union a compact neighborhood of $A_i \cap U_{ij}$. The known PL (and compact) version of 2.1 shows that M_{ij} is a PL handlebody on $M_{i-1} \cap U_{ij}$. So adding to M_{i-1} , we get $M_i = M_{i-1} \cup M_{i,1} \cup M_{i,2} \cup \ldots$ a TOP handlebody on M_{i-1} . This completes the inductive construction of $M = M_0 \subset M_1 \subset \ldots \subset M_m = W$ as required to prove 2.1 when $\partial W = \phi$. ∎

The proof of 2.1 when $\partial W \neq \phi$ is word for word the same if $m \geqslant 7$. However if $m = 6$, the Product Structure Theorem does not apply unless

a_{i-1}) *$\partial(U_i \cap M_{i-1})$ is a PL submanifold of U_i near ∂U_i .*

So we will assume a_{i-1}) inductively and also

b_{i-1}) *$M_{i-1} - \mathring{M}$ is a clean submanifold of $W - \mathring{M}$.*

Further, we arrange prior to the induction that , for each chart U_j , $U_j \cap M$ is a clean PL submanifold of U_j . (Just work from the outset with such charts only) .

Now the above construction works and produces $M_i' = M_{i-1} \cup M_{i,1} \cup M_{i,2} \cup \ldots$ which contains $M \cup A_0 \cup \ldots \cup A_i$ and satisfies b_i) when each M_{ij} is chosen with the similar property. Next we build $M_i \subset M_i'$ verifying a_i) as well – roughly by peeling M_i' away from ∂U_{i+1}.

Consider $N_i \equiv U_{i+1} \cap M_i' \subset U_{i+1}$. Choose a clean PL submanifold X of $\partial U_{i+1} - \mathring{M}$ lying in N_i and containing a neighborhood in $\partial U_{i+1} - \mathring{M}$ of $(A_0 \cup \ldots \cup A_i)$.

Choose a narrow clean collaring

$$h : [\partial U_{i+1} \cap (N_i - \mathring{M})] \times I \to N_i - \mathring{M}$$

that is PL near X and has image whose intersection with intW is closed in intW. Define

$$M_i = M_i' - h\{(\partial U_{i+1} \cap N_i - M - X) \times [0,1)\} .$$

Then M_i satisfies a_i), b_i) and if h is sufficiently narrow, M_i contains $M \cup A_0 \cup \ldots \cup A_i$. But M_{i-1} is a handlebody on M so M_i' clearly is also one. Then M_i is one because clearly $(M_i', M) \approx (M_i, M)$, (see [$Si_4$, Lemmas in §3] for assistance). ∎

The last paragraph above is trivial but a bit confusing, so we illustrate for dimension $m = 3$ in Figure 2-a. W is defined by $x \geqslant 0$ or rather U_{i+1} is, since U_{i+1} is all we see of W; M is $x \geqslant 0$, $z \leqslant 0$, i.e. what lies below the xy plane; M_i' is M union the curved snowpile on the xy-plane up against the yz wall. X lies beyond the piecewise linear path in the yz wall; the slab sitting on the xy plane on the dark base is what is deleted from M_i' to get M_i.

Figure 2-a .

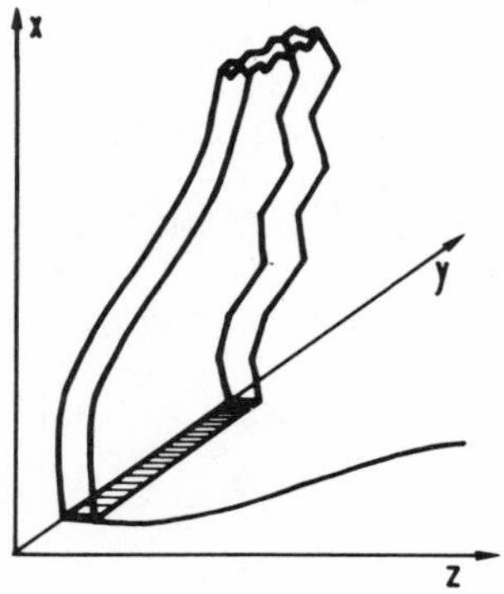

Handlebodies have CW spines. †

This basic observation was made circa 1970 by M. Cohen with a helping hand from B. Sanderson. As we shall not in fact use this result we are content to discuss a compact and non-relative version.

THEOREM 2.2 .

Let W^m *be a compact TOP m-manifold that is a handlebody the handles being added in order of increasing* ‡ *dimension.*

Then W^m *is homeomorphic to the mapping cylinder Map(f) of a continuous map* $f : \partial W \to X$ *to a finite CW complex* X *with one k-cell for each k-handle of W ,* $0 \leq k \leq m$.

X is called a *spine* of W . In case $\partial W = \phi$, one has $X = W$, i.e. W itself is a CW complex. Thus every closed TOP manifold of dimension ≥ 6 is a CW complex. Is in fact every TOP manifold a CW complex? Contrast this with the more delicate question whether every TOP manifold is a *simplicial* complex, this latter being unsettled in 1975 even for closed manifolds of high dimension, cf. [Si9] ,[Matu] .

Proof of 2.2.

Proceed by induction on the maximum handle index, using a direct and elementary argument suggested by these diagrams:

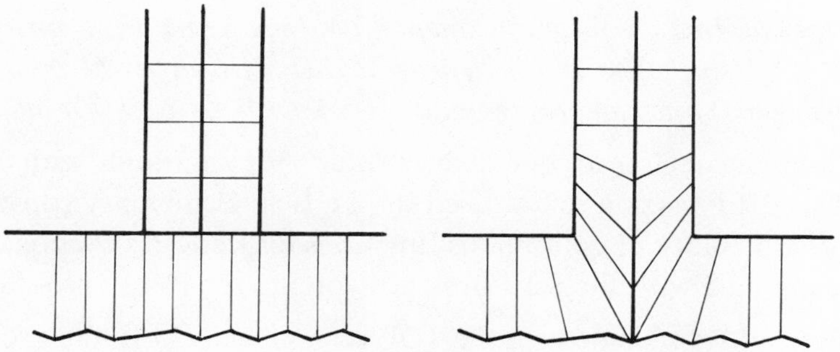

■

† The existence of *polyhedral* spines is discussed in $[\mathrm{Ped}_1]$.

‡ Not an essential restriction; see [Edw5] and any exposition of handlebody theory.

§3. MORSE FUNCTIONS. †

Consider a CAT (= DIFF or TOP) m-manifold M and a continuous map $f : M \to [a, b] \subset R$. We call f a CAT **Morse function** (compare Morse [Ms_2]) if for each point $p \in M$, *either* f is a CAT submersion into [a, b] near p, *or* $p \in \text{int}\, M$ and there exists a CAT embedding $h : R^m \to M$ with $h(0) = p$ so that for some k, $0 \leqslant k \leqslant m$, $fh(x_1, x_2, \dots, x_m) = f(p) - (x_1^2 + \dots + x_k^2) + (x_{k+1}^2 + \dots + x_m^2)$ for all $(x_1, \dots, x_m)$ near the origin $0 = (0, \dots, 0) \in R^m$. In the latter case p is called a **critical point** of index k. Clearly the critical points are isolated, so only finitely many lie in a given compactum. We will say f is Morse **near** a subset $C \subset M$ if $f|U : U \to R$ is Morse for some open neighbourhood U of C.

Our main result in a convenient relative form is the

EXISTENCE THEOREM 3.1. (CAT = DIFF or TOP)

Let $f : M \to R$ be a continuous map of a CAT open manifold to R, that is Morse on an open neighborhood U of a closed subset C of M. Let $D \subset M$ be closed and $V \supset D$ an open neighborhood of D in M. If CAT = TOP suppose $m \geqslant 6$.

One can find a continuous map $g : M \to R$ equal to f near $C \cup (M - V)$ so that g is Morse near $C \cup D$. Given $\epsilon : M \to [0, \infty]$ positive near D one can arrange that $|g(x) - f(x)| \leqslant \epsilon(x)$ for all $x \in M$.

Collaring devices let one deduce results for manifolds with boundary.

The DIFF version (cf. Morse [Ms_1].) is an elementary consequence of approximability of continuous functions by smooth functions [Mu_1, §4] combined with the Sard-Brown theorem, see [Mi_8, proof of 2.7].

From it we shall deduce the TOP version using both the Product Structure Theorem and the Concordance Implies Isotopy Theorem. The heart of the proof is:

PROPOSITION 3.2.

Let $f : M^m_\Sigma \to R$ be a TOP Morse function defined on an a DIFF manifold without boundary, of dimension $m \geqslant 6$.

† Cf. [Ok].

There exists a DIFF structure Σ' on M concordant to Σ so that $f : M_{\Sigma'} \to R$ is a DIFF Morse function.

Suppose $f : M_\Sigma \to R$ is already DIFF Morse on an open neighborhood U of a closed set $C \subset M$. Let $D \subset M$ be compact and $V \supset D$ be open. Then there exists a DIFF structure Σ' on M concordant to Σ rel $C \cup (M-V)$ such that $f : M_{\Sigma'} \to R$ is DIFF Morse near $C \cup D$.

To get this particularly clearcut result we shall require some assistance from a local contractibility principle for homeomorphisms.

PROOF OF THEOREM 3.1 (assuming 3.2). By the usual finite chart-by-chart induction, it suffices to prove the case where $M = R^m$ and D is compact. In this case 3.2 offers a CAT structure Σ' on $U \subset R^m$ concordant to the standard one, so that $f : U_{\Sigma'} \to R$ is DIFF Morse. We extend Σ' to all M by the Concordance Extension Theorem [I, 4.2], then apply the DIFF version of 3.1 to conclude. ■

PROOF OF PROPOSITION 3.2.

Clearly the second statement implies the first; so let us prove the second.

The second statement is clearly a consequence of two special cases applied successively:

(a) *D is a (finite !) set of critical points.*

(b) *f has no critical points in $D - C$.*

And the case (b) reduces by enlarging C and cutting back M and V, to the case

(b′) *f has no critical points at all.*

Proof of case (a).

The definition of TOP Morse function shows that there exists a finite disjoint union B of closed discs in $V-C$ so that $D \subset \mathring{B}$, and there exists a structure Σ'' on $\mathring{B}$ so that $f : \mathring{B}_{\Sigma''} \to R$ is DIFF Morse. The Stable Homeomorphism Theorem [Ki_1] shows that $\Sigma'' | \mathring{B}$ is concordant to $\Sigma | \mathring{B}$ so that the Concordance Extension Theorem [I, 4.2] provides a concordance $\Sigma \simeq \Sigma'$ rel $C \cup (M-V)$ with $f : M_{\Sigma'} \to R$ DIFF Morse near $C \cup D$. Alternatively, use of [Ki_1] and [I, 4.2] can be replaced by solution of one 0-handle problem for each point of $D-C$, by the method of [KS_1]. ■

Concordance Implies Sliced Concordance Lemma 3.2.1.

Consider $(M\times I)_\Gamma$ *a conditioned* CAT *(= DIFF or PL) concordance of* CAT *structures* $\Gamma : \Sigma \simeq \Sigma'$ *, on a* TOP *m-manifold* M^m , $m \geq 5$, *without boundary. Then, as* CAT *structure,* Γ *is concordant* rel $M\times\partial I$ *to a structure* Γ' *that is sliced over* I *in the sense that the projection* $p_2 : (M\times I)_{\Gamma'} \to I$ *is a* CAT *submersion; thus* Γ' *is a conditioned and sliced* CAT *concordance* $\Sigma \simeq \Sigma'$ *. Furthermore, if*† Γ *is a product along* I *near* $C\times I$ *, with* C *closed in* M *, then the asserted concordance* $\Gamma \simeq \Gamma'$ *can be* rel $C\times I$.

Proof of 3.2.1 . Apply to Γ the Concordance Implies Isotopy Theorem [I, §4] (rel $C\times I$ if necessary) to get a sliced and conditioned concordance $\Theta : \Gamma \simeq \Sigma\times I$. This Θ is the image of $(\Sigma\times I)\times I$ by the isotopy involved; and we can require that Θ be constant on the initial half [0,½] of its interval. Let H be $(\mathrm{id}|M)\times h$ where h is a homeomorphism of the square $I\times I$ fixing $I\times 0$ and mapping $1\times[½,1]$ linearly onto the top $I\times 1$. Then $H(\Theta)$ is the concordance required $\Gamma \simeq \Gamma'$ rel $M\times\partial I$, at least if CAT = PL ; if CAT = DIFF , we must go on to unbend (see §4.3) the corners of $H(\Theta)$ along $M\times h(0,1)$ and inversely create corners along $M\times(1,1)$, much as in [I; §4, case 3] . ■

Figure 3-a

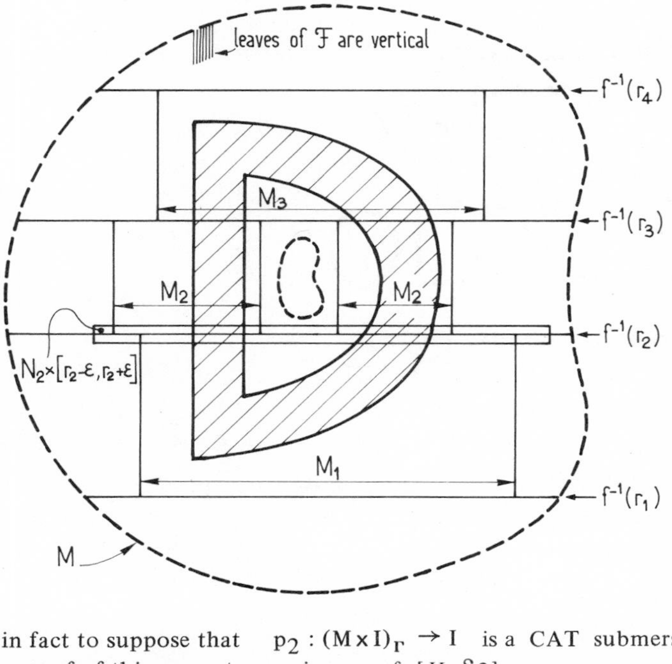

† It suffices in fact to suppose that $p_2 : (M\times I)_\Gamma \to I$ is a CAT submersion near $C\times I$; but the proof of this seems to require use of [II, §2] .

We are now ready to give the remaining

Proof of case (b').

To gain clarity *we assume at first that* $C = \phi$ *and* $V = M$; afterwards we will fill in the details when $C \neq \phi$ or $V \neq M$. Allowing $V \neq M$ will pose no problem at all since the concordances we find will all have compact support and so extend by the constant concordance from V to all of M.

Applying the local contractibility principle for homeomorphisms, we find a TOP foliation $\mathfrak{F}$ on M transverse to the fibers of f, with model R^{m-1}, and with 1-dimensional leaves that are imbedded in R by f – see [Si_{12}, §6.25].

For a subset $S \subset f^{-1}(s)$, $s \in R$, we often speak of a subset $i: S \times [a, b] \hookrightarrow M$, $a \leqslant s \leqslant b$. This inclusion map (if it exists) is the unique continuous imbedding i such that for $x \in S$ and $t \in [a, b]$ one has $i(x, s) = x$, $fi(x, t) = t$, and $i(x \times [a, b])$ lies in the leaf of $\mathfrak{F}$ through (x, s). For compact S, the inclusion i clearly does exist if a, b are sufficiently near s, (since it exists locally).

We can find, using compactness of D, numbers $r_1 < r_2 < \ldots < r_k$, and compacta $M_1, \ldots, M_{k-1}$ so that

(i) $M_i \subset f^{-1}(r_i)$, *and* $M_i \times [r_i, r_{i+1}]$ *exists in* M,

(ii) $\bigcup_{i=1}^{k-1} M_i \times [r_i, r_{i+1}]$ *is a neighborhood of* D *in* M.

Next we can find compacta $N_i \subset f^{-1}(r_i)$, $i = 1, \ldots, k$, so that the interior $\mathring{N}_i$ in $f^{-1}(r_i)$ contains $(M_i \times r_i) \cup (M_{i-1} \times r_i)$. There then exists $\epsilon > 0$ so small that each $N_i \times [r_i - \epsilon, r_i + \epsilon]$ exists in M and $2\epsilon < r_{i+1} - r_i$, for all i. See Figure 3-a opposite.

We apply the local Product Structure Theorem [I, §5.2] to each $(\mathring{N}_i \times (r_i - \epsilon, r_i + \epsilon))_\Sigma$, $i = 1, \ldots, k$, to get a DIFF structure Σ^* on M with the properties

1) Σ^* *is a product along* $(r_i - \epsilon, r_i + \epsilon)$ *near* $(M_i \times r_i) \cup (M_{i-1} \times r_i)$, *and (hence)* $f: M_{\Sigma^*} \to R$ *is a DIFF submersion near* $(M_i \times r_i) \cup (M_{i-1} \times r_i)$,

2) $\Sigma^* = \Sigma$ *outside a compactum.*

We next apply the lemma 3.2.1 proved above, together with the Concordance Extension Theorem of [I], to $(\overset{\circ}{M}_i \times [r_i, r_{i+1}])_{\Sigma^*}$; we get a concordance $\Sigma^* \cong \Sigma'$ rel $\bigcup_i (\overset{\circ}{M}_i \times \{r_i, r_{i+1}\})$ constant outside a compactum in $\bigcup_i(\overset{\circ}{M}_i \times [r_i, r_{i+1}])$ so that $f : M_{\Sigma'} \to R$ is a DIFF submersion near $\bigcup_i D_i \times [r_i, r_{i+1}] \supset D$ where D_i is the compact projection on $\overset{\circ}{M}_i$ of $D \cap (\overset{\circ}{M}_i \times [r_i, r_{i+1}])$. ∎

When $C \neq \phi$ we must make the following additions to the proof. We want the foliation $\mathfrak{F}$ to be DIFF near C. To achieve this, construct a vector field ξ on $U \supset C$, where f is DIFF, so that $f_* \xi = \frac{\partial}{\partial x}$. Integrating ξ we obtain a DIFF foliation $\mathfrak{F}_0$ DIFF transverse to f on U. Now construct $\mathfrak{F}$ so that $\mathfrak{F} = \mathfrak{F}_0$ near C, using [Si_{11}, §6.25].

Let $C^* = C \cap (\bigcup_{i=1}^{k-1} M_i \times [r_i, r_{i+1}])$. By refining the partition $r_1, \ldots, r_k$ if necessary (on adding r' between r_i and r_{i+1} we can let $M_{r'} = M_i \times r'$) we can find compacta C_i in M_i such that

(iii) $C^* \subset \bigcup_{i=1}^{k-1} C_i \times [r_i, r_{i+1}]$,

(iv) f *is* DIFF *near* $\bigcup_{i=1}^{k-1} C_i \times [r_i, r_{i+1}]$.

Next we choose ϵ (above) so small that f is DIFF near the set denoted $\{(C_{i-1} \times r_i) \cup (C_i \times r_i)\} \times [r_i - \epsilon, r_i + \epsilon]$ forgiving an obvious abuse of notation. The local Product Structure Theorem [I, §5.2] is applied relative to this set to assure that

3) Σ^* *is concordant rel* C *to* Σ

and that we can apply the Concordance Implies Isotopy Theorem [I, §4.1] rel $\bigcup_i C_i \times [r_i, r_{i+1}]$ getting a concordance $\Sigma^* \cong \Sigma'$ rel $C \subset \bigcup_i C_i \times [r_i, r_{i+1}]$. This finishes the proof of 3.2. ∎

Our discussion of topological Morse functions would not be complete without mention of gradient-like fields. For TOP we define them in an integrated form.

DEFINITION 3.3. Let $f : M \to [a, b]$ be a TOP Morse function on a TOP manifold M^m. A **gradient-like field** $\mathfrak{F}$ for f is a TOP foliation†

† Defined by an atlas of submersions into R^m_+.

by lines transverse to the level surfaces of f, defined on the complement of the critical points of f, such that about each critical point $p \in \text{int}M$ of f, there are co-ordinates $(x_1, \ldots, x_m)$ so that $f(\vec{x}) = f(p) - (x_1^2 + \ldots + x_\lambda^2) + (x_{\lambda+1}^2 + \ldots + x_m^2)$, and $\mathfrak{F}$ is given near p by the integral curves of the vector field $v(\vec{x}) = (-x_1, \ldots, -x_\lambda, x_{\lambda+1}, \ldots, x_m)$. If p is a non-critical point of f, there are co-ordinates $(x_1, \ldots, x_m)$ near p such that $f(\vec{x}) = x_1$ and, on the leaves of $\mathfrak{F}$, only x_1 varies.

If $\mathfrak{F}_0$ is a gradient-like field for $f|U$ where U is an open neighborhood of C closed $\subset M^m$ (any m), the local contractibility theorem for homeomorphisms provides a gradient-like field $\mathfrak{F}$ for f equal $\mathfrak{F}_0$ near C, see [Si_{12}, §6.25]. One first builds $\mathfrak{F}$ explicitly near the (isolated!) critical points, then extends.

Using TOP gradient-like fields, one can carry through for the topological case the elementary discussion of Morse functions as they relate to cobordisms, surgeries, handles etc. – following the discussion in [Mi_8, §§ 1–3] almost word for word.

3.4 THE FUNCTIONAL TOPOLOGICAL SMALE THEORY.

To conclude this section we indicate how the techniques of the functional Smale-theory of [Mi_8] or [CeG] can be adapted to TOP Morse functions in dimension $m \geqslant 6$. Taken together with the simple type theory of §4 and §5 below, these suffice to transcribe for TOP manifolds the s-cobordism theorem [Ke_1], the boundary theorem of [Si_1] [Ke_2] and the splitting theorem of Farrell & Hsiang [FH]. The splitting theorem of Cappell [Cap] requires Appendix C as well.

Rearrangement of critical levels of a TOP Morse function on a compact triad to get a 'nice' Morse function (of Smale) as in [Mi_8, §4] is possible as soon as a small ambient isotopy can eliminate all intersections of locally flat 'left hand' spheres S_L^v with 'right hand' spheres S_R^w when $v + w < m-1$ in a noncritical level N^{m-1}. This is easily done for $m-1 \geqslant 5$ by the stable homeomorphism theorem – compare [Si_{10}, §7.2]; TOP general position *always* applies, cf. [Ed_5].

Creation of a pair of critical points [Mi_8; §8] is a local matter and poses no problems.

Elementary change of basis among critical points of index λ, $2 \leqslant \lambda \leqslant m-2$, [$Mi_8$, §7] requires at worst the stable homeomorphism theorem once again, in getting a 'linking' isotopy to deform the gradient-like field .

Here is a discussion for $m \geq 6$ of techniques for *eliminating a pair of critical points* p and q of index λ and $\lambda + 1$ with $f(p) < f(q)$ and no critical values between . Consider $N = f^{-1}(c)$, where $f(p) < c < f(q)$. We assume f is proper so that the chart about q (normalizing f) can be extended (using the gradient lines of $\mathfrak{F}$) to a chart U_q containing the left hand disc $D_L^{\lambda+1}$ of increasing trajectories from N to q with $\partial D_L^{\lambda+1} = S_L \subset N$, so that $f|U_q$ and $\mathfrak{F}|U_q$ are DIFF . Similarly from U_p about p , containing the right hand disc $D_R^{n-\lambda}$ with $\partial D_R = S_R \subset N$. Using the Stable Homeomorphism Theorem repeatedly, one can build a small isotopy of N^{m-1} and use it to change $\mathfrak{F}$ (and U_p) as in [Mi_8 , §4.7] so that

1) *The structures of U_p and U_q agree near $S_R \cup S_L$.*
2) *S_L is DIFF transverse to S_R at $S_R \cap S_L$ (for the common structure) .*

By 1) , U_p and U_q provide a chart $U \supset D_R^{n-\lambda} \cup D_L^{\lambda+1}$ making $f|U$ and $\mathfrak{F}|U$ DIFF . By 2) , S_L and S_R intersect at isolated points in U . If there is *only one* intersection point , f can now be changed on a compactum in U to eliminate the critical points p and q , by the purely DIFF cancellation procedure of [Mi_8 , §5] . ■

The above cancellation result makes it important to be able to *elimate superfluous intersection points of* $S_R(p)$ *and* $S_L(q)$ *by Whitney's procedure* . Suppose given a pair (a, b) of points of $S_R \cap S_L$ having appropriate intersection number ± 1 [Mi_8 , §6] and a loop γ consisting of a DIFF arc A_R running a to b in S_R and a DIFF arc A_L running b to a in S_L such that γ is contractible in N . See figure 3-b below .

Assertion . *Whitney's procedure [Whn_2] can be adapted to eliminate the intersection points a, b via an isotopy of id|N – under the usual technical restrictions :* $m - 1 \geq 5$; $\dim S_R \geq 3$; *and* $\pi_1(N - S_R) \to \pi_1 N$ *is injective when* $\dim S_L \leq 2$.

Here are instructions for adapting the DIFF Whitney procedure in [Whn_2] or [Mi_8] , somewhat different from those in [Si_{10} , §7] .

The DIFF procedure unaltered produces for us a DIFF imbedding

$$\varphi : D - \mathring{D}_o \to U \cap N$$

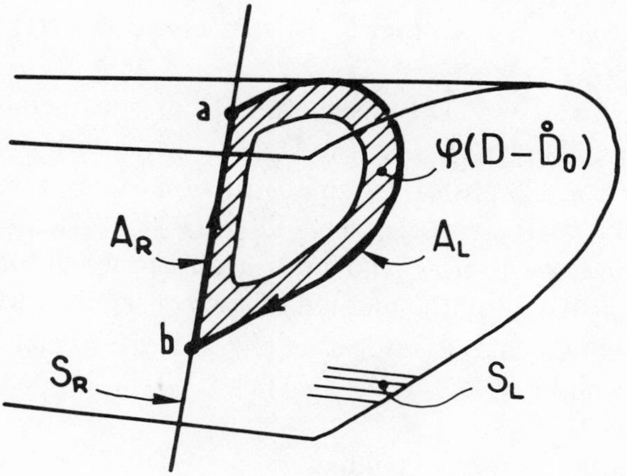

Figure 3-b

of the Whitney model 2-disc D (of [Mi$_8$, §6.7]) *minus* the interior of a central standard 2-disc D_o . And we can at least find a continuous map

$$g : D_o \to N - (S_L \cup S_R)$$

equal φ on ∂D_o .

Now apply any one of three techniques (a) , (b) , (c) mentioned below to find an open (m – 1) – ball $B \subset N - (S_L \cup S_R)$ and an approximation g′ to g equal g on ∂D_o so that $B \supset g'(D_o)$.

(a) Regard D_o as the cone on ∂D_o and use Newman's chart-by-chart version of engulfing [Ne] cf. [Si$_5$] , to make a small ball about g (cone point) swell up to engulf $g'(D_o)$ for some approximation g′ to g equal g on ∂D_o .

(b) Apply Homma's convergence technique [Glu] to find a locally flat topological embedding $g' : D_o \to N - (S_L \cup S_R)$ equal g on ∂D_o . Then an elementary stretching argument [Lac] produces a (m – 1)–ball $B \supset g'(D_o)$.

(c) Replace Homma's method in (b) by a chart-by-chart procedure using general position and a smoothing theory or the Product Structure Theorem to advance from chart to chart .

Applying DIFF general position (the Sard-Brown theorem [Mi_7]) in U, we now adjust g' so that $g'(D_o)$ meets $\varphi(D - \mathring{D}_o) \subset N$ in $\varphi(\partial D_o) = g'(\partial D_o)$.

Next we find a new DIFF structure Σ on the open ball B so that near $\varphi\partial D_o = g'\partial D_o = g'D_o \cap \varphi(D - \mathring{D}_o)$ it agrees with the structure of $N \cap U$. This is accomplished by an obstruction theory for smoothing using $\pi_i(TOP/O) = 0$, $i = 0,1$, see [KS_1][IV][V] [Si_{10}]. More simply one can solve a zero- and a one-handle problem for two handles with core in $\varphi(\partial D_o)$, by the method of [KS_1] applied to DIFF.†

Now $(N \cap U)$ and B_Σ together give a DIFF structure Σ' on a small neighborhood V of $\varphi(D - \mathring{D}_o) \cap g'(D_o)$ in N extending that of U near $\varphi(D - \mathring{D}_o)$.

Finally alter $\varphi \cup g'$ extending φ to a DIFF embedding $\varphi' : D \to V_{\Sigma'}$. Then Whitney's DIFF process applies to eliminate the pair (a, b) of intersection points. ∎

Remark: The argument outlined in [Si_{10}, §7.3], although tougher, is more elementary in that it does not use the surgery concealed in [KS_1].

* * *

The 1-parameter theory of generic functions due to J. Cerf has (in 1975) not yet been reworked in the TOP context. Nevertheless, the TOP version of its main theorem ([Ce_7], [HaW]) has been established via a reduction to the smooth case, see [BaLR, pp. 148-9] (E. Pederson's appendix): *Theorem: Let* V^n, $n \geqslant 5$, *be a connected compact TOP manifold such that* $\pi_1 V$ *acts trivially on* $\pi_2 V$. *Then the components of the space of automorphisms of* $V^n \times I$ *fixing* $(\partial V^n) \times I \cup V^n \times 0$ *pointwise, are in bijective correspondence with the elements of an algebraically defined abelian group* $Wh_2(\pi_1 V) \oplus Wh_1(\pi_2 V; \pi_1 V)$ *that is* 0 *whenever* $\pi_1 V = 0$. In case V admits a DIFF structure, the reduction is a simple application of the Concordance Implies Isotopy Theorem (Essay I) and its 1-parameter version (see Essays II, V). In general one seeks to replace V by a subhandlebody of index $\leqslant 3$, which does admit a DIFF structure. (When $n = 5$, Pederson *hypothesises* that V^n is a handlebody.

† For 0- and 1-handles, a PL solution easily gives a DIFF solution; so [KS_1] could be applied directly.

§4. THE SIMPLE HOMOTOPY TYPE OF A TOPOLOGICAL MANIFOLD†

We shall define a prefered simple type for a topological manifold M by properly imbedding M in a high dimensional euclidean space R^n with closed normal disc bundle D ; the (local) Product Structure Theorem of [I , §5.2] applied near ∂D in R^n gives D a triangulation ; then by definition the preferred simple type of M is the one making $i : M \hookrightarrow D$ a simple equivalence . If this procedure is carried out again yielding $i^* : M \hookrightarrow D^*$ we observe using the Concordance Implies Isotopy Theorem of [I , §4] that (after stabilisation) there is a PL isomorphism $f : D \to D^*$ so that fi is proper homotopic to i^* . This implies that our preferred simple type is well-defined . The proofs are very easy ; most space is devoted to recollecting basic facts about simple types , triangulations , normal bundles and imbeddings .

First the basic facts about simple homotopy types (from [Chn_2] and [Si_7]) .

A simplicial map $f : X \to Y$ of finite (unordered) simplicial complexes is an **elementary expansion** if f is injective and $Y - f(X)$

† Since this essay was written T. Chapman has succeeded in defining the simple type more generally for any locally compact metric space X such that $X \times [0, 1]^\infty$ is a manifold with model the Hilbert cube $[0, 1]^\infty$, see [Ch_1] [Ch_2] [Si_{14}] . (For example X can be a TOP manifold or locally a finite CW complex ; it seems that X can be any locally compact ANR , see [Mil_3][Wes][Ed_6]).This flows from Chapman's triangulation theorem and Hauptvermutung for Hilbert cube manifolds M : one has $M = K \times [0, 1]^\infty$ for some locally finite simplicial complex K ; and if $K_1 \times [0, 1]^\infty \to K_2 \times [0, 1]^\infty$ is a homeomorphism (or even proper and cell-like [Ch_2]) , then the proper homotopy equivalence $K_1 \to K_2$ determined is a simple equivalence of complexes . Several other interesting finite dimensional arguments have since appeared , each showing at least that a homeomorphism $K_1 \to K_2$ of locally finite simplicial complexes is a simple equivalence [Ed_2] [Ed_4] [Ch_3] . The argument [Ed_2] [Si_{14}] of R. D. Edwards is based on this section .

consists of exactly two open simplices, one being a face of the other.

In the category $\mathcal{C}$ of injective simplicial maps (inclusions) of locally finite unordered simplicial complexes we must specify the subcategory $\mathcal{E}$ of **expansions**. It is the least family in $\mathcal{C}$ containing isomorphisms and elementary expansions that is closed under 1) finite composition, 2) disjoint union, 3) pushout. Closed under pushout means that if $X = Y \cup Z$ in $\mathcal{C}$ and $Y \cap Z \hookrightarrow Y$ is in $\mathcal{E}$ then $Z \hookrightarrow X$ is also in $\mathcal{E}$, see [Si_7]. A map $f : X \to Y$ of $\mathcal{C}$ where X is compact clearly belongs to $\mathcal{E}$ if and only if it is a finite composition of isomorphisms and elementary expansions.

A proper homotopy equivalence $f : X \to Y$ of locally finite simplicial complexes is **simple** if it is proper homotopic to a finite composition $X = X_1 \to X_2 \to \ldots \to X_n = Y$ where each map is an expansion or a proper homotopy inverse to an expansion. Clearly each expansion is a proper homotopy equivalence – i.e. a homotopy equivalence in the category of proper continuous maps. It is an elementary fact that a PL homeomorphism is a simple equivalence.

It is enlightening to recall (from Whitehead) that when X and Y are properly simplicially imbedded in R^n and $n \geqslant 2 \max\{\dim X, \dim Y\} + 1$, then X and Y have PL isomorphic regular neighborhoods if and only if X and Y are simple homotopy equivalent.

A **simple type** for an arbitrary (metrizable) space X is represented by a proper homotopy equivalence $f : X \to Y$ to a locally finite simplicial complex Y (if such f exist). Another such equivalence $f' : X \to Y'$ represents the same simple type if there exists a simple proper homotopy equivalence $s : Y \to Y'$ so that sf is proper homotopic to f'.

A space equipped with a given simple type will be called a **simple space**. Every locally finite simplicial complex X has a canonical simple type represented by the identity map $X \to X$; so X is a simple space. A proper homotopy equivalence $f : X \to Y$ of simple spaces is called **simple** if when $g : Y \to Z$ represents the given simple type of Y, then $gf : X \to Z$ represents that of X. When X and Y are locally finite simplicial complexes this is equivalent to the earlier definition of a simple proper homotopy equivalence.

THEOREM 4.1.

Every separable metrizable topological manifold M has a preferred *simple type — so that M is a simple space. It can be defined by the following CAT(= DIFF or PL) rule: Imbed any topological closed disc bundle over M as a (clean) CAT codimension zero submanifold of an euclidean space and select a CAT*[†] *triangulation T of it; then the inclusion map $M \to T$ represents the preferred simple type.*

Other equivalent definitions of simple type will be discussed in §5.

COROLLARY 4.1.1. *Any homeomorphism $h: M \to M'$ of separable metrizable topological manifolds is a proper simple homotopy equivalence.*

Proof of 4.1.1. from 4.1. This is seen in the nature of the rule defining the preferred simple type. ∎

We can establish the preferred simple type of 4.1 using only DIFF or only PL arguments; for the readers convenience, the specifically DIFF arguments are marked off by square brackets ⟦ ⟧ and/or double stars **. ⟦If we use DIFF as our working tool, Whitehead's C^∞ triangulation theory is of course used, in particular to prove:

****LEMMA 4.2.** *If T and T′ are two (Whitehead) DIFF triangulations of a DIFF manifold M_Σ possibly with corners, then the identity map $T \to T'$ is a simple homotopy equivalence; in fact it is isotopic to a PL homeomorphism.*

The proof is in [Wh_1] [Mu_1, §10]. ∎

****4.3. Straightening corners.**

Let $B^2(\theta) = \{(r\cos\varphi, r\sin\varphi) \in R^2 \mid 0 \leq r \leq 1,\ 0 \leq \varphi \leq \theta\}$. If M_Σ has corners and $N = B^2(\pi/2) \times C \subset M_\Sigma$ is DIFF tubular neighborhood of a manifold C of corners we can **straighten** (or **unbend**) **corners** along C altering Σ to Σ^* on $B^2(\pi/2) \times C$ defining Σ^* so that $\rho \times (\mathrm{id}|C): (B^2(\pi/2) \times C)_{\Sigma^*} \to B^2(\pi) \times C$ is a diffeomorphism where $\rho: (r\cos\varphi, r\sin\varphi) \mapsto (r\cos 2\varphi, r\sin 2\varphi)$ doubles angles.

****LEMMA 4.3 (for DIFF with data above)**

If T^* is a DIFF triangulation of M_{Σ^*}, the identity map $T \to T^*$ is a simple equivalence.

† By DIFF triangulation we mean a Whitehead C^∞ triangulation [Mu_1].

Proof of 4.3. By 4.2 it suffices to prove this for *suitable* choices of DIFF triangulation T and T* of Σ and Σ^* respectively. DIFF triangulate C arbitrarily; DIFF triangulate $B^2(\pi/2)$ (resp. $B^2(\pi)$) by bisecting it into two 2-simplices at $\pi/4$ (resp. at $\pi/2$). We let T and T* give the corresponding product triangulations on $B^2(\pi/2) \times C \subset M$, and we let T and T* both give the same DIFF triangulation T_o of $(M - N)_\Sigma$. Then the inclusions $T_o \to T$ and $T_o \to T^*$ are expansions, proving that $\mathrm{id} : T \to T^*$ is simple. (In fact it is not hard to see a topological isotopy of it to a simplicial isomorphism). ■]]

An essential tool will be the theorem of M. Hirsch and B. Mazur stating that *a bundle with fiber euclidean space contains a disc bundle after stabilization, which is up to isotopy unique after (more) stabilization*. Its compactifying rôle is crucial; recall that the rôle of torus in [I, §3] was similarly compactifying. So it is appropriate to recall the precise statement required and indicate the proof.

Consider a locally trivial bundle E with fiber R^n and zero section, over an ENR X; let X be identified with the zero-section so that $X \subset E$ and the projection $p : E \to X$ is a retraction that completely specifies the bundle. We say E has fiber $(R^n, 0)$ since the group of E as a Steenrod bundle is the group (with compact-open topology) of all self homeomorphisms of $(R^n, 0)$, i.e. of R^n fixing 0. We will use similar terminology for disc bundles.

PROPOSITION 4.4 (see [Hi$_6$], also [KuL$_{1,2}$]).†

A) (Mazur) *The stabilized $(R^{n+1}, 0)$ bundle $E \times R$ over X contains a $(B^{n+1}, 0)$ bundle $D \subset E \times R$ over X. This D is understood to come from a reduction of E to the group of homeomorphisms h of R^{n+1} so that $hB^{n+1} = B^{n+1}$ and $h(0) = 0$.*

B) (Hirsch) *If E itself contains two $(B^n, 0)$ bundles D_1, D_2, then $D_1 \times B^1$, $D_2 \times B^1$ are isomorphic $(B^{n+1}, 0)$ bundles. In fact the $(R^{n+1}, 0)$ bundle $[1,2] \times (E \times R)$ over $[1,2] \times X$ contains a $(B^{n+1}, 0)$ bundle D′ extending $k \times D_k \times B^1$ over $k \times X$, $k = 1,2$ – called a concordance from $D_1 \times B^1$ to $D_2 \times B^1$ in $E \times R$.*

Proof of A).

Let Σ^{n+1} be the sphere $R^{n+1} \cup \infty$. There is clearly a unique $(\Sigma^{n+1}, 0, \infty)$ bundle $\bar{S}$ so that the complement of the ∞-section is $E \times R$. The closure in $\bar{S}$ of $E \times [-1, \infty)$ is a $(B^{n+1}, 0)$ bundle $\bar{D}$.

† Concerning reduction of (the group of) a bundle like E, see [Hus].

This $\overline{D}$ contains a disc bundle lying in $E \times R$. Indeed, by Alexander's theorem [Al$_1$], any $(B^{n+1}, 0)$ bundle reduces to the group of homeomorphisms h of B^{n+1} so that $h(x) = |x| h(x/|x|)$ for $x \neq 0$. So reduced, $\overline{D}$ contains a $(B^{n+1}, 0)$ bundle in $E \times R$ of any radius in $(0, 1)$. ∎

Proof of B).

Let S be the $(\Sigma^n, 0, \infty)$ bundle associated to E, where $\Sigma^n = R^n \cup \infty$. By Alexander's theorem, there is a reduction ρ_1 of S to the group of those homeomorphisms h of Σ^n fixing $0, \infty$ with $h(x) = |x|\, h(x/|x|)$ for $x \neq 0, \infty$, a reduction such that D_1 is the unit disc bundle.

We *assert* that there is an isotopy h_t, $0 \leq t \leq 1$, of $\mathrm{id}|\overline{S} = h_0$ respecting fibers and fixing the zero-section, such that $h_1(D_1 \times B^1) = \overline{D}$, where $\overline{S}$ and $\overline{D}$ were defined for A). Then, using the reduction of $\overline{D}$ mentioned for A) we have a $(B^{n+1}, 0)$ bundle $\frac{1}{2}\overline{D}$ of radius ½ and can easily deduce from h_t an isotopy h'_t of $\mathrm{id}|\overline{S} = h'_0$ respecting fibers and zero section so that $h'_1(D_1 \times B^1) = \frac{1}{2}\overline{D}$ and $h'_t(D_1 \times B^1) \subset E \times R$, for all t. This h'_t gives a concordance from $D_1 \times B^1$ to $\frac{1}{2}\overline{D}$ in $E \times R$. Similarly we get one from $\frac{1}{2}\overline{D}$ to $D_2 \times B^1$, so from $D_1 \times B^1$ to $D_2 \times B^1$ as required.

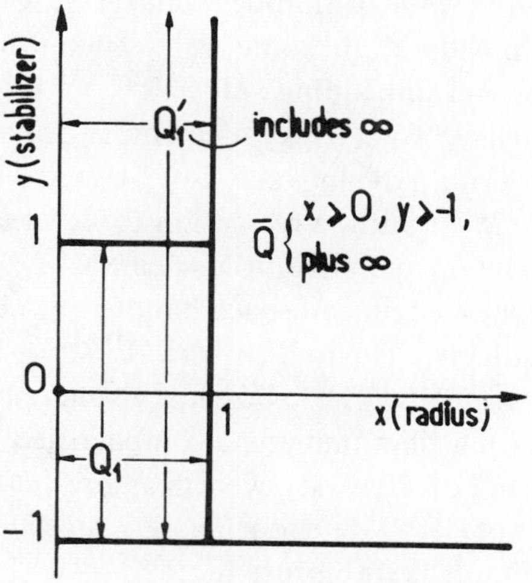

Figure 4-a *(Illustration for the paragraph overleaf)*

To construct h_t note that ρ_1 gives an obvious quotient map

$$q_1 : R_+ \times \delta D_1 \longrightarrow E \qquad \text{(polar coordinates)}$$

where δD_1 is the boundary sphere bundle of D_1, and by natural extensions quotient maps $R_+ \times R \times \delta D_1 \to E \times R$ and

$$q : \{(R_+ \times R) \cup \infty\} \times \delta D_1 \longrightarrow \bar{S} ,$$

so that with the notation of Figure 4-a $\bar{D}$ is the quotient of $\bar{Q} \times \delta D_1$ and $D_1 \times B^1$ of $Q_1 \times \delta D_1$. We leave the reader to produce an isotopy g_t, $0 \leqslant t \leqslant 1$ of the identity of the Gauss sphere $R^2 \cup \infty$ so that $g_1(Q_1) = \bar{Q}$, $g_t(Q_1) \subset (R_+ \times R) \cup \infty$, $g_t(0) = 0$, and $g_t \times (\mathrm{id} \,|\, \delta D)$ passes to the quotient by q defining the asserted isotopy h_t. (Just isotop Q_1 onto Q_1', then Q_1' onto $\bar{Q}$, taking care to respect (y-axis) $\cup \infty$ throughout !) ■

PROOF OF 4.1 .

Given M, let $M' = \{M \cup \partial M \times [0, 1)\} / \{\partial M = \partial M \times 0\}$ be M with an open collar tacked on the boundary . Find an imbedding $i : M' \to R^n$, n large , with a closed locally flat disc bundle $D' \subset R^n$, and let $D \subset R^n$ be the restriction of D' to M. We arrange that D is a clean submanifold of R^n.

[We recall simple ways of finding i and D'. Using an atlas and a partition of unity imbed M' in some R^a. Then using a proper map $M' \to R$ we get a proper imbedding $M' \to R^{a+1} = R^a \times R$. We identify M' now with its image . By a tricky elementary argument† [Mi_5] [Hi_5] . M' has a normal microbundle in some R^b, $b > a + 1$. This is an open neighborhood of M' with a retraction to M' that is a submersion near M', all fibers being open manifolds near M'. This microbundle (indeed any) contains a euclidean space bundle E' by the Kister-Mazur theorem [Kis] [KuL_1] . Then $E' \times R \subset R^{b+1} \equiv R^n$ contains a disc-bundle D' over M' whose boundary sphere bundle $\partial D'$ is a copy of E' with each fiber one-point compactified , see 4.4 . As D' is a clean submanifold of $E' \times R$, we can assure that D and D' are clean submanifolds of R^n by choosing E' so small that its closure in the normal microbundle is its closure in $R^{n-1} = R^b$] .

† The argument of Hirsch is explained in [IV , Appendix A] ; Milnor's argument is different .

Note that the manifold boundary ∂D is bicollared in R^n and apply the Product Structure Theorem [I, §5.2] to obtain a CAT structure Σ on R^n concordant to the standard one so that ∂D_Σ and D_Σ are CAT submanifolds. Then fix a CAT triangulation of D_Σ as a simplicial complex T. If CAT = PL this means that the identity map $\theta : T \to D_\Sigma$ is PL ; existence of T is an elementary fact [Hu_2]. ⟦If CAT = DIFF, this means that $\theta : T \to D_\Sigma$ is smooth C^∞ non-singular on each closed simplex ; existence of T is Whitehead's theorem [Mu_1].⟧

By definition the canonical simple type for M is given by the inclusion

$$M \xrightarrow{i} T .$$

By the CAT Concordance Implies Isotopy Theorem [I, §4.1], there is a small isotopy of $id | R^n$ carrying D_Σ onto a CAT submanifold of R^n. Thus the definition in the statement of 4.1 is seen to be equivalent.

Note that we have already proved

Local finiteness theorem 4.1.3. *Every metrizable topological manifold is proper homotopy equivalent to a locally finite simplicial complex.* ■

Preliminary to showing that the chosen simple type is canonical (i.e. independent of choices) here are two remarks :

(1) *The final choice of T is irrelevant*. (Recall that a PL isomorphism is simple ⟦and for CAT = DIFF use 4.2⟧.)

(2) *Our choice of simple type is unaffected by the following stabilization* which we describe first for CAT = PL. Replace R^n by $R^n \times R$; D by $D \times B^1$; Σ by $\Sigma \times R$; T by $T \times B^1$ (with standard subdivision, B^1 having vertices $-1, 0, 1$). This holds because $T \times 0 \to T \times B^1$ is an expansion. And the stabilized construction is again of the sort described. ⟦If CAT = DIFF this stabilization is to be *corrected* as follows : (a) Alter $\Sigma \times R$ on R^{n+1} by a concordance, unbending the corners of $(\Sigma | D) \times B^1$ using a tube about $\partial D \times \partial B^1$ respecting D – thus getting Σ^* for instance. (b) Replace $T \times B^1$ by any DIFF triangulation T^* of $(D \times B^1)_{\Sigma^*}$. Note that $T \times B^1$ is not a DIFF triangulation of Σ^*. However the identity map $T \times B^1 \to T^*$ is a simple equivalence by 4.3 so the original reasoning now applies.⟧

Now suppose the construction of $i : M' \to R^n$, D', D, Σ, T has been carried out in two ways indicated by subscripts 1 and 2. We must show that

$$M \xrightarrow{i_1} T_1 \qquad , \qquad M \xrightarrow{i_2} T_2$$

define the same simple type.

After stabilization as described in (2) above, we can arrange that $R^{n_1} = R^{n_2} = R^n$, that i_1, $i_2 : M' \to R^n$ are proper, that there is a clean proper imbedding

$$j : M' \times [1, 2] \to R^n \times [1, 2]$$

equal $i_1 \times 1$ on $M' \times 1$ and equal $i_2 \times 2$ on $M' \times 2$, and also that there is a normal disc bundle B' to $j(M' \times [1, 2])$ in $R^n \times [1, 2]$ extending $D_1' \times 1$ and $D_2' \times 2$. We will give more details for construction of j and B' presently. Clearly, we can arrange that, as a bundle, B' is a product along $[1, 2]$ near $R^n \times 1$ and $R^n \times 2$.

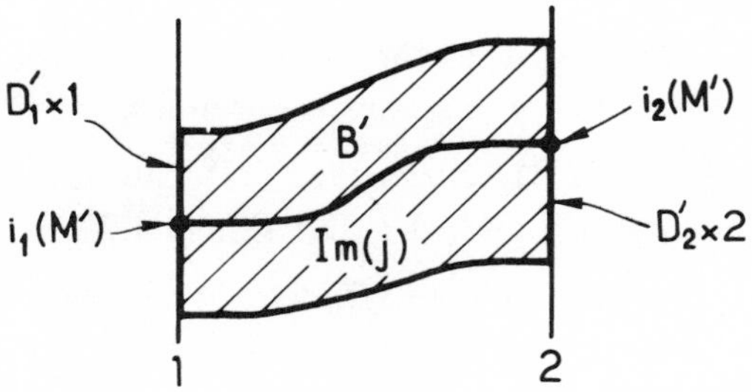

Figure 4-b.

There exists a CAT structure θ on $R^n \times [1,2]$ extending $\Sigma_1 \times 1$ and $\Sigma_2 \times 2$, also a product along $[1,2]$ near both $R^n \times 1$ and $R^n \times 2$. Then the restriction B of B' to $j(M \times [1, 2])$ is CAT in $(R^n \times [1, 2])_\theta$ near there. So the (local) Product Structure Theorem [I, §5.2] lets us alter θ in $R^n \times (1, 2)$ so that B is (everywhere) a CAT submanifold B_θ, which cannot in general be a CAT *bundle*.

Using the bundle homotopy theorem, identify B to $D_1 \times [1, 2]$ sending $j(M \times [1, 2])$ to $i_1(M) \times [1, 2]$ in the standard way,

$D_1 \times 1$ to itself, and $D_2 \times 2$ to $D_1 \times 2$. Then the Concordance Implies Isotopy Theorem applied to $(D_1 \times [1,2])_\theta$ provides us with a CAT isomorphism h making the triangle

$$\begin{array}{ccccc} & \overset{i_1}{\nearrow} & (D_1)_{\Sigma_1} & = & (D_1 \times 1)_\theta \\ M & & \downarrow h & & \\ & \underset{i_2}{\searrow} & (D_2)_{\Sigma_2} & = & (D_1 \times 2)_\theta \end{array}$$

proper homotopy commutative, because, as a map $D_1 \to D_1$, h is TOP isotopic to the identity. In view of preliminary remark (1), we have proved that the choices do not affect the simple type assigned to M – i.e. the type is canonical. ∎

We conclude with

Notes on construction of j and B′ (cf. Figure 4-b).

j is a concordance from i_1 to i_2, constructed after stabilization via preliminary remark (2), much as we suggested i_1 and i_2 be constructed.

The concordance B′ from D'_1 to D'_2 is a bit harder to find. Hirsch's elementary *relative* normal bundle existence theorem [Hi_5], [IV] coupled with the Kister-Mazur theorem [KuL_1], does provide, after stablization, a concordance F say from E_1 to E_2, viz. a normal euclidean space bundle to $j(M' \times [1,2])$ in $R^n \times [1,2]$ restricting to $E_k \times k$ over $i_k(M')$, $k = 1,2$, here E_k being the bundle in which D'_k was found. Just as one found D′ in E one can find a disc bundle B″ in F after one stabilization, that is a clean submanifold of $R^n \times [1,2]$. This B″ can, as usual, be a product along [1, 2] near 1 and 2. If $B'' | (i_1 M' \times 1)$ is concordant to $D'_1 \times 1$ as a disc bundle in $E_1 \times 1$ and similarly in $E_2 \times 2$, then clearly B″ can be corrected near 1 and 2 to give B′ as required. Now 4.4 asserts this after one more stabilization; so then we get B′. ∎

§5. SIMPLE TYPES DECOMPOSITIONS AND DUALITY

This concluding section gives the essential links between our definition of the canonical simple type of a TOP manifold and TOP handlebody theory (or Morse function theory) and TOP surgery. More specifically

(a) We prove that various definitions of simple type for a topological manifold - using (arbitrary !) triangulations, handle decompositions, or Morse functions - all give the same result whenever they apply.†

(b) We show that, with the canonical simple type, any compact topological manifold (or triad) is a simple Poincaré duality space - i.e. roughly speaking the Poincaré duality cap product comes from a simple chain homotopy equivalence.

We continue to give two alternative arguments using DIFF and PL methods respectively.

The oldest way to define the simple type of a manifold M involved choosing *if possible* a CAT (= DIFF or PL) structure Σ on M, then a CAT triangulation $M_\Sigma \approx T$ by a simplicial complex T. That this gives the canonical simple type for M is a special case of

PROPOSITION 5.1. If D is *any* closed disc bundle over the TOP manifold M and D has a CAT (= DIFF or PL) structure and a CAT triangulation; then $M \hookrightarrow D$ gives M the canonical simple type of §4.

Proof: CAT imbed D properly in euclidean space with a closed cleanly imbedded normal CAT disc bundle D* (see [$\mathrm{Hi}_{5,6}$] for PL) and CAT triangulate D* so that the projection $p : D^* \to D$ is a piecewise-linear map. ⟦ for CAT = DIFF assure this by building the Whitehead C^∞ triangulation [Mu_1] [Wh_1] of D* using only charts on D* presenting p as a linear projection.⟧ Now $M \hookrightarrow D^*$ gives by

† Where triangulations are concerned T. Chapman's or R. D. Edwards' more recent results are stronger (see §4). We recommend to the reader Chapman's brief proof [Ch_3] that a proper cell-like mapping of simplicial complexes is a simple equivalence.

definition the canonical simple type. But the inclusion $D \hookrightarrow D^*$ is simple, as an induction up through the skeleta of D readily shows (alternatively the PL retraction $p: D^* \to D$ is collapsible [Chn_1]). Hence $M \hookrightarrow D$ also gives the canonical simple type. ■

It is important in handlebody theory to have a rule that assigns a simple type to a manifold X for each suitably well behaved filtration of X. We present a simple geometric one that will apply to both handlebody decompositions and arbitrary triangulations.

Let us operate in the category $\mathfrak{T}$ of *locally compact metrizable spaces and proper continuous maps*. (A map is proper if the preimage of each compactum is compact). Thus homotopy means proper homotopy; the word *equivalence* means proper homotopy equivalence.

An inclusion map $X \hookrightarrow Y$ is a *cofibration* if any map $f_o: X \times [0,1] \cup Y \times 0 \to Z$ extends to a map $Y \times [0,1] \to Z$. We are here working in $\mathfrak{T}$, to be sure. But note that if an extension $f: Y \times [0,1] \to Z$ of a proper map f_o exists so that f is continuous but not proper, then one can deduce from it a proper extension using the fact (easily proved) that f is necessarily proper on a sufficiently small closed neighborhood of $X \times [0,1] \cup Y \times 0$ in $Y \times [0,1]$.

Recall that the set $\mathcal{S}(X)$ of simple types on a locally finite simplicial complex is an abelian group and that the rule $X \mapsto \mathcal{S}(X)$ is functorial on proper homotopy classes of maps. An equivalence $f: X \to Y$ of simplicial complexes represents an element $[f] \in \mathcal{S}(X)$, also denoted $[Y, X]$ if f is an inclusion. Both addition and functoriality come from astutely forming amalgamated sums [Si_7] [Chn_2]: if $f: X \hookrightarrow Y$ and $g: X \hookrightarrow Z$ with $W = Y \cup Z$ and $X = Y \cap Z$, then $g_*: \mathcal{S}(X) \to \mathcal{S}(Z)$ sends $[f] = [Y, X]$ to $[W, Z]$; if $g: X \hookrightarrow Z$ is also an equivalence one defines $[f] + [g] \equiv [Y, X] + [Z, X]$ to be $[W, X] \equiv [Y \cup Z, X]$.

We extend this trivially to a functor $\mathcal{S}: \mathfrak{T}_o \to$ {Abelian groups} where $\mathfrak{T}_o$ is the category of proper homotopy classes of maps between spaces X such that there exists some (unspecified) proper equivalence $f': X \to X'$ to a locally finite simplicial complex. We simply define $\mathcal{S}(X) = \mathcal{S}(X')$. This makes sense because, if $f'': X \to X''$ is another such equivalence, there is a uniquely determined isomorphism $g_*: \mathcal{S}(X') \cong \mathcal{S}(X'')$ where $gf' \simeq f''$. Here $\simeq$ indicates proper homotopy.

Beware that the group $\mathcal{S}(X)$ is *not* naturally identified to the set of simple types on X unless X is a simple space. For X in $\mathfrak{I}$ or $\mathfrak{I}_o$, the simple types form a merely affine set (possibly empty), and clearly it does not vary functorially with X in $\mathfrak{I}$. For $X \in \mathfrak{I}_o$, our group $\mathcal{S}(X)$ can be defined more intrinsically as a group whose elements are ordered pairs of simple types on X.

DEFINITION 5.2. A **spaced simple block decomposition** of a space X in $\mathfrak{I}$, or an **s-decomposition** for short, consists of two things. *First* is a finite filtration of X

$$X_0 \subset X_0^+ \subset X_1 \subset X_1^+ \subset X_2 \subset X_2^+ \subset \ldots \subset X_n \subset X_n^+ = X$$

by closed subsets $X_i \subset X_i^+$ (equal ϕ if $i < 0$) enjoying the properties:

(a) $X_i \hookrightarrow X_i^+$ is an equivalence.

(b) writing $B_i = \mathrm{Cl}(X_i - X_{i-1}^+)$ (Cl indicating closure in X), and $\partial_- B_i = B_i \cap X_{i-1}^+$, the inclusion $\partial_- B_i \hookrightarrow X_i$ is a cofibration.

Second is a prescription of simple type for each "block" B_i and each $\partial_- B_i$.

LEMMA 5.3. *Suppose $Z \in \mathfrak{I}$ is a union $Z = Z_1 \cup Z_2$ of closed subsets so that $Z_o = Z_1 \cap Z_2 \hookrightarrow Z$ is a cofibration. If Z_o, Z_1, Z_2 are simple spaces, then there is an equivalence of triads $f : (Z; Z_1, Z_2) \to (Z'; Z_1', Z_2')$ to a simplicial triad in $\mathfrak{I}$, giving simple equivalences $f : Z_k \to Z_k'$, $k = 0,1,2$, where $Z_o' = Z_1' \cap Z_2'$. The resulting simple type for Z is independent of choices.*

Proof of 5.3. Form a homotopy commutative diagram

$$(5.3.1) \qquad \begin{array}{ccccc} Z_1' & \xleftarrow{j_1} & Z_o' & \xrightarrow{j_2} & Z_2' \\ {\scriptstyle f_1}\uparrow & & {\scriptstyle f_o}\uparrow & & {\scriptstyle f_2}\uparrow \\ Z_1 & \longleftarrow & Z_o & \longrightarrow & Z_2 \end{array}$$

where f_1, f_o, f_2 represent the simple types of Z_1, Z_o, Z_2 respectively. By simplicial approximation and a mapping cylinder device we can arrange that j_1 and j_2 are simplicial inclusions of simplicial complexes. Form Z' from the sum of Z'_1 and Z'_2 identifying the copies of Z'_o. Adjusting f_1 and f_2 using evident cofibration properties we can arrange that f_1 and f_2 agree with f_o on Z_o and so give a map $f : Z \to Z'$. This is a (proper homotopy) equivalence; indeed it is an equivalence of triads $(Z ; Z_1, Z_2) \to (Z' ; Z'_1, Z'_2)$ – by a well known argument† [Di] [Si_2, Appendix III]. Now f defines the wanted simple type. We insert

REMARK 5.3.2. In the homotopy equivalence of triads $f : (Z' ; Z_1, Z_2) \to (Z' ; Z'_1, Z'_2)$ let us look for the originally given representative of the simple type of Z_1, say $f_{1,0} : Z_1 \to Z'_{1,0}$. We see that Z'_1 can be a simplicial mapping cylinder of a map to $Z'_{1,0}$ – namely j_1 in (5.3.1) made simplicial – so $Z_{1,0} \hookrightarrow Z'_1$ is an expansion. And $f | Z_1$ can be $f_{1,0}$. ∎

That the simple type given by $f : Z \to Z'$ is independent of choices follows from the *sum theorem* of [Si_7]. This states that, if $g : (Z' ; Z'_1, Z'_2) \to (Z'' ; Z''_1, Z''_2)$ is an equivalence of tripples of complexes, then in $\mathcal{S}(Z')$ one has

(5.4) $$[g] = i_{1*}[g_1] + i_{2*}[g_2] - i_{0*}[g_o]$$

where g_k representing $[g_k] \in \mathcal{S}(Z'_k)$ is the restriction of g to an equivalence $Z'_k \to Z''_k$, and $i_k : Z'_k \to Z'$ induces $i_{k*} : \mathcal{S}(Z'_k) \to \mathcal{S}(Z')$. Thus g is a simple equivalence if g_1, g_2 and g_0 are. This completes the proof of 5.3. ∎

REMARK 5.4. *Notice that the sum formula carries over automatically to triads of simple spaces where all the subspaces are closed and their inclusions cofibrations.* ∎

We now deduce

THEOREM 5.5.

In a s-decomposition as above, each X_i and each X_i^+ has a preferred simple type.

Proof of 5.5. By hypothesis $X_0 = B_0$ has a preferred simple type. Suppose inductively that X_{k-1} has one, where $n \geqslant k \geqslant 1$. Give to

† It transcribes without change into the proper category.

X_{k-1}^+ the simple type making $X_{k-1} \to X_{k-1}^+$ a simple equivalence . Now apply 5.3 to the tripple

$$(X_k ; X_{k-1}^+ , B_k) \quad , \quad \partial_- B_k = X_{k-1}^+ \cap B_k ,$$

to define the simple type of X_k . This completes the induction to make all X_i , X_i^+ simple spaces . ■

REMARK 5.5.1. Suppose in the above theorem that X is a simplicial complex which is *simplicially* s-decomposed in the sense that
(a) each X_i and X_i^+ is a subcomplex and $X_i \hookrightarrow X_i^+$ is a simple equivalence .
(b) B_i and ∂B_i are subcomplexes and are assigned their natural type . Then it is clear that this theorem assigns to X its natural type (as complex) . ■

DEFINITIONS 5.6. Consider $f : X \to \hat{X}$ a map in $\mathfrak{T}$ of spaces with s-decompositions with the same number $n + 1$ of blocks , such that for each i , $0 \leqslant i \leqslant n$, the map f gives by restriction a homotopy equivalence of triads $(X_i ; X_{i-1}^+ , B_i) \to (\hat{X}_i ; \hat{X}_{i-1}^+ , \hat{B}_i)$. Then we call f a **blocked equivalence** of s-decompositions .

The s-decomposition of X is called **complete** if inclusion induces the zero map $\mathcal{S}(B_i) \to \mathcal{S}(X)$ for each i .

THEOREM 5.7 (with data of 5.6) . *The class of f in* $\mathcal{S}(X)$ *is* $b_0 + (b_1 - b_1') + (b_2 - b_2') + \ldots + (b_n - b_n')$, *where* b_i *is the image in* $\mathcal{S}(X)$ *of the class in* $\mathcal{S}(B_i)$ *of the map* $f : B_i \to \hat{B}_i$ *and* b_i' *is similarly the image in* $\mathcal{S}(X)$ *of the class of* $f : \partial_- B_i \to \partial_- \hat{B}_i$.

Proof of 5.7 . Use induction on m , applying the sum theorem 5.4 extended to simple spaces . ■

COROLLARY 5.7.1. *If an s-decomposition of a space X is complete , the simple type it gives X (via 5.5) is independent of the simple types chosen for the blocks* B_i *and for* $\partial_- B_i \subset B_i$. ■

PROPOSITION 5.8. *Let X be a space in* $\mathfrak{T}$ *with an s-decomposition . There always exists a simplicially s-decomposed simplicial complex* $\hat{X}$ *(see 5.7) and a blocked equivalence* $f : X \to \hat{X}$.

Proof of 5.8. If this is true for decompositions of length n , it follows for length $n + 1$ by applying Lemma 5.3 and Remark 5.3.2 . ■

DEFINITION 5.9. An s-decomposition of a TOP manifold M^m : $M_0 \subset M_0^+ \subset M_1 \subset M_1^+ \subset \ldots \subset M_n \subset M_n^+ = M$ is called a **TOP s-decomposition** if :

a) Each M_i and M_i^+ is a clean TOP m-submanifold of M .

b) Each 'space' $S_i = \mathrm{Cl}(M_i^+ - M_i)$ is a clean submanifold of M_i^+ , and $(S_i , \partial_- S_i) \approx \partial_- S_i \times ([0, 1], 0)$ where $\partial_- S_i = S_i \cap M_i$. Here $\approx$ indicates homeomorphism . (Note that S_i may well be empty .)

c) Each 'block' $B_i = \mathrm{Cl}(M_i - M_{i-1}^+)$ is a clean submanifold of M_i . And both B_i and $\partial_- B_i = B_i \cap M_{i-1}^+$ have the canonical simple type of §4 , (which is necessarily true if $(B_i , \partial_- B_i)$ is a disjoint union of handles) .

Example 5.9.1. From a TOP Morse function $f : (M^m ; V, V') \to ([0, 1] ; 0, 1)$ and a gradient-like field on a compact triad (see §3) , a familiar procedure [Mi_8 , §3] produces a TOP s-decomposition of M with M_o a collar of V , and with one block B_i for each critical value of f , each $(B_i , \partial_- B_i)$ being a disjoint union of handles , copies of $(B^k , \partial B^k) \times B^{m-k}$ for various k .

Example 5.9.2. Using §2 one obtains an s-decomposition for any TOP manifold M^m , $m \geqslant 6$. An s-decomposition of M is complete as soon as each component of each block lies in a contractible co-ordinate chart (see 2.1.1) .

THEOREM 5.10. *Every TOP s-decomposition of a TOP manifold M yields (via 5.5) the canonical simple type for M defined in §4 .*

Proof of 5.10. Assume indicutively that this holds for such decompositions that have $\leqslant n$ blocks . (It holds trivially for $\leqslant 1$ block) . Let M have n blocks ; then by inductive hypothesis M_{n-1}^+ gets canonical type .

Choose an embedding of M^m in some euclidean space of high dimension with normal cleanly embedded disc-bundle D . We find a CAT (= DIFF or PL) structure Σ on $D|M_n$ arranging , by the (local) Product Structure Theorem [I , §5.2] , that $(D|M_{n-1}^+)$ is a clean CAT submanifold of $(D|M_n)_\Sigma$.

Then we find a CAT triangulation of the CAT triad $(D|M_n ; D|M_{n-1}^+ , D|B_n)_\Sigma$. This clearly determines by inclusion the canonical simple type for all parts of the triad $(M_n ; M^+_{n-1}, B_n)$, even for the intersection $\partial_- B_n$. Thus the canonical type for M_n agrees with the type given by 5.5.

The same is true for $M^+_n = M$ because the inclusion $M_n \to M^+_n$ is a simple equivalence for the canonical simple types. Indeed $M_n \hookrightarrow M^+_n$ is proper homotopic to a homeomorphism. This completes the proof of 5.10. ■

COMPLEMENT 5.10.1. *Suppose a TOP manifold M has an s-decomposition, which is not TOP, but gives a TOP s-decomposition of $M \times B^r$ by producting with a disc B^r, $r \geqslant 0$. Then this s-decomposition of M again gives the canonical simple type of M.*

Proof of 5.10.1. The inclusion $M \xrightarrow{\times 0} M \times B^r$ is a simple equivalence both if we use the canonical types of §4 and again if we use the s-decomposition types of 5.5 (see 5.7). Thus the result follows from 5.10 applied to $M \times B^r$. ■

† **THEOREM 5.11.** *An arbitrary triangulation of a TOP manifold M as a simplicial complex gives to M the canonical simple type of §4.*

Proof of 5.11.

Let $M^{(i)}$ be the i-skeleton of M for a fixed triangulation of M, and define an s-decomposition of M by setting $M_i = M_i^+$ equal the union of all closed simplices of M'' which meet $M^{(i)}$. Here $M'' = (M')'$ denotes a second derived subdivision of M.

Since $M_0 \subset M_1 \subset M_2 \subset \ldots$ is a filtration of M by subcomplexes of M'', the simple type of the filtration is by 5.5.1 the simple type defined by $\mathrm{id} : M \to M''$ or again by $\mathrm{id} : M \to M$.

But this s-decomposition also gives the canonical simple type, by complement 5.10.1, since we shall prove the

ASSERTION. $M_0 \times B^3 \subset M_1 \times B^3 \subset \ldots \subset M_m \times B^3$ *is an unspaced TOP s-decomposition of* $M \times B^3$.

The proof is broken into two steps.

† Omit on first reading, as much stronger results are now available, eg. [Ch$_3$]. However this is a part of R. D. Edward's proof that a homeomorphisms of polyhedra is a simple equivalence [Si$_{14}$] [Ed$_2$].

Note that B_i has one component H_σ for each i-simplex σ of M. It is the closed star neighborhood $St(\hat{\sigma}, M'')$ of the center $\hat{\sigma}$ of σ (the vertex of M' in the interior of σ).

Step 1. *The component $(H_\sigma, \partial_- H_\sigma)$ of $(B_i, \partial_- B_i)$, corresponding to an i-simplex σ of M, is homeomorphic to $(B^i, \partial B^i) \times cL$. Here $B^i = $ (i-disc), and cL is a copy of the closed cone on the link $Lk(\sigma, M)$ of σ in M. Also B_i has collared frontier in $M - \mathring{M}_{i-1}$.*

Step 2. *$B_i \times B^3$ is a manifold and $\partial_- B_i \times B^3$ is a clean submanifold of its boundary.*

Proof of step 2. We verify this for any one component of $(B_i, \partial_- B_i)$, which we express as $(B^i, \partial B^i) \times cL$ using step 1. We know that $(\text{int} B^i) \times \mathring{c}L$, with $\mathring{c}$ indicating open cone, is an open subset of M. Thus $\mathring{c}L \times R^i$, $L \times R^{i+1}$, $L \times R^{i+1} \times [0, 1)$, $cL \times R^{i+1}$, $\partial B^i \times cL \times R^2$, $\partial B^i \times cL \times B^3$ are successively seen to be manifolds, from which step 2 follows. ■

For completeness we give

Proof of step 1. For any simplicial complex M this step succeeds and the result is in fact PL. In the first derived M' of M we have the join

$$N_\sigma = (\partial\sigma)' * \hat{\sigma} * L$$

where L is the subcomplex with typical simplex $\langle \hat{\sigma}_1, \hat{\sigma}_2, \ldots, \hat{\sigma}_k \rangle$ spanned by the (bary-) centers of simplices with $\sigma < \sigma_1 < \sigma_2 < \ldots < \sigma_k$. Here $<$ indicates "a subcomplex of". Recall that L is isomorphic to the 1st derived of $Lk(\sigma, M)$ by pseudo-radial projection sending the vertex $\hat{\sigma}_i$ of L to $\hat{\tau}_i$, where $\sigma_i = \sigma * \tau_i$. As N_σ is a neighborhood of H_σ in M it will suffice to examine $N_\sigma \cap (M_i, M_{i-1})$. This is naturally (PL) homeomorphic to the pair $(A \cup B, A)$ in N_σ described as follows. Use barycentric co-ordinates of the join N_σ to express a typical point p of N_σ as

$$p = u \cdot x + v \cdot \hat{\sigma} + w \cdot z \in (\partial\sigma)' * \hat{\sigma} * L \equiv N_\sigma$$

with $x \in (\partial\sigma)'$, $z \in L$ and u, v, w in [0, 1] with $u + v + w = 1$. Then define

$$A = \{p \mid u \geqslant 1/2\} \quad ; \quad B = \{p \mid w \leqslant 1/4\}.$$

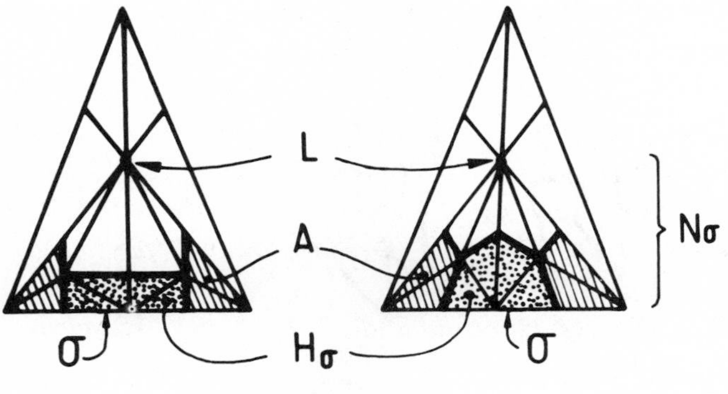

Figure 5-a

See Figure 5-a (left). In fact the two pairs $N_\sigma \cap (M_i, M_{i-1})$ and $(A \cup B, A)$ are identical when (following Zeeman [Ze, Chap. 3]) we make special (non-standard) choices for the centers $\hat{\tau}$ of simplices τ of $N_\sigma < M'$ to define the second derived – namely, so that if $\hat{\tau} = u \cdot x + v \cdot \hat{\sigma} + w \cdot z$, then (a) $u = 0, 1/2$ or 1, and (b) $w = 0$ or $1/4$ when $v \neq 0$. (In general, the PL homeomorphism comes from shift of centers, see Figure 5-a comparing right and left).

With the special choices of centers $H_\sigma = A \cap B = \{p \mid u \leqslant 1/2, w \leqslant 1/4\}$, $\partial_- H_\sigma = \{p \mid u = 1/2, w \leqslant 1/4\}$, $\partial_+ H = \{p \mid u \leqslant 1/2, w = 1/4\}$, and $N_\sigma \cap (M - \mathring{M}_{i-1}) = \{p \mid u \leqslant 1/2\}$. Step 1 follows; a suitable homeomorphism $H_\sigma \to c(\partial\sigma) \times cL \approx B^i \times cL$ is given by $u \cdot x + v \cdot \hat{\sigma} + w \cdot z \leftrightarrow (2u \cdot x, 4w \cdot z)$, and a suitable (PL) collar of $\partial_+ H_\sigma$ in $M - \mathring{M}_{i-1}$ is $\{p \mid u \leqslant 1/2, 1/4 < w \leqslant 1/2\}$. ∎

Theorem 5.11 is now proved. ∎

Next is the result about s-decompositions necessary for the TOP s-cobordism theorem.

THEOREM 5.12.

Let $M_0 \subset M_0^+ \subset M_1 \subset M_1^+ \subset \ldots \subset M_{m+1}^+ = M^m$ be a TOP s-decomposition of a compact connected TOP m-manifold M such that $(B_i, \partial_- B_i)$ is a disjoint union of $(i-1)$-handles for $i \geqslant 1$.

If $j : M_0 \hookrightarrow M$ *is a homotopy equivalence, its Whitehead torsion*† $\tau(j) \in Wh(\pi_1 M) \cong \mathcal{S}(M)$ *is the torsion of the finitely based acyclic complex over* $Z[\pi_1 M]$ *derived from the induced filtration* $\tilde{M}_0 \subset \tilde{M}_1 \subset \ldots \subset \tilde{M}^m$ *of the universal covering* $\tilde{M}$ *of* M .

Proof of 5.12. By 5.8 , there is an s-decomposed finite complex Z and a blocked equivalence $f : M \to Z$, giving , by 5.7 , a simple equivalence $M_i \to Z_i$ for each i . Then the induced map

$$f_* : \mathbf{C}_*(\tilde{M}, \tilde{M}_0) \to \mathbf{C}_*(\tilde{Z}, \tilde{Z}_0)$$

of complexes associated to the filtrations of the universal covers is an isomorphism commuting with the action of covering translations . As f is blocked , it maps a prefered basis on the left to a prefered basis on the right . [Recall that , for each handle component (H, ∂_H) of $(B_{i+1}, \partial\_B_{i+1})$, one forms a basis element for $\mathbf{C}_*(\tilde{M}, \tilde{M}_0)$ as follows : One chooses at will a lifting $(\bar{H}, \partial\_\bar{H})$ in $\tilde{M}$, and at will an integral homology generator on $(\bar{H}, \partial\_\bar{H}) \simeq (B^i, \partial B^i)$, giving an element of $H_i(\tilde{M}_{i+1}, \tilde{M}_i^+) = H_i(\tilde{M}_{i+1}, \tilde{M}_i) \equiv \mathbf{C}_i(\tilde{M}, \tilde{M}_0)$. This is the basis element for (H, ∂_H) . Similarly in $\mathbf{C}(\tilde{Z}, \tilde{Z}_0)$.]

Now the torsion $\tau(j)$ of $j : M_0 \hookrightarrow M$ equals the torsion of $Z_0 \hookrightarrow Z$, which Milnor's algebraic subdivision theorem [Mi9 , §5.2] proves to be that of $\mathbf{C}_*(\tilde{Z}, \tilde{Z})$. As the latter is based isomorphic to $\mathbf{C}_*(\tilde{M}, \tilde{M}_0)$, the proof is complete . ■

We close this final section by explaining why Poincaré duality expresses itself at the chain level by an algebraically simple chain homotopy equivalence . Complete proofs would involve the commodious machinery of simple homotopy theory and thus require an extravagant amount of space . We we will give proofs in outline , as in [Wa] , leaving the reader to fill in the details ; some points are clarified by Maumary [Mau] .

Consider a TOP *compact* manifold triad $(M ; M_-, M_+)$, $\partial M = M_- \cup M_+$, $M_- \cap M_+ = \partial M_\pm$. There always exists a finite simplicial

† The parallel statement and proof apply to Reidemeister representation torsions [Mi9 , §8] .

complex triad $(X ; X_- , X_+)$ and by 5.3 a simple homotopy equivalence of triads $f : (M ; M_- , M_+) \to (X ; X_- , X_+)$. It is understood here that M , M_- , M_+ and $M_- \cap M_+ = \partial M_\pm$ have the canonical type of §4. The map f gives a simple equivalence of all four of these.

THEOREM 5.13. *In this situation* $(X ; X_- , X_+)$ *is a (simple) Poincaré duality triad in the sense of Wall [Wa , p.23].*

As this is independent of the choice of $(X ; X_- , X_+)$, see [Wa , p.24] , it is convenient to say that (M , M_- , M_+) *itself is a simple Poincaré duality triad*.

Proof of 5.13.

Let D be a normal disc bundle to M embedded as a clean CAT (= DIFF or PL) submanifold of some R^n, using the (local) Product Structure Theorem [KS_5 , 5.2] we arrange that the three manifolds $D_+ \equiv D \,|\, M_+$, $D_- \equiv D \,|\, M_-$ and the sphere bundle $\dot{D}$ are clean CAT submanifolds of ∂D (possibly with corners if CAT = DIFF). Then we choose a CAT triangulation of D giving CAT triangulations of the submanifolds $\dot{D} , D_+ , D_-$. This makes D a simplicial complex.

The proof centers on a homotopy commutative diagram of chain homotopy equivalences (with shift of dimension), over the group ring $\Lambda = Z[\pi_1 M]$

$$
\begin{array}{ccccc}
C^*(M, M_-) & \xrightarrow{r^*} & C^*(D, D_-) & \xrightarrow{[D]\cap} & C_*(D, \dot{D} \cup D_+) \\
(\star)\ [M]\cap \Big\downarrow & & & \searrow^{i_*[M]\cap} & \Big\downarrow \cap U \\
C_*(M, M_+) & & \xleftarrow[\quad r_* \quad]{} & & C_*(D, D_+)
\end{array}
$$

As $(X; X_-, X_+)$ can be $(D; D_-, D_+)$ our task is to show that $i_*[M]\cap$ is always a simple homotopy equivalence.

With the one important difference that we use singular theory, we adopt the conventions of Wall [Wa , Chap. 2] . For example , for *any pair* (X, Y) with X mapping naturally to M (as does D), $C_*(X, Y)$ is the complex of singular chains of $\tilde{X}$ modulo that of $\tilde{Y}$. Here $\tilde{X}$ is the covering of X induced from the universal covering $\tilde{M} \to M$. Similarly for $\tilde{Y}$. This complex $C_*(X, Y)$ is a free right Λ complex via the free right action of $\pi_1 M$ on $\tilde{X}$ as covering translations. $C^*(X, Y)$ is $\mathrm{Hom}_\Lambda(C_*(X, Y), \Lambda)$, converted from a left to a right Λ module using the involution of Λ given by $g \leftrightarrow w_1(g)g^{-1} \in \Lambda$, $g \in \pi_1 M$. Recall that $w_1 : \pi_1 M \to Z_2$ is the 1st Stiefel-Whitney class of M. $C^*(X, Y)$ is thought of as a graded group with C^i in dimension $-i$ and with differential $\delta : C^{i+1} \leftarrow C^i$ equal to $(-1)^i \mathrm{Hom}(d, \Lambda)$, $d : C_{i+1} \to C_i$ (the sign $(-1)^i$ is essential to make the cap products $[M]\cap$ and $[D]\cap$ chain maps). $[M]$ is a chain representing the fundamental homology class of M with w_1-twisted integer coeficient bundle Z_M. Similarly $[D]$, but the coeficients are untwisted: $Z_D = Z$. U represents the cohomology Thom class of the bundle $D \to M$ lying in $H^k(D, \dot{D}; Z_M)$, $k+m = n$.

Each complex of (☆) is a *simple complex*, i.e. it is equipped with a *simple structure* represented by a chain homotopy equivalence over Λ to a free complex over Λ with a finite Λ-basis. This equivalence is well defined only up to chain homotopy and composition with a simple equivalence. We are imitating the notion of *simple space* given in §4; we derive the notion of a simple equivalence of simple complexes.

For example $C_*(D, D_+)$ is naturally chain homotopy equivalent to the similarly defined chain complex $\mathbf{C}_*(D, D_+)$ of cellular rather than singular theory. See Wall [Wa_1, Lemma 1], or better, use a cellular approximation to the natural map $(SD, SD_+) \to (D, D_+)$ from geometrically realized singular complexes. And $\mathbf{C}_*(D, D_+)$ has a preferred finite Λ-base consisting of oriented lifts to $\tilde{D}$ of the simplices of $D-D_+$.

As $(X; X_-, X_+) \equiv (D; D_-; D_+)$, the simple structure on $C_*(M, M_-)$ is the unique one so that r_* is a simple equivalence. Similarly r^* is simple.

The cap product $[D]\cap$ is simple because it is known to be simple in cellular homology theory. One proof (Wall [Wa, 2.1]) uses dual cells and special diagonal approximations, another uses a handle induction† in $(D; D_-, D_+)$.

† This sort of handle induction could perhaps be used in $(M; M_-, M_+)$ to prove 5.13 somewhat more directly in case M has a handle decomposition on (a collar of) $M_\pm$; we know it has one always for $\dim M \geqslant 6$, but not always for $\dim M < 6$ [Si_8].

If we can prove that the Thom equivalence $\cap U$ is simple , it will follow that $[M]\cap$ is simple . To check the simplicity of $\cap U$ is easy if M admits a handle decomposition on a (collar of) M_+ . Since the result is clear for just one handle , it can then be established conveniently by induction on the number of handles . When M does *not* admit such a decomposition , we can reduce to the case where it does (§2, §3) , by producting everything with a disc to reach dimension $\geqslant 6$. Alternatively reduce to this case by pulling back D over a copy M′ of D ; with this approach (M, M_+) gets replaced by a CAT pair (M', M'_+) , and §§2 and 3 are not needed .

Thus Poincaré duality is simple . ∎

Appendix A
ON AVOIDING SURGERY BY USE OF STABLE STRUCTURES

A.1. It is regretable that the very basic results of this essay depend via the Product Structure Theorem [KS_5 , §5] on the sophisticated non-simply-connected surgery used in [Ki_1]. One aim of this appendix is to show that this unfortunate dependence disappears if we reformulate all our results in the realm of STABLE topological manifolds of Brown and Gluck. Thus the basics of the theory of STABLE manifolds are relatively elementary. What at present requires the sophisticated surgery is the fact [Ki_1] that in high dimensions STABLE manifolds are precisely TOP manifolds. Another aim is to show that this surgery is entirely unnecessary to well-define the simple type of a TOP manifold as in §4.

A.2. Recall that a homeomorphism $h : U \to U'$ of open subsets of R^m is called STABLE if for each point $x \in U$, there is an open neighborhood U_x of x in U so that $h|U_x$ is isotopic through open embeddings to a restriction of a linear isomorphism $R^m \to R^m$. When such an isotopy exists, it can be made to fix x.

The topological isotopy extension theorem [EK] shows that the above definition is equivalent to the more complex definition of Brown and Gluck [BrnG].

Then the STABLE homeomorphisms of open subsets of R^m form a pseudo-group [KN], so we derive the notion of a STABLE m-manifold without boundary.

If $h_t : M \to M'$, $0 \leqslant t \leqslant 1$, is an isotopy through open embeddings where M and M' are STABLE manifolds, and h_0 is a STABLE isomorphism onto its image in M', then h_1 is likewise STABLE. The proof is straightforward.

Hence if $h : M \to M'$ is a homeomorphism of connected STABLE manifolds (without boundary) that is STABLE on some open subset, then h itself is STABLE. This is the **cardinal fact** about STABLE structures.

STABLE m-manifolds *with* boundary can be defined similarly (as in [I , §5]) by using R^m_+ in place of R^m .

There is an equivalent definition as follows . A STABLE structure on a manifold M with boundary ∂M consists of a pair (Σ,σ) consisting of a STABLE structure Σ on intM and a STABLE structure σ on ∂M such that , for one (or any) collar $\partial M \times [0,1]$ of $\partial M = M \times 0$ in M , the STABLE structures Σ and $\sigma \times (0,1)$ coincide on $\partial M \times (0,1) \subset \text{int} M$. The pair (Σ,σ) together with the collar determine an unique STABLE structure Σ' in the first described sense , with model R^m_+ . And Σ' clearly gives back Σ and σ by restriction . The choice of collar does not affect Σ' , since collars are (locally) unique up to isotopy [I, Appendix A] .

We now give boldly the recipe for translating the results of this essay into the realm of STABLE manifolds : *In the definitions and hypotheses , as well as in the conclusions of each theorem , each manifold and homeomorphism which is mentioned or could naturally be mentioned is to be supposed STABLE . Every condition of mutual compatibility which can be naturally expressed for the STABLE structures is to be insisted upon .*

In practice this recipe should lead to a perfectly unique maximally STABLE version of each notion and result . The reader is however cautioned not to overlook manifolds given implicitly which should be supposed STABLE . Let us illustrate by describing a STABLE Morse function $f : M^m \to R$ where M is a STABLE manifold without boundary . Let C be the set of critical points . For each point $y \in R$, $f^{-1}(y) - C$ is a "naturally expressed" manifold , so it is supposed to have a *given* STABLE structure . (This is extra equipment necessarily given with f !) . These structures are naturally subject to a compatibility condition : for each $x \in M{-}C$, there must exist U open in $R^{m-1} \times R = R^m$ and a STABLE embedding $h : U \to M{-}C$ so that $x \in hU$, and $fh = p_2 : R^m \to R$, and so that , for each $y \in R$, the restriction of h to $U \cap (R^{m-1} \times y) \to f^{-1}(y) - C$ is STABLE . Thus $f|M{-}C$ is a STABLE submersion in the best sense . About each critical point p for f there is a STABLE chart $h : R^m \to M$ with $h(0) = p$ so that $fh(x) = f(p) + Q(x)$ for x near 0 , where $Q(x_1, \ldots, x_m) = \pm x_1^2 \pm x_2^2 \pm \ldots \pm x_m^2$. Further , hear 0 , h is a STABLE embedding of each smooth surface $Q^{-1}(\text{constant})$ into a level surface of $f|(M{-}C)$.

The STABLE results are in each case† to be proved like the corresponding TOP result in §§1–5. Whenever the Product Structure Theorem was applied for the TOP result the STABLE version of it [I, §5] (using no surgery!) can be substituted when we rely on the compatible system of STABLE structures that is carried (and developed) throughout the STABLE proof.

A.3. The TOP h-cobordism theorem without surgery?? It seems plausible to us that there is a still undiscovered proof, without difficult surgery, of at least that part of the Stable Homeomorphism Theorem (for dimensions $\geqslant 5$) [Ki_1] called the Annulus Theorem, which in dimension n is equivalent to surjectivity of $\pi_0(\mathrm{TOP}_{n-1}) \twoheadrightarrow \pi_0(\mathrm{TOP}_n)$. To support this prejudice, note that the surjectivity $\pi_i(\mathrm{TOP}_{n-1}) \twoheadrightarrow \pi_i(\mathrm{TOP}_n)$, $0 < i \leqslant n \geqslant 6$, requires only the DIFF or PL s-cobordism theorem and some geometry. (See [Si_{10}, §5] for example).

Anyone fortunate enough to discover such an elementary proof (for $n \geqslant 6$) could for example show (cf. [BrnG]) that any simply connected h-cobordism is STABLE and so acquire an elementary proof of the TOP h-cobordism theorem.

Simple type without surgery. We turn now to the matter of well defining the simple homotopy type of a TOP manifold (in §4) without reliance on surgery‡. The surgery intervened in the two applications of the Product Structure Theorem of [I]: one application for constructing the simple type, one for proving its well-definition. We can clearly replace these two applications by applications of the STABLE Product Structure Theorem (independent of surgery) as soon as we can prove

PROPOSITION A.4. *Let M be a TOP manifold of which an open subset $M_0 \supset \mathrm{int}\, M$ has a given STABLE structure. If $\partial M \subset M$ gives a π_1 surjection (of connected spaces), one can find (by elementary means) a unique STABLE structure on $M \times [-1, 1]$ extending the given STABLE structure on $M_0 \times [-1, 1)$.*

† There is one exception. The discussion of elimination of double points by Whitney's method should follow [Si_{10}, §7.3] not our treatment in §3, which relied on surgery.

‡ T. A. Chapman's recent more general results [Ch_1] [Ch_2] [Si_{14}] certainly do not require surgery either.

Indeed, in revising the proof of 4.1 one applies this once to D and to B. An allowable stabilization of each by $\times [-1,1]$ is introduced.

The above proposition follows from the next proposition applied to $\partial(M \times [-1,1])$. To see the implication recall the *cardinal fact* that STABLE structures on a connected open manifold coinciding near some point coincide throughout.

PROPOSITION A.5. *Let M be a connected open TOP manifold and let $M_0 \subset M$ be a connected open subset with a given STABLE structure. If $\pi_1 M_0 \to \pi_1 M$ is a surjection, one can find (by elementary means) a unique STABLE structure on M that extends that of M_0.*

Some preliminary remarks (from [BrG]) are required for the proof. Given any topological manifold M without boundary there is a canonical covering space $p : \widetilde{M} \to M$ where $\widetilde{M}$ is equipped with a STABLE structure. $\widetilde{M}$ is in a sense the space of STABLE structures on M; to construct it, form the disjoint sum of all U_σ where U is an open subset of M and σ is a STABLE structure on U, then identify $x \in U_\sigma$ to another such point $x' \in U'_{\sigma'}$ whenever, as points of M, x equals x', and σ equals σ' near x in M. To see that the natural map $p : \widetilde{M} \to M$ is a covering, use the cardinal fact again. (By the STABLE Homeomorphism Theorem of [K_1], we know, for $\dim M \neq 4$, that $\widetilde{M} = M$ and p is the identity. But we avoid using this theorem and treat $p : \widetilde{M} \to M$ as a possibly nontrivial covering.)

Proof of A.5. A STABLE structure on an open subset U of M is the same as a continuous section of $p : \widetilde{M} \to M$ defined over U. Thus in the proposition we are given a section s defined over M_0. Given $x \in M$, let $\lambda : [0,1] \to M$ be a path joining the base point $x_0 = \lambda(0) \in M_0$ to $\lambda(1) = x$. There is a lifting $\widetilde{\lambda}$ of λ to $\widetilde{M}$ unique once we lift x_0 to $s(x_0)$. Let $s(x) = \widetilde{\lambda}(1)$ define $s : M \to \widetilde{M}$. To see that $s(x)$ is independant of the choice of λ, note that a different choice λ' gives a loop "$\lambda^{-1} \circ \lambda'$" in M which deforms to a loop in M_0; but as the loop in M_0 lifts to a loop in $\widetilde{M}$, so does $\lambda^{-1} \circ \lambda'$. This s is the required continuous section. ∎

Appendix B. STRAIGHTENING POLYHEDRA IN CODIMENSION $\geqslant 3$.

This appendix briefly discusses (but does not prove) a strong codimension $\geqslant 3$ straightening theorem due to R. T. Miller, J. L. Bryant and others. It was first attained via simplicial techniques largely unrelated to these essays! We made use of the version for manifolds in our treatment of imbedded TOP microbundle transversality for manifolds in §1 of this essay.

Consider a polyhedron P topologically imbedded as a closed subset of a TOP manifold M. We say that P is **locally tame** near $C \subset M$ if C is covered by open charts U of M with PL *manifold* structure, such that $P \cap U \hookrightarrow U$ is piecewise linear from the PL structure of P to that of U.

THEOREM B.1. [Bry_2] [BS, Theorem 3] [Mil_1]

Let M^m be a PL manifold with metric d and $P \subset M$ a topologically imbedded polyhedron closed and locally tame in M. Provide that $m - \dim P \geqslant 3$, and $\partial M = \phi$. Then there exists a small homeomorphism $h: M \to M$ such that $P \xrightarrow{h} M$ is PL.

One can also find a small isotopy h_t, $0 \leqslant t \leqslant 1$, of h to $id|M$, (small means that for given continuous $\epsilon: M \to (0, \infty)$, one can find h and h_t so that $d(h_t(x), x) < \epsilon(x)$ for all $x \in M$ and all $t \in [0, 1]$). Further, if $P \hookrightarrow M$ is PL near C closed in P, then the isotopy can fix a neighborhood of C in M.

We add a few words about first proof available. An astute convergence technique of T. Homma 1962, [Ho], cf. [G1], succeeds for $2\dim P + 2 \leqslant m$. So we can assume $m \geqslant 5$ henceforth.

The case where P is a PL manifold was solved next. The Stable Homeomorphism Theorem of Kirby [Ki_1] 1968, reinforced by E. Connell's PL approximation theorem for STABLE homeomorphisms [Cnl_1] 1963 (or by [I, §4.1]), provided small ambient isotopies making $P \hookrightarrow M$ PL on any given open ball. Using the version with majorant approximation of the PL local unknotting theorem for codimension ≥ 3 proved in R. T. Miller's thesis 1968 [Mil_1] , one easily puts together these local solutions to get a global one , cf. Rushing [Ru] .

The proof when P is a non-manifold was reduced by J. Bryant (and J. Cobb) [Bry_2] 1970 to the manifold case .

There are several variants and alternatives to this proof , see [Ru] [KS_2] [RS_6] [Edw_1] for information ; torus methods have succeeded , but only for the case of manifolds . From our point of view it would be desirable for some future exposition to (a) suppress the surgery entering via [Ki_1] † , and (b) suppress the delicate PL topology used in [Mil_1] and [Bry_2] . For manifolds , the result should follow *gracefully* from the PL and TOP unknotting of spheres in codimension ≥ 3 , and topological geometry .

We have taken the trouble to mention the general version of B. 1 (treating polyhedra) since it is particularly useful - for example in adapting to TOP manifolds the classical PL general position principles [Hu_2 , Chap. 5] for mappings of polyhedra to PL manifolds . We refer to [SGH] for a typical example of this applied in TOP engulfing . Of course for B. 1 to apply , the polyhedra must have codimension $\geq$ 3 . The intuitive idea behind these adaptions is that , so far as locally tame sub-polyhedra of codimension ≥ 3 are concerned , the charts of a TOP manifold are by B. 1 *as good as* PL compatible . This naive idea (an old one) is capable of making many other PL results topological .

† This is accomplished by very exacting PL geometry in [Mil_2] .

Appendix C. A TRANSVERSALITY LEMMA USEFUL FOR SURGERY.

In smooth surgery of dimension ≥ 5 manifolds, as presented for instance in [Wa], the surgery obstructions are met and their invariance ascertained through making certain smooth immersions suitably transverse to themselves and to each another. We shall present lemmas which permit one to carry over the same procedure to topological manifolds. The point is that the manifolds immersed in the TOP manifolds undergoing surgery are standard *smooth* manifolds such as $S^k \times R^n$, $B^k \times R^n$, etc. immersed with codimension zero†; and the expected intersection dimensions of the 'cores' $S^k \times 0$, $B^k \times 0$, etc. are ≤ 2 (even ≤ 1 it seems). This will permit us to adjust the immersions producing a smooth open subset of the target manifold so that the immersions become smooth open immersions into this open set. Then the procedures of smooth surgery apply.

The tools we need for this smoothing are

– (a) a relatively new topological general position principle from [Ed_5] [Su] [Ur].

– (b) handle smoothing of index ≤ 2 (as in [KS_1], cf. [Si_{10}, §5]) corresponding to the vanishing of π_i(TOP/DIFF), $i \leq 2$.

Now (a) could be replaced for surgery applications by the Miller-Bryant straightening theorem B.1 above; however we prefer to introduce the reader to this new and extremely useful topological principle. As explained to us by R. D. Edwards, it requires nothing stronger than simple engulfing in the sense of J. Stallings.

The reader will recall that (b) requires the full force of DIFF surgery; but that is a reasonable price to pay for TOP surgery.

Finally some hints for those readers who would prefer to set up TOP surgery without using this appendix. First recall that the surgery obstructions are encountered with immersions for which the expected 'core' intersections are of dimension 0. This case is easily dealt with

† i.e. $k + n$ is the dimension of the target and the immersion is open; it is thus equally a submersion.

using only the Stable Homeomorphism Theorem of [Ki_1] (with [I, §4.1]) ; indeed we have already done as much in §3.4 and [Si_{10}, §7]. Thus invariance of the surgery obstructions is the only real problem . But there is a stronger (and perhaps more difficult...?) result known , the invariance of surgery obstructions for Poincaré spaces . See [J] [Qn] [Hod] [LLM] for the multiple approaches to this . In the simply connected case , there is a brief homological argument [BrH] for invariance of surgery obstructions .

The promised sequence of lemmas has to do with the following situation : V^w is a TOP manifold without boundary ; $M \subset V^w$ is a closed subset† and $g : V^w \to W^w$ is an open immersion (= submersion) into a TOP manifold W such that $g|M$ is proper

Recall that one says M is **k-LC imbedded***$(k \geqslant 1)$ in V if $V - M$ is k–LC in V , i.e. if for each point $x \in M$ and each neighborhood N of x in V , there exists a smaller neighborhood $N' \subset N$ of x so that each continuous map $S^k \to (N' - M)$ is null-homotopic in $N - M$. Next recall $\mathbf{LC^k}$ means $j - LC$, for $j = 0,1,\ldots,k$. These are local properties.

One says that a closed set $M \subset V$ has **Štan'ko homotopy dimension of imbedding** m in V and one writes‡ $\operatorname{dem}M = m$ in case $-1 < \dim M < w$ and M is LC^{w-m-2} imbedded in V but not better . If $\dim M = -\infty$‡‡ or w we set $\operatorname{dem}M = \dim M$. Here dim is the usual covering dimension of M [Na] .

Observe that dem = dim for a subpolyhedron of euclidean space ; $\operatorname{dem} \leqslant k$ closed subsets share some key properties of $\dim \leqslant k$ subpolyhedra .

Note that $\operatorname{dem}M \leqslant w - 1 \Leftrightarrow$ M is nowhere dense ; and that $\operatorname{dem}M \leqslant w - 2 \Leftrightarrow \dim M \leqslant w - 2$. If M is of $\operatorname{dem} \leqslant w - 1$ and LC^1 imbedded in V , then $\operatorname{dem}M = \dim M \leqslant w - 3$. This is seen with the help of Alexander duality applied in small open discs of V , together with the basic fact that , when M is nowhere dense in V , it is LC^k if and only if for each open $V_o \subset V$, the inclusion $V_o - M \hookrightarrow V_o$ is $(k + 1)$-connected , cf. [EW , Theorem 1] . For examples of non-LC^1 embeddings of Cantor sets and arcs in high dimensions see [B1] [AnC] .

* Perhaps a better term is k–LCC = k-colocally connected = having k-locally connected complement .

† M will be supposed to be a manifold only in our final application C.6 .

‡ Beware that Štan'ko writes $\operatorname{dem}_\pi M$, in [$\check{S}t_1$] .

‡‡ We need the convention that $\dim \emptyset = -\infty$ (not -1) .

Given the facts set out above, one easily verifies that any *closed subset* of a dem $\leq k$ set in V is of dem $\leq k$; also as V is complete metrizable, a *closed countable union* of closed dem $\leq k$ sets is of dem $\leq k$.

GENERAL POSITION LEMMA C.1 .

Consider an open immersion $g : V^w \to W^w$ *of TOP manifolds*, $w \geq 5$, *without boundary. Let* M *be a closed set of dem* m *in* V *such that* $g|M$ *is proper. Let* $C \subset M$ *be a closed subset such that* $g|C$ *is in general position in the sense that the closed subset of* V

$$S(g|C) = \{x \in C \mid \exists\ y \in C,\ y \neq x,\ g(y) = g(x)\}$$

is of dem $\leq 2m - w$. *Finally let* $D \subset M$ *be compact.*

Then there exists a regular† *homotopy* $g_t : V^w \to W^w$, $0 \leq t \leq 1$, *of* $g_0 = g$ *fixing* $g|C$ *and with compact support in* V, *to an immersion* $g_1 : V^w \to W^w$ *which is in general position (as above) on* $C \cup D$. *Further* g_t *can be as near to* g *as we please.*

Proof of C.1. In view of the relative nature of this result we can assume that $g|D$ is injective. This uses the basic fact (proved readily by contradiction) that if $g|D$ is injective then g' is injective near D for every immersion $g' : V^w \to W^w$ sufficiently near to g .

In case $g|D$ is injective, we choose an open neighborhood V_o of $D - C$ in V , having compact closure in V , so that $D - C$ is closed in V_o, and $g|V_o$ is injective. Then we use [Ed_5] [Su] [Ur] (the noncompact general position theorem involving dem for embedded closed sets is further explained in C.3 below) to obtain a small isotopy h_t, $0 \leq t \leq 1$, of $id|g(V_o)$ such that $h_1 g(D - C)$ meets the dem $\leq m$ set gC in general position, i.e.

$$\text{dem}\{x \in D-C \mid \exists\ y \in C,\ h_1 g(x) = g(y)\} \leq \text{dem}\, C + \text{dem}\, D - w\ .$$

If h_t is chosen (majorant) sufficiently near $id|g(V_o)$, it extends as the identity outside $g(V_o)$ to an isotopy h_t of $id|W$. Then setting $g_t = h_t g$ on V_o and $g_t = g$ elsewhere in V we have a regular homotopy of g as required. ■

† A regular homotopy is here just a homotopy through open immersions.

We will want to a sharper result if $M \subset V$ is a manifold with several components of *various* dimensions. Then we need a finer notion of general position as follows. Suppose $M = M_1 \cup \ldots \cup M_k$, where each M_i is closed in V^w and of dem $\leqslant m_i$.

C.2. Complement to C.1.

The general position lemma holds true when the following meaning is assigned to general position for $g|C$ (and similarly for $g|(C \cup D)$): For each ordered pair of integers i, j in $[1,k]$ ($i = j$ allowed) we suppose, writing C_i for $C \cap M_i$, that:

$$S_{ij}(g|C) \equiv \{x \in C_i \mid \exists\, y \in C_j,\ y \neq x,\ g(y) = g(x)\}$$

is of dem $\leqslant m_i + m_j - w$.†

Proof of Complement. We can assume that $m_1 \leqslant m_2 \leqslant \ldots \leqslant m_k$, and that $M_1 \subset M_2 \subset \ldots \subset M_k = M$. Then it further suffices to deal with the case where $D = D_i$ ($\equiv D \cap M_i$) for some i and $D_{i-1} \subset C$. This case is dealt with much as before but in k steps. After the jth step we delete the image of M_j from V, so that at each step we use just the same absolute (noncompact, epsilon) general position result. The details are left to the reader, see [Ur]. ∎

C.3. Background (from [Ed_5] [Su] [Ur]).

Here is an indication of how two closed subsets X, Y of a coordinate chart (or PL manifold) V_o^w, $w \geqslant 5$, are pushed to meet in general position. This is just what is needed for C.1 and C.2 above. For any continuous $\epsilon : V_o \to (0, \infty)$, (called a majorant) one can find a closed polyhedron P in V_o with $\dim P \leqslant \operatorname{dem} X \equiv x$ so that X lies in the ϵ-neighborhood of P and can be ambient isotopically moved arbitrarily near P by ϵ-isotopies; P will be called an ϵ-spine of X in V_o. (The notion is due to J. Bryant [Bry_1] and Štan'ko [$\check{S}t_1$]). This P can be the x-skeleton of a sufficiently fine PL triangulation of V_o; because the dual skeleton P* (of dimension $w - x - 1$) can be PL pushed a little to miss X, by the device (R. D. Edwards') of simplex-by-simplex engulfing‡ of P* with $V_o - X$, using Stallings' engulfing and connectivity facts we have mentioned above C.1.

† One can add that $w = 4$ is permissible provided each m_i is $\geqslant 2$; the technique explained in C.3 below applies unchanged.

‡ This argument fails if $\operatorname{dem} X \leqslant 1$, i.e. if $\dim P^* \geqslant w - 2$; but the low dem permits another argument [Bry_1] [Ed_5]. This low dem argument breaks down for $w \leqslant 4$: for $w = 3$ the general position result is false as stated, cf. [McR] [Ur].

Similarly Y has an ϵ-spine Q^y, $y = \text{dem} Y$. We observe that, if f and g are ϵ-pushes† of X and Y respectively sufficiently near to P and Q, then $fX \cap gY$ has $P \cap Q$ as ϵ-spine. Also $\dim(P \cap Q) \leqslant x + y - w$ if P and Q are in PL general position, as we may suppose. Composing such pushes for infinitely many such pairs of ϵ_n-spines, where the $\epsilon_n : V_o \to (0, \infty)$ are majorants tending to zero and carefully chosen by induction‡, one obtains ϵ-pushes f_∞, g_∞ so that $f_\infty X \cap g_\infty X$ has ϵ_n-spines of dimension $\leqslant x + y - w$ for all n. Then $f_\infty X \cap g_\infty Y$ is easily seen to be as highly LC^k imbedded as its ϵ_n-spines, which implies that $\text{dem}(f_\infty X \cap g_\infty Y) \leqslant x + y - w$. ■

SMOOTHING LEMMA C.4.

Consider an open immersion $g : V^w \to W^w$ *of a* CAT (= DIFF *or* PL) *manifold without boundary* V^w *to a* TOP *manifold* W^w. *Let* $C \subset V$ *be a closed subset such that* g *is 'weakly smooth' near* C *in the sense that for some open neighborhood* V_C *of* C *in* V, g *imposes a* CAT *structure* Ω *on* gV_C *making* $V_C \xrightarrow{g} (gV_C)_\Omega$ *a* CAT *immersion. Let* $D \subset V$ *be a compact set such that* $S \equiv \{x \in D \mid \exists\ y \in C \cup D,\ y \neq x,\ g(y) = g(x)\}$ *is of dim* $\leqslant 2$. *Finally provide that* $w \geqslant 5$.

Then there exists a CAT *concordance* $\Gamma : \Sigma \sim \Sigma'$ *rel* C *with compact support, from the given* CAT *structure* Σ *of* V *to a new* CAT *structure* Σ' *such that* $g : V_{\Sigma'} \to W$ *is 'weakly smooth' near* $C \cup D$.

Proof of C.4. As this lemma is strongly relative, we can assume that g is injective on an open neighborhood V_D of D in V. Let $V_o = gV_D$ and Σ_0 be the structure on V_o making $V_D \to (V_o)_{\Sigma_0}$ a diffeomorphism. Then it suffices to find a concordance $\Gamma_o : \Sigma_0 \sim \Sigma'_0$ rel gC with compact support to a structure Σ'_0 coinciding with Ω near gS. Indeed the wanted concordance Γ can then be defined to have compact support V_D and make $(V_D \times I)_\Gamma \xrightarrow{g} (V_o \times I)_{\Gamma_o}$ a diffeomorphism.

† We mean homeomorphisms of V_o that are ϵ-isotopic to $\text{id} | V_o$.

‡ To arrange that f_∞, g_∞ are *homeomorphisms*, one can apply to the 1-point compactification of V_o the fact that for any metric compactum (C, d_c), the space of homeomorphisms $C \to C$ has complete metric $d(h, h') = \sup\{d_c(h(x), h'(x)) + d_c(h^{-1}(x), h'^{-1}(x)) \mid x \in C\}$.

The existence of Γ_o follows from the next lemma using substitutions we shall indicate presently :

LEMMA C.5 .

Let A be a closed subset of (covering) dimension $\leqslant 2$ in a TOP manifold W without boundary and of dimension $\geqslant 5$. Let U_α, V_β be open neighborhoods of A in W equipped with CAT (= DIFF or PL) structures α and β. Suppose $\alpha = \beta$ near a closed set $B \subset A$.

Then there exists a concordance $\Gamma : \alpha \sim \alpha'$ rel B to a CAT structure α' that equals β near A.

Proof of Lemma C.5. According to the classification theorem [IV, 10.1 - 10.9] , the obstructions to finding Γ lie in the Čech cohomology groups $\check{H}^i(A, B ; \pi_i(TOP/CAT))$. These vanish for $i \geqslant 3$ since $\dim A \leqslant 2$; and they vanish for $i \leqslant 2$ since $\pi_i(TOP/CAT) = 0$ for $i \leqslant 2$ by [KS_1] , cf. [Si_{10}, §5] . ■

Remark. - In case $\text{dem} A \leqslant 2$, one can (with due care) prove this more directly using an ϵ-spine P^2 of A in U_α and solving CAT handle problems of index $\leqslant 2$, as the following sketch suggests for $\text{dem} A = \dim P = 1$. ■

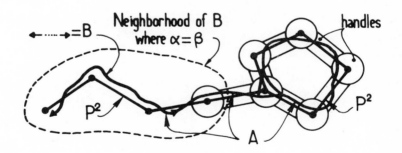

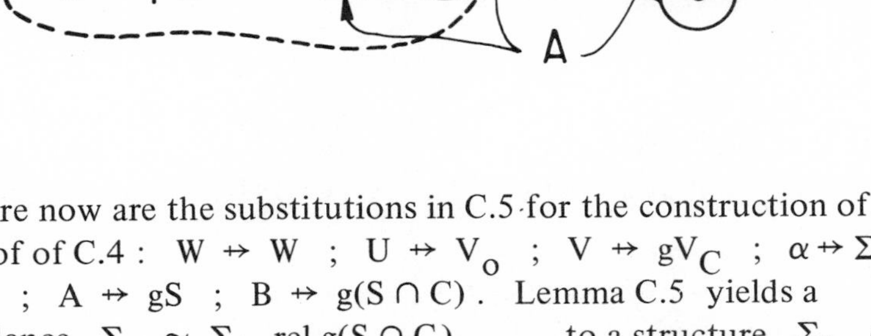

Here now are the substitutions in C.5 for the construction of Γ in the proof of C.4 : $W \leftrightarrow W$; $U \leftrightarrow V_o$; $V \leftrightarrow gV_C$; $\alpha \leftrightarrow \Sigma_o$; $\beta \leftrightarrow \Omega$; $A \leftrightarrow gS$; $B \leftrightarrow g(S \cap C)$. Lemma C.5 yields a concordance $\Sigma_o \sim \Sigma_1$ rel $g(S \cap C)$, to a structure Σ_1 equal Ω near gS . By the Concordance Extension Theorem [I, §4.2] we deduce another concordance $\Gamma_o : \Sigma_o \sim \Sigma'_o$ rel $g(C \cap V_D)$ and with compact support , to Σ'_o equal Ω near gS , as required to prove C.4 .

■

Applying the topological general position lemma C.1 & C.2 then the smoothing lemma C.4 together with the (relative) Concordance Implies Isotopy Theorem [I, §4.1], and finally a DIFF transversality theorem (essentially the Sard-Brown theorem [Mi_7]) one routinely deduces :

TRANSVERSALITY LEMMA FOR SURGERY C.6 .

Consider an open immersion $g : V^w \to W^w$ *of a DIFF manifold without boundary* V^w *to a TOP manifold* W^w. *Let* $M \subset V$ *be a DIFF submanifold (with components possibly of various dimensions)*† , *so that* $g|M$ *is proper. Let* $D \subset M$ *be compact, and let* $C \subset M$ *be a closed subset.*

Suppose that, for some open neighborhood V_C *of* C *in* V, *the restriction* $g|V_C$ *is weakly smooth in the sense that it imposes a DIFF structure* Ω *on* gV_C *making* $V_C \xrightarrow{g} (gV_C)_\Omega$ *a DIFF immersion.*

Suppose also that $g|C$ *is in DIFF transversal position in the sense that, if* $g(x) = g(y)$ *for* $x \neq y$ *both in* C, *then the images under the differential* Dg *of the tangent planes* $T_x M$ *and* $T_y M$ *are linear transversal.*

Finally provide that $w \geq 5$, *and that for all* $x \in C \cup D$ *and* $y \in D$, *one has* $\dim T_x M + \dim T_y M - w \leq 2$.‡

Then there exists an arbitrarily small regular homotopy $g_t : V^w \to W^w$, $0 \leq t \leq 1$, *of* $g = g_0$ *rel* C *with compact support in* V, *such that, on some neighborhood* $V_{C \cup D}$ *of* $C \cup D$ *in* V, g_1 *is weakly smooth, and* $g|C \cup D$ *is in transversal position, in the sense explained above.* ■

For manifolds with boundary one deduces a parallel result exploiting suitable collarings of the boundary. (Beware a dimension ≥ 5 condition for ∂V).

This lemma helps one establish for TOP manifolds just as for DIFF manifolds not only the basic theorems of Wall's book [Wa] but also many recent geometrically proven theorems on surgery such as the splitting theorems of Farrell & Hsiang [FH_1] and of Cappell [Cap_1].

† Against our usual conventions.

‡ *Generalization :* If $g|D$ is an imbedding, $g_t|D$ remains an imbedding throughout any small regular homotopy of g. Thus, it suffices in this case to assume $\dim T_x M + \dim T_y M - w \leq 2$ merely for $x \in C$ (not $C \cup D$), and $y \in D$.

It can do no harm to describe how a TOP surgery is done (cf. [Ls] [Wa_1]), at least in the absolute case. One is faced with a degree one map $f: W^w \to X^w$ from a closed TOP w-manifold W to a closed manifold or Poincaré space X ; and as extra data one has a TOP microbundle ξ and a stable microbundle map $\varphi: \tau(W) \to \xi$ covering f . The pair (f, φ) is called a *normal map* . Given an element $x \in \pi_{k+1}(f)$, $k < w$, we shall see that a surgery on x is possible precisely if a certain regular homotopy class of immersions $S^k \times R^n \to W^w$, $n = w - k$, contains an imbedding. (A. Haefliger introduced these immersions in the early 60's.)

The element $x \in \pi_{k+1}(f)$ is represented by the comutative square at left, which we cover by a commutative square of stable microbundle maps at right :

$$\begin{array}{ccc} W & \xrightarrow{\ f\ } & X \\ \partial g \uparrow & & \uparrow g \\ \partial H = S^k \times R^n & \hookrightarrow & B^{k+1} \times R^n = H \end{array} \qquad \begin{array}{ccc} \tau(W) & \xrightarrow{\ \varphi\ } & \xi \\ \partial\gamma \uparrow & & \uparrow \gamma \\ \tau(\partial H) & \xrightarrow{\ nat\ } & \tau(H) \end{array}$$

Given g , the map γ exists since the handle H is contractible; then γ determines $\partial\gamma$ uniquely. (By convention, for a manifold M with boundary, the tangent bundle $\tau(M)$ is $\tau(M_+)$, where M_+ is the open manifold obtained by adding an open collar $(\partial M) \times (-\infty, 0]$ along $\partial M = \partial M \times 0$; and $\tau(\partial M) \xrightarrow{nat} \tau(M)$ is defined using a collaring of ∂M – see [V, §4] .) The stable map $\partial\gamma$ is nevertheless , up to microbundle homotopy, in a unique class of non-stable maps, because $\pi_k TOP \cong \pi_k TOP_w$ for $k < w$, by a stability theorem [KS_1] [Si_{10}, §5.3] [V, §5] . Thus, by TOP immersion theory [Ls] [Las_2] [V, App. A] , we can arrange that ∂g is an open immersion while $\partial\gamma$ is its differential. By a parallel argument this process determines a *unique* regular homotopy class of immersions ∂g (except for reflection in R^{w-k}).

In case ∂g is an imbedding, surgery on x is done as follows: To $W \times [0,1]$, attach the compact handle $B^k \times B^n \subset H$, $n = w - k$, by imbedding $(\partial B^k) \times B^n$ into into $W \times 1$ using $\partial g \times 1$, to get Z say. Referring to the right hand square, observe that, as soon as φ is made a product along the first stabilising factor, φ and γ together **will determine a stable** microbundle map $\Phi: \tau(\mathrm{int} Z) \to \xi$, over the map $F: Z \to X$ determined by f and g . (Pause to check this!) Certainly Φ can then be modified to extend as $\Phi: \tau(Z_{\dot{+}}) \to \xi$, leaving it unchanged near $W \times 0$. The *surgered normal map* (f', φ') is then got by restriction over $\partial(Z - W \times 0)$, from the *normal cobordism* (F, Φ) .

A key question to decide, for $k \leqslant w/2$, is whether immersion ∂g is regularly homotopic to an imbedding. It always is by general position (C.1) if $k < w/2$. For $k = w/2$ our transversality lemma (C.6) leaves $\partial g \mid S^k \times 0$ with just isolated double points of smooth transversal intersection. A suitable way of counting these (relative to a fixed fundamental class) gives Wall's self-intersection number $\mu(\partial g)$; and a further application of C.6 , with 1-dimensional self-intersections, reveals that $\mu(\partial g)$ is regular homotopy invariant, i.e. $\mu(\partial g)$ is $\mu(x)$. For $w \geqslant 5$, Whitney's process (see [IV, § 3.4] [Si_{10}, § 7.3]) permits a regular homotopy of ∂g to an imbedding precisely if $\mu(x) = 0$.

These $\mu(x)$ determine a nonsingular quadratic form on $\pi_{k+1}(f)$ when f is k-connected, which, modulo hyperbolic forms, is Wall's obstruction (for $w \geqslant 5$) to surgering (f, φ) repeatedly to obtain a simple homotopy equivalence.

Essay IV

STABLE CLASSIFICATION OF SMOOTH AND PIECEWISE LINEAR MANIFOLD STRUCTURES

by

L. Siebenmann

Section headings in Essay IV

§0. INTRODUCTION

Let M be a (metrizable) topological (= TOP) manifold of dimension ≥ 6 (or ≥ 5 if $\partial M = \phi$). In Essay I we have used results of handlebody theory to establish the Product Structure Theorem [I, §5.1] relating smooth C^∞ (= DIFF) or piecewise-linear (= PL) manifold structures on M to those on $M \times R$, where R = {real numbers}. It says, roughly, that the classifications up to isotopy are the same, a bijection of isotopy classes being given by crossing with R.† Recall that two CAT structures (CAT = DIFF or PL) Σ and Σ' on M are isotopic if there is a TOP isotopy (= path of homeomorphisms) $h_t : M \to M$, $0 \leq t \leq 1$, from $h_0 = \mathrm{id}|M$, the identity map of M, to $h_1 : M_\Sigma \to M_{\Sigma'}$ a homeomorphism giving a CAT isomorphism from M with the structure Σ to M with structure Σ'.

Our goal in this essay is to use our Product Structure Theorem and some well-established ideas of earlier smoothing theories to prove a serviceable version of the

CLASSIFICATION THEOREM (see §10).

Let M be a (metrizable) topological manifold of dimension ≥ 6 (or ≥ 5) if the boundary ∂M is empty). There is a one-to-one correspondence between isotopy classes of PL [respectively DIFF] manifold structures on M and vertical homotopy classes of sections of a (Hurewitz) fibration over M. It is natural for restriction to open subsets of M.

This fibration is the pull-back, by the classifying map $M \to B_{TOP}$ for the tangent microbundle $\tau(M)$ of M, of a fibration $B'_{PL} \to B_{TOP}$ [resp. $B'_O \to B_{TOP}$] homotopy equivalent to the usual 'forgetful' map $B_{PL} \to B_{TOP}$ [resp. $B_O \to B_{TOP}$], of stable classifying spaces. The fiber, called TOP/PL [resp. TOP/O], has homotopy groups $\pi_i(\mathrm{TOP/PL})$ equal zero if $i \neq 3$ and equal Z_2 if

† Strangely enough, this is false in some cases where M has dimension 3 or 4, see [I, §5], [Si_8].

$i = 3$, see [KS_1] , [resp . $\pi_i(TOP/O) = \Theta_i = \Gamma_i$ if $i > 4$, and $\pi_i(TOP/O) = \pi_i(TOP/PL)$ if $i \leqslant 6$] , cf . [Si_{10}] or §10 .

The version of this theorem for PL structures was announced in [KS_1] . The proof outlined there proceeded handle by handle and used results of handlebody theory , surgery and immersion theory . But as we suggested there , this was more extravagant than necessary . It has turned out that one can purify the arguments of [KS_1] by following either one of two radically different general methods . One method (pioneered by Cerf , Haefliger , Lashof , and Morlet [Ce_0] [Ha_1] [Las_1] [$Mor_{2,3,4}$]) is to rely heavily in one way or another on ideas of immersion theory ; we shall exploit this method in Essay V getting more precise (unstable , sliced) classification theorems , at the cost of introducing numerous semi-simplicial spaces and fibrations .

As we have stated, the course we follow here relies on the Product Structure Theorem , hence chiefly on handlebody theory . We were encouraged to adopt it by some extremely direct and natural applications in [III] of this Product Structure Theorem to TOP manifolds , which , we hope , assure its role as a landmark in the theory .

An analogue of the Product Structure Theorem for compatible DIFF structures on PL manifolds , called the Cairns-Hirsch theorem has become well known since 1961 [Ca] [Hi_2] [Hi_4] . It allowed a classification of compatible DIFF structures quite analogous to the theorem of this article .† The question of existence was first decided by Milnor [$Mi_{3,4,5}$] , who introduced microbundles for this purpose . The corresponding uniqueness question proved technically more delicate ; it was worked out by Hirsch and Mazur and by Lashof and Rothenberg . (See Morlet [Mor_1] for an outline of the theory as of 1963 . See also [MazP] [HiM_1] [LR_1] [Wh] and [HiM_2] .) After this work , it became quite clear to the well-informed that a similar classification for DIFF structures on TOP manifold would be feasible if and when our Product Structure Theorem could be proved . This made us hesitate to write down the details . But once started we made the exposition both leisurely and elementary . We are fortunate that no technical bugbear as nasty as the functorial triangulation of vector bundles [HiM_1] [LR_1] ever intervenes .

We now briefly sketch the proof to come .

† This classification was constructed against the background of Thom's conjectures [Th_2] and an obstruction theory of J. Munkres [Mu_3] .

Milnor's method for introducing a CAT (= DIFF or PL) structure on the TOP manifold M involves the observation that when M is embedded in euclidean space R^n as retract of a neighborhood N by $r : N \to M$, then the pull-back $r^*\tau(M)$ over N of the tangent microbundle $\tau(M)$ of M contains a copy of $M \times R^n$ as an open neighborhood of embedded M. There results, given the Product Structure Theorem, a mapping from concordance classes of CAT microbundle structures on $r^*\tau(M)$ or on $r^*\tau(M) \oplus \epsilon^s$, $s \geq 0$, to isotopy classes of CAT structures on M; we call this mapping the *smoothing rule*.

We proceed to prove in §§4-6 that this smoothing rule is a bijective correspondence provided that $M \hookrightarrow N$ is a homotopy equivalence (which is trivial to arrange once M is known to carry a CAT structure). The stable relative existence theorem for normal microbundles makes this bijectivity fairly easy to verify; for the sake of completeness we explain M. Hirsch's proof of this theorem [Hi_5] in Appendix A. (It was first established by R. Lashof and M. Rothenburg in [LR_1], for a similar purpose.) It remains then to recall, in §8 and §9, the classification of stable concordance classes of CAT microbundle structures on $r^*\tau(M)$ by sections of a fibration as described above. Then the full classification theorem is presented in §10.

Hirsch and Mazur initially followed a slightly different route to the classification of CAT structures on M (see [HiM_1]), which we quickly see to be equivalent in §7. It reveals that the isotopy classes of CAT structures on a CAT manifold M naturally form an abelian group.

Some features of our exposition are intended to make the results more readily applicable. We take care to describe how to pass efficiently both from a manifold structure to a bundle section, and from bundle section to a manifold structure. And at every stage we give a relative version.

Since we will discuss certain non-manifolds in connection with topological and piecewise bundles, we adopt the following broad meanings for TOP and PL in this essay. TOP indicates the category of continuous maps of topological spaces; PL indicates the category of piecewise linear maps of (locally compact) polyhedra. However, DIFF (as always) indicates the category of C^∞ smooth maps of finite dimensional smooth manifolds.

Our basic conventions about manifolds remain those of Essay I , §2 . Thus manifolds mean metrizable manifolds with boundary (possibly an empty boundary) , having the same dimension in each component . DIFF manifolds usually do not have corners ; but in a few situations , where they obviously should have , we allow them without mention . For example if Γ is a DIFF structure on $M \times I$, $I = [0,1]$, where M is a TOP manifold with $\partial M \neq \phi$, then $(M \times I)_\Gamma$ by assumption has corners at $\partial M \times 0 \cup \partial M \times 1$ of the same sort as $M_\Sigma \times I$ has . Recall that Γ is a *concordance* Σ_0 to Σ_1 if $\Gamma | M \times \{i\} = \Sigma_i \times \{i\}$ for $i=0,1$.

§ 1. RECOLLECTIONS CONCERNING MICROBUNDLES

The reader should familiarise himself with the rudiments of Milnor's notion of **microbundle** before proceeding ; see [Mi_5] . A n-microbundle ξ over a space X is a diagram $\xi : X \xrightarrow{i} E \xrightarrow{p} X$ (X = base ; i = zero section ; E = total space ; p = projection) so that $pi = id\,|\,X$, and in E near each point of $i(X)$, p looks like the projection $X \times R^n \to X$. Often , without special apology , we will identify X with $i(X)$ and regard X as a subspace of $E(\xi)$.

According to the *Kister-Mazur theorem* [Kis] [KuL_1] [SGH_1] there is always a neighborhood U of $i(X)$ in E such that $\xi_0 : X \xrightarrow{i} U \xrightarrow{p|U} X$ is a locally trivial bundle with fiber R^n and zero section i . And ξ_0 is unique up to isomorphism .

A **micro-isomorphism** $\xi_1 \to \xi_2$ of n-microbundles over X consists of a neighborhood U of the zero section $i_1(X)$ in the total space $E(\xi_1)$ of ξ_1 and a topological imbedding h onto a neighborhood of $i_2(X)$ in $E(\xi_2)$, making this diagram commute .

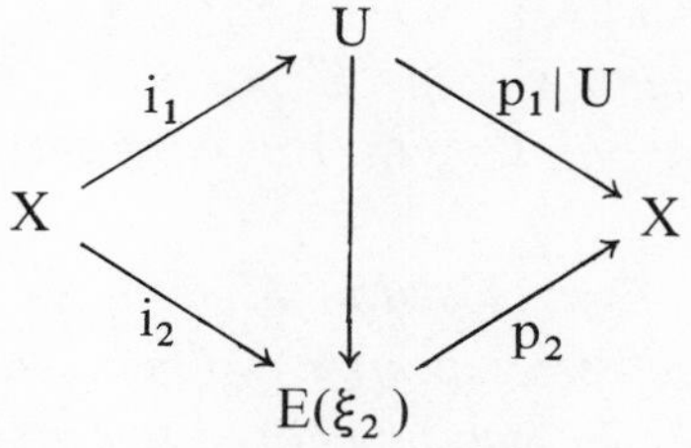

Then ξ_1 and ξ_2 are called **(micro) isomorphic** (= isomorphic in the category of germs of micro-isomorphisms) . If h is an inclusion of a subspace it is called a micro-identity and ξ_1 , ξ_2 are called **micro-identical** . This is written $\xi_1 \doteq \xi_2$. Two micro-isomorphisms $h : \xi_1 \to \xi_2$ and $h' : \xi_2 \to \xi_3$ can be composed to give a micro-isomorphism $h'h : \xi_1 \to \xi_3$. But note that the domain of definition of $h'h$ is in general smaller than that of h .

The microbundles defined above are called topological or TOP microbundles .

The notion of a PL n-microbundle micro-isomorphism and micro-identity are defined like the topological entities above but one works in the category of piecewise linear maps of polyhedra (= Hausdorff spaces equipped with a maximal piecewise linearly compatible atlas of charts to locally finite simplicial complexes , cf. [Hu_2, §3]) . It is convenient to allow that a topological microbundle is a PL microbundle if it is one after deletion of part of the total space outside of a neighborhood of the zero section .

Similarly define DIFF microbundles starting from the category of DIFF manifolds and DIFF maps .

If X is a CAT submanifold of Y and ξ is a CAT microbundle whose total space $E(\xi)$ is a neighborhood of X in Y with inherited CAT structure , we say that ξ is a CAT **normal microbundle** to X in Y . (See Appendix A .)

The Kister-Mazur theorem mentioned above is valid in all three categories . A stronger version holding for DIFF states that DIFF microbundles are essentially equivalent to vector bundles . However , we retain DIFF microbundles for the sake of uniformity .

PROPOSITION 1.1

Every DIFF microbundle $\xi : M \xrightarrow{i} E(\xi) \xrightarrow{p} M$ *over a DIFF manifold* M *contains a neighborhood* E' *of* $i(M)$ *which can be regarded as a smooth vector bundle with the same zero-section and projection .*

Furthermore suppose C *is closed in* M *and* E'_0 *is such a vector bundle neighborhood of the zero section of* $\xi | U$, *where* U *is an open neighborhood of* C . *Then* E' *can be chosen to coincide (as vector bundle) with* E'_0 *over a neighborhood of* C . *And if* E'' *is another choice for* E' *there is an isotopy* $h_t : E' \to E$ *of* $E' \hookrightarrow E$ *through DIFF imbeddings yielding such vector bundle neighborhoods , to a vector bundle isomorphism* $E' \to E''$.

Proof of 1.1 . The uniqueness clause is a well known portion of the uniqueness theorem for DIFF tubular neighborhoods , cf. [Lan] . But it yields the existence assertion by dint of a chart-by-chart construction . ∎

An alternative proof is suggested by Milnor [Mi_5 , §2.2] . ∎

If $f: X \to Y$ is a map and ξ, η are CAT n-microbundles over X, Y a **microbundle map** $\mathbf{f}: \xi \to \eta$ covering f can be defined to be a CAT map $U \to E(\eta)$ of a neighborhood U of X in $E(\xi)$ that restricts to f on $X \subset E(\xi)$, that respects fibers, and that gives a CAT open embedding to each fiber. (See [Mi_5, §6].) For example one has a canonical CAT microbundle map $f_{\#}: f^*\eta \to \eta$ when f is a CAT map. The fiber product rule defining $f^*\eta$ shows this is essentially the only example. Indeed, if $\mathbf{f}$ is any CAT microbundle map covering f, there is a unique micro-isomorphism $\varphi: \xi \to f^*\eta$ so that $f_{\#}\varphi = \mathbf{f}$. Beware that quite a variety of symbols such as $f_{\#}$ or φ or $\mathbf{f}$ may stand for a microbundle map.

We will constantly use the basic **homotopy theorem** [Mi_6, §6] for microbundles. (It is perhaps worth noting that the Kister-Mazur theorem mentioned above reduces this result to the parallel, better known result for locally trivial CAT bundles with fiber euclidean space.) It states that any CAT microbundle ζ over a product $I \times X$ (X being Hausdorff paracompact) admits a CAT micro-isomorphism $f: \zeta \to I \times \xi$, where $0 \times \xi$ is just a copy of $\zeta|(0 \times X)$. The standard *relative version* (which is easily deduced!) states that given C closed in X and a CAT micro-isomorphism $f': \zeta|(I \times U) \to I \times (\xi|U)$, where U is an open neighborhood of C in X, it is possible to choose f to agree† with f' over a neighborhood of C. A *sharply relative version*‡ states that if C is a closed and a CAT

† The domains of definition of f and f' in any fiber may differ nevertheless.

‡ To prove this sharply relative version, use the more basic *Observation : To make f agree with given f'' over C it suffices to find, for each point $x \in C$ an open neighborhood V_x in X and a CAT micro-isomorphism $f_x : \zeta|(I \times V_x) \to I \times (\xi|V_x)$ so that f_x and f'' agree over $I \times (V_x \cap C)$*. To verify this observation, reread the proof of the homotopy theorem in [Mi_5, §6]. To then apply the observation to get the given sharply relative version choose V so that $\xi|V_x$ is trivial and there is a CAT retraction $r_x : V_x \to V_x \cap C$. Next, as $\xi|V_x$ is trivial, there exist CAT microbundle maps α and β covering $I \times r_x$, and themselves retractions :

$$\begin{array}{ccc} \zeta|(I \times V_x) & \xrightarrow{f_x} & I \times (\xi|V_x) \\ \alpha \downarrow & & \downarrow \beta \\ \zeta|(I \times (V_x \cap C)) & \xrightarrow{f''} & I \times (\xi|V_x \cap C). \end{array}$$

Now there is a canonical CAT micro-isomorphism f_x making the square commute and clearly f_x agrees with f'' over $I \times (V_x \cap C)$.

retract of a neighborhood in X, then given a CAT micro-isomorphism $f'' : \zeta|(I \times C) \to I \times (\xi|C)$ it is possible to choose f to agree with f'' over C.

Here is a useful *reformulation* of the microbundle homotopy theorem, which explains its name. If $\mathbf{f} : \xi \to \gamma$ is a CAT microbundle map over $f : X \to B$ say, and $F : I \times X \to B$ is a CAT homotopy of f, i.e. $p_2 \circ (0 \times f) = F|(0 \times X)$, then there exists a CAT **microbundle homotopy** over F $\quad \mathbf{F} : I \times \xi \to \gamma$ of $\mathbf{f}$, i.e. $\mathbf{F}$ is a CAT microbundle map over F with $p_2 \circ (0 \times \mathbf{f}) = \mathbf{F}|(0 \times \xi)$. Here p_2 generically denotes projection to second factor. If F is rel C, then $\mathbf{F}$ can be rel C, i.e. be equal to $p_2 \circ (I \times \mathbf{f})$ *near* $I \times C$ (the relative version). If $F = p_2 \circ (I \times f)$ *on* $I \times C$ and C is a CAT neighborhood retract in X, then $\mathbf{F}$ can coincide with $p_2 \circ (I \times \mathbf{f})$ *over* $I \times C$ (a sharply relative version).

If ξ is a TOP microbundle over a polyhedron X, a **PL structure on** ξ, also called a **reduction of** ξ **to** PL **microbundle**, or a **PL structure on** ξ, is simply a PL microbundle ξ' whose base is the same polyhedron X and whose underlying TOP microbundle is ξ. If ξ is a TOP microbundle over a DIFF manifold X, we make a similar definition with DIFF in place of PL.

A **stable** CAT structure on ξ will mean a CAT structure η on $\xi \oplus \epsilon^s$, $s \geqslant 0$, where $\epsilon^s : X \to X \times R^s \to X$ is the standard trivial microbundle. If η' is a CAT structure on $\xi \oplus \epsilon^t$, $t \geqslant s$, then we write $\eta \doteq \eta'$ if $\eta \oplus \epsilon^{t-s} \doteq \eta'$. This extends the (CAT) micro-identity relation $\doteq$ to an evident equivalence relation $\doteq$ on stable CAT structures on ξ.

Let ξ be a TOP microbundle over a CAT object X (CAT = DIFF or PL). A **concordance** between CAT structures ξ_0, ξ_1 on ξ is a CAT structure γ on $I \times \xi$ such that the restriction $\gamma|i \times X$ is CAT micro-identical to $i \times \xi_i$ for $i = 0,1$. (In symbols $\gamma|i \times X \doteq i \times \xi_i$, for $i = 0,1$.) If C is closed in X and there is a neighborhood U of X such that $\gamma|(I \times U) \doteq I \times (\xi_0|U)$, then γ is called a *concordance rel* C. Concordance rel C is an equivalence relation, written

$$\xi_0 \simeq \xi_1 \quad \text{rel } C.$$

The microbundle homotopy theorem provides a CAT micro-isomorphism $h : \gamma \to I \times \xi_0$ that is the identity near $I \times C$. Hence

$\xi_0 \triangleq \xi_1$ rel C means that there exists a neighborhood U of X in $E(\xi_1)$, and an isotopy $h_t : U \to E(\xi)$, $0 \leq t \leq 1$, of the inclusion through micro-isomorphisms of ξ to itself to a CAT micro-isomorphism $\xi_1 \to \xi_0$; further, h_t can fix a neighborhood of C. Compare the Concordance Implies Isotopy Theorem.

Next define **stable concordance** rel C of stable CAT structures ξ_0, ξ_1 on ξ to mean $\xi_0 \doteq \zeta_0 \triangleq \zeta_1 \doteq \xi_1$ ($\triangleq$ being CAT concordance rel C) for stabilizations ζ_0 of ξ_0 and ζ_1 of ξ_1.

If η is a stable CAT structure on $\xi|V$ for some neighborhood V of C in the base we will write

$$\mathrm{TOP/CAT}\ (\xi \text{ rel } C, \eta)$$

for *the set of stable concordance classes rel C of stable CAT structures ξ' on ξ that coincide near C with η* in the sense that $\xi'|U \doteq \eta|U$ for some neighborhood U of C. If the choice of η is evident we omit η from the symbol; and if $C = \phi$ we omit rel C, η getting simply TOP/CAT(ξ).

The **tangent** CAT **microbundle** $\tau(M)$ $(= \tau_M)$ of a CAT manifold M without boundary is defined by Milnor to be

$$\tau(M): M \xrightarrow{\Delta} M \times M \xrightarrow{p_1} M \quad \text{where } \Delta(x) = (x, x) \text{ and } p_1(x, y) = x.$$

§2. ABSORBING THE BOUNDARY

Consider a TOP manifold M^m of dimension m, possibly with boundary. Suppose that a CAT structure Σ_0 is given on a neighborhood of a closed subset C of M. For m large we aim to settle, first in bundle theoretic terms, then in homotopy theoretic terms, the problem of finding a CAT structure Σ on M that agrees with Σ_0 near C, and of classifying all such structures up to concordance rel C if one such exists. The symbol $\mathcal{S}_{CAT}(M \text{ rel } C, \Sigma_0)$ will denote the set of concordance classes rel C of such structures. It is abbreviated by omitting Σ_0 if the choice of Σ_0 is evident (e.g., if M is a CAT manifold), and by omitting C if $C = \phi$.

The following consequence of the Product Structure Theorem shows that, if $m \geqslant 6$, it will suffice to accomplish this task when $\partial M = \phi$.†

PROPOSITION 2.1

Let M' be the open topological manifold obtained from M by attaching to M the collar $\partial M \times [0, \infty)$ identifying x in ∂M to $(x, 0)$ in $\partial M \times [0, \infty)$. Extend Σ_0 to a CAT structure Σ_0' on a neighborhood of C in M' that is a product along $[0, \infty)$ in $\partial M \times [0, \infty)$.

Provided that $m \geqslant 6$ (or $m = 5$ and $\partial M \subset C$), there is a natural bijection

$$\theta : \mathcal{S}_{CAT}(M \text{ rel } C, \Sigma_0) \to \mathcal{S}_{CAT}(M' \text{ rel } C, \Sigma_0')$$

natural for restriction to open subsets of M.

Remark. Naturality of θ here means that θ is defined by a rule such that if $\hat{M}$ is open in M, and

$$\hat{C} = \hat{M} \cap C, \quad \hat{\Sigma}_0 = \Sigma_0 | \hat{M}, \quad \hat{\Sigma}_0' = \Sigma_0' | \hat{M}'$$

then one has a commutative square

† Readers content with the case $\partial M = \phi$ should pass directly on to §3.

$$\begin{array}{ccc} \mathcal{S}_{\mathrm{CAT}}(M \text{ rel } C, \Sigma_0) & \xrightarrow{\theta} & \mathcal{S}_{\mathrm{CAT}}(M' \text{ rel } C, \Sigma'_0) \\ \text{restriction} \downarrow & & \downarrow \text{restriction} \\ \mathcal{S}_{\mathrm{CAT}}(\hat{M} \text{ rel } \hat{C}, \hat{\Sigma}_0) & \xrightarrow{\theta} & \mathcal{S}_{\mathrm{CAT}}(\hat{M}' \text{ rel } \hat{C}, \hat{\Sigma}'_0) \end{array} .$$

Most propositions of this essay will enjoy a similar naturality for restriction .

Proof of 2.1 . To define θ at the level of representing structures consider a CAT structure Σ in the class $x \in \mathcal{S}_{\mathrm{CAT}}(M \text{ rel } C, \Sigma_0)$. To represent $\theta(x)$ choose a CAT structure Σ' on M' that coincides with Σ on M and is a product along $[0, \infty)$ on $\partial M \times [0, \infty)$. Clearly , Σ' is unique if CAT = PL , and it coincides with Σ'_0 near C . If CAT = DIFF we specify Σ' more precisely by choosing a DIFF imbedding $h_\Sigma : \partial M_\Sigma \times (-\infty, 0] \hookrightarrow M_\Sigma$ extending the identity $\partial M_\Sigma \times 0 = \partial M_\Sigma$ and such that the resulting imbedding $h'_\Sigma : \partial M_\Sigma \times (-\infty, \infty) \to M'$ is DIFF near $(\partial M \cap C) \times 0$ as a map to the structure Σ'_0 . Then we let Σ' be the unique DIFF structure on M' so that h'_Σ is a DIFF imbedding and $\Sigma' | M = \Sigma$. Thus $\Sigma' = \Sigma'_0$ near C and Σ' is a product on $\partial M \times [0, \infty)$.

If Σ is concordant rel C to Σ_1 it is easily verified that Σ' is concordant rel C to Σ'_1 (similarly constructed from Σ_1) . Thus θ is well defined . Naturality of θ is evident .

To show that θ is *onto* consider any CAT structure Σ' representing $y \in \mathcal{S}_{\mathrm{CAT}}(M' \text{ rel } C, \Sigma'_0)$. Choose a TOP collaring $h : \partial M \times (-\infty, 0] \to M$ of ∂M so that the resulting bicollaring $h' : \partial M \times (-\infty, \infty) \to M'$ of ∂M in M' is a CAT imbedding near $(C \cap \partial M) \times 0$. Applying the Product Structure Theorem (local version) to h' one gets a concordance of Σ' rel C to a structure Σ'_1 that is a product along $[0, \infty)$ near $\partial M = \partial M \times 0$ in $\partial M \times [0, \infty) \subset M'$.

Using the map $F : I \times \partial M \times [0, \infty) \to I \times \partial M \times [0, \infty)$, defined by $F(t, x, u) = (t, x, (1-t)u)$ one can form a CAT concordance rel M from Σ'_1 to Σ'_2 with $\Sigma'_2 | (\partial M \times [0, \infty)) = (\Sigma'_2 | \partial M) \times [0, \infty)$. Clearly the class of Σ'_2 , which is still y , lies in the image of θ .

The *injectivity* of θ requires a similar application of the Product Structure Theorem . We leave the details to the reader . It would suffice here that $m \geqslant 5$. ∎

§3. MILNOR'S CRITERION FOR TRIANGULATING AND SMOOTHING

For the analysis we fix a topological imbedding† of M^m in a euclidean space R^n, $n >> m$, that is a CAT locally flat imbedding near C; for simplicity we now regard M as a subspace of R^n. As M is an ANR there exists an open neighborhood N of M in R^n that admits a retraction $r : N \to M$ onto M (so that $r|M = \mathrm{id}$). We can and do arrange that r is a CAT map to Σ_0 on the preimage of a neighborhood of C. Then $r^*\tau(M)$ restricted to a neighborhood of C in N has a CAT microbundle structure that we call ξ_0.

THEOREM 3.1

Consider the data: M an TOP manifold without boundary of dimension $m \geqslant 5$; $C \subset M$ a closed subset; Σ_0 a CAT structure defined on a neighborhood of C in M; $r : N \to M$ a neighborhood retraction as introduced above; ξ_0 the CAT structure on the restriction of $r^*\tau(M)$ to a neighborhood of C, derived from Σ_0 as above.

Suppose that the topological microbundle $r^*\tau(M)$ over N admits after stabilizing a CAT structure ξ coinciding near C with ξ_0. Then M admits a CAT structure Σ coinciding with Σ_0 near C. Conversely, if Σ exists, $r^*\tau(M)$ admits such a CAT structure, even without stabilizing.

Proof of 3.1. First we dispose of the converse using

3.2 The pull-back rule π. Given a CAT structure Σ on M homotop r to a CAT map $\hat{r}: N \to M_\Sigma$. Then, applying the microbundle homotopy theorem to a homotopy from r to $\hat{r}$, obtain a TOP micro-isomorphism from $r^*\tau(M)$ to $\hat{r}^*\tau(M)$. Now $\hat{r}^*\tau(M_\Sigma)$ is a CAT microbundle over N (since $\hat{r}$ is CAT), which is topologically identical to $\hat{r}^*\tau(M)$. Hence the above isomorphism endows $\hat{r}^*\tau(M)$ with a CAT microbundle structure ξ.

† It is not necessary to assume that this imbedding is locally flat, or even proper. Hence for $n \geqslant (m+1)m$ such an imbedding is easily deduced from a covering of M^m by open subsets imbeddable in R^m such that the nerve of the covering has dimension $\leqslant m$. See [Mu_1, §2.7], or [II, §1.1].

the above isomorphism endows $r^*\tau(M)$ with a CAT microbundle structure ξ.

If $C \neq \phi$ and Σ equals Σ_0 near C we take some care in making ξ agree near C with ξ_0. Let $H : I \times N \to M$ be a homotopy from r to $\hat{r}$ fixing a neighborhood of C. Then $H^*\tau(M)$ is identical to $I \times r^*\tau(M)$ near $I \times C$. Using the microbundle homotopy theorem, extend this identification from a neighborhood of $I \times C$ to a micro-isomorphism $H^*\tau(M) \cong I \times r^*\tau(M)$ over $I \times N$. Over $1 \times N$ it gives a micro-isomorphism from $H^*\tau(M) | 1 \times N =$ $1 \times \hat{r}^*\tau(M)$ to $1 \times r^*\tau(M)$. Use this micro-isomorphism $\hat{r}^*\tau(M) \to r^*\tau(M)$ to define the structure ξ on $r^*\tau(M)$ as the image of the structure $\hat{r}^*\tau(M_\Sigma)$ on $\hat{r}^*\tau(M)$. The construction $\Sigma \mapsto \xi$ is called the *pull-back rule* and is denoted π. ■

It remains to suitably construct a CAT structure on M from a CAT structure on $r^*\tau(M) \oplus \epsilon^s$, $s \geqslant 0$, using what we call (even for CAT = PL):

3.3 The smoothing rule σ. Since $\tau(M)$ is $M \overset{\Delta}{\to} M \times M \overset{p_1}{\to} M$ with $\Delta(x) = (x, x)$ and $p_1(x, y) = x$, the induced microbundle $r^*\tau(M)$ over N has total space $Er^*\tau(M) = \{(y, r(y), x) \in N \times M \times N\}$ while its projection is $(y, r(y), x) \mapsto y \in N$, and its zero section is $i : y \mapsto (y, r(y), r(y))$. There is a natural homeomorphism $g : Er^*\tau(M) \to M \times N \subset M \times R^n$ given by $(y, r(y), x) \mapsto (x, y)$ which sends $i(M)$ to $\Delta(M) \subset M \times N$. Composing with the imbedding $j : M \times N^n \to M \times R^n$, $j(x, y) = (x, y-x)$ we get an imbedding

$$h = jg : Er^*\tau(M) \to M \times R^n$$

that sends $i(M)$ to $M \times 0$ by $i(x) \mapsto (x, 0)$. Via h we can think of $Er^*\tau(M)$ as an open subset of $M \times R^n$ with $i(M) = M \times 0$. Clearly h gives a CAT imbedding of a neighborhood of $i(C)$ in $E(\xi_0) \subset Er^*\tau(M)$.†

To proceed with the construction suppose that $r^*\tau(M)$ is endowed with a structure of CAT microbundle (without stabilizing), which agrees near $i(C)$ in $Er^*\tau(M)$ with ξ_0. Then a neighborhood of $i(M)$ in $Er^*\tau(M)$ has (a fortiori!) a CAT **manifold** structure

† This pleasantly explicit description was suggested by M. Brown [Brn_3]. Milnor's description is slightly more general. He uses only the fact that N is a CAT manifold with trivial tangent bundle.

which h sends to a CAT manifold structure Θ on a neighborhood of $M \times 0$ in $M \times R^n$ coinciding with $\Sigma_0 \times R^n$ near $C \times 0$. The Product Structure Theorem (local version) produces a concordance of Θ rel $C \times 0$ to a CAT structure that is a product of the form $\Sigma \times R^n$ near $M \times 0$. This Σ is the wanted structure on M.

For the general case suppose that $r^*\tau(M) \oplus \epsilon^s$ over N has a CAT structure ξ. As $E(r^*\tau M \oplus \epsilon^s) = E(r^*\tau M) \times R^s$, the imbedding $h \times (\mathrm{id} | R^s)$ gives us a CAT structure near $M \times 0 \times 0$ in $M \times R^n \times R^s$ from which the Product Structure Theorem produces as above a structure Σ on M. This construction $\xi \nrightarrow \Sigma$ is called the *smoothing rule* and denoted σ. ■

The proof of 3.1 is now complete. ■

We record the following proposition whose proof is straightforward.

PROPOSITION 3.4 (for data of 3.1).

The pull-back rule π and the smoothing rule σ yield two well-defined mappings

$$\mathrm{TOP/CAT}(r^*\tau(M) \text{ rel } C, \xi_0) \underset{\pi}{\overset{\sigma}{\rightleftarrows}} \mathcal{S}_{\mathrm{CAT}}(M \text{ rel } C, \Sigma_0).$$

Each is natural for restriction to open subsets of M. ■

The meaning of naturality was explained under 2.1.

§4. STATEMENTS ABOUT BIJECTIVITY OF THE SMOOTHING RULE AND THE PULL-BACK RULE.

We now complement Theorem 3.1 by classifying CAT structures on M when one is known to exist.

CLASSIFICATION THEOREM 4.1 (for data of 3.1)

If $M \hookrightarrow N$ is a homotopy equivalence, the smoothing rule

$$\sigma : \mathrm{TOP/CAT}(r^*\tau M \text{ rel } C, \xi_0) \to \mathcal{S}_{\mathrm{CAT}}(M \text{ rel } C, \Sigma_0)$$

is bijective (= one to one and onto).

Remark. Once a CAT structure Σ is known on M one can choose $M \to R^n$ to be a CAT imbedding of M_Σ, thus making trivial the choice of the open set N so that $M \hookrightarrow N$ is a homotopy equivalence .†

THEOREM 4.2 (for same data) . *The composition* $\sigma\pi$ *of the pull-back rule* π *with the smoothing rule* σ *is always the identity on* $\mathcal{S}_{CAT}(M \text{ rel } C, \Sigma_0)$.

The proofs of 4.1 and 4.2 appear in §6 below .

Here is an important corollary of 4.1 and 4.2 .

THEOREM 4.3 . *Suppose* M *is a* CAT *manifold without boundary of dimension* $\geqslant 5$ *and* $C \subset M$ *is closed . Then there is a bijective correspondence*

$$\sigma_o : \mathrm{TOP/CAT}(\tau(M) \text{ rel } C) \to \mathcal{S}_{\mathrm{CAT}}(M \text{ rel } C)$$

natural for restriction to open subsets of M . *Its inverse can be described by a pull-back rule like that of 3.2 .*

Proof of Theorem 4.3 from 4.1 and 4.2 . Consider the triangle

$$\begin{array}{ccc} \mathrm{TOP/CAT}(r^*\tau(M) \text{ rel } C) & \xrightarrow{\sigma} & \mathcal{S}_{\mathrm{CAT}}(M \text{ rel } C) \\ & \text{restriction} \searrow \quad \nearrow \sigma_o & \\ & \mathrm{TOP/CAT}(\tau(M) \text{ rel } C) & \end{array}$$

where the data of 3.1 are so chosen that $M \hookrightarrow R^n$ is a CAT imbedding and $r : N \to M$ is a CAT deformation retraction . Then the restriction map indicated is easily seen to be a bijection.‡ *We merely define* σ_o *to make the triangle commute .*

Theorem 4.2 shows σ_o^{-1} can be defined by the following pull-back rule . Let Σ represent $x \in \mathcal{S}_{\mathrm{CAT}}(M \text{ rel } C)$. Find a homotopy from $\mathrm{id}|M$ fixing a neighborhood of C to a CAT map $i' : M \to M_\Sigma$, and using the microbundle homotopy theorem , derive a micro-isomorphism $h : i'^*\tau(M_\Sigma) \to \tau(M)$ that is CAT near C . Then

† However, it is an elementary fact, related to the Kister-Mazur theorem [Si_{13}] , that such an N exists whenever $M \hookrightarrow N$ is a topologically locally flat imbedding .

‡ This requires some care when $C \neq \phi$; indeed , a neighborhood extension property for microbundle maps is wanted . The bijectivity is anyhow implied by Proposition 9.2 below .

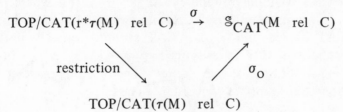

$\sigma_o^{-1}(x)$ is represented by the CAT structure ξ given to $\tau(M)$ by h.

This interpretation of σ_o^{-1} shows first that σ_o is independent of the imbedding $M \hookrightarrow R^n$ and the retraction r and second that σ_o is natural for restriction to open subsets.

This completes the proof of 4.3 granting 4.1 and 4.2. ■

4.4 Supplementary Remarks.

(i) If Σ is a CAT structure on M equal the standard one near C the relative simplicial (or smooth) approximation theorem and a pull-back rule define a natural bijection φ : TOP/CAT(τ(M) rel C) → TOP/CAT($\tau(M_\Sigma)$ rel C). And $\sigma_o \varphi = \sigma_o$ in view of the pull-back rule defining σ_o^{-1}. *Thus 4.3 is meaningful and valid if M is merely a topological manifold with a CAT structure specified near C.*

(ii) This suggests that one define a CAT microbundle structure on a TOP microbundle ξ^n over any (paracompact) base space X in a manner independent of any eventual CAT structure on X. Indeed it can be defined as a TOP microbundle map $\xi^n \to \gamma^n_{CAT}$ where γ^n_{CAT} over $B_{CAT(n)}$ is a fixed universal CAT n-microbundle (cf. §8 below). If X is a CAT object, and η is a CAT structure on ξ^n in the original sense choose a CAT classifying map $\eta(= \xi) \to \gamma^n_{CAT}$ to get a corresponding structure in the new sense. The resulting correspondence gives a bijection of appropriate concordance classes.

§5. THE MAIN LEMMA FOR BIJECTIVITY OF THE SMOOTHING RULE

The proof of Theorem 4.1 will require the following proposition which we shall deduce from Hirsch's relative stable existence and uniqueness theorem for normal microbundles given in Appendix A .

PROPOSITION 5.1 (CAT = DIFF or PL) .

Consider the data : N *a* CAT *manifold without boundary ;* $\zeta : N \hookrightarrow E(\zeta) \overset{p}{\to} N$ *a* TOP *microbundle over* N *;* Θ *a* CAT *manifold structure defined on a neighborhood of the zero-section* N *in the total space* $E(\zeta)$ *;* $C \subset N$ *a closed subset such that* Θ *endows the restriction of* ζ *to a neighborhood of* C *with a* CAT *microbundle structure .*

Then there exists an integer s *and a concordance* rel C *of the restriction of* $\Theta \times R^s$ *to a neighborhood of* N *in* $E(\zeta \oplus \epsilon^s)$*, to a new structure that makes* $\zeta \oplus \epsilon^s$ *a* CAT *microbundle .*†

Proof of 5.1 : Clearly we can assume Θ is defined on all of $E(\zeta)$. Find an integer t and an embedding

$$h : I \times N \to I \times E(\zeta) \times R^t$$

so that h is a product along $I = [0, 1]$ near $\{0, 1\} \times N \cup I \times C$, and $h | (1 \times N)$ is a CAT embedding $1 \times N \to 1 \times E(\zeta)_\Theta \times R^t$, while $h | (0 \times N)$ is the inclusion $0 \times N \to 0 \times E(\zeta) \times R^t$. (See Figure 5-a .)

Figure 5-a .

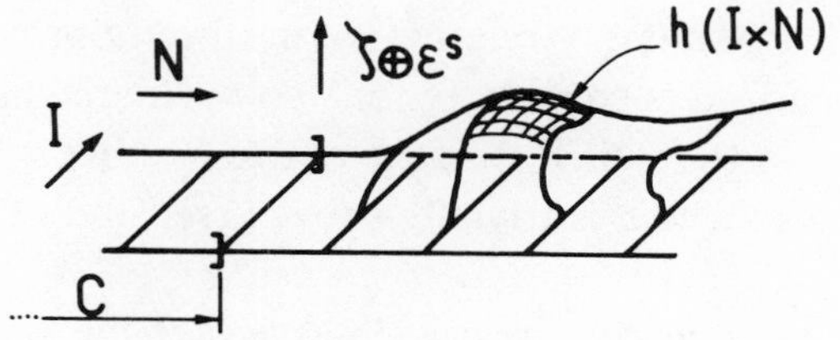

† This integer s can be made to depend on dimN only . We shall use this fact in §9.4 . It is established by the proof below in view of the parallel remark in Appendix A (footnote to A.1 and A.2) .

(This is easily done if t is sufficiently large, using relative approximation of continuous maps by CAT maps, and a covering of N by finitely many co-ordinate charts, cf. the 2nd last step of proof of 4.1 in [III]).

The stable relative existence theorem for CAT and TOP normal microbundles given in Appendix A assures that for some integer $s \geqslant t$, there exists a normal microbundle η to $h(I \times N)$ in $I \times E(\zeta) \times R^s$ that is equal to $I \times (\zeta \oplus \epsilon^s)$, near $0 \times N \cup I \times C$ and that gives a normal CAT microbundle to $h(1 \times N)$ in $1 \times E(\zeta)_\Theta \times R^s$. Indeed the CAT version of Appendix A first gives $\eta | h(1 \times N)$; then granted an inflation of s the TOP version defines η on all of $h(I \times N)$.

The microbundle homotopy theorem now provides a neighborhood U of N in $E(\zeta) \times R^s$ and an open imbedding

$$H : I \times U \to E(\eta) \subset I \times E(\zeta) \times R^s$$

that is the identity on $0 \times U$ and a product along I near $I \times C$, and that presents a micro-isomorphism $1 \times (\zeta \oplus \epsilon^s) \to \eta | h(1 \times N)$. This micro-isomorphism gives the desired new CAT structure on $\zeta \oplus \epsilon^s$; the desired concordance $(I \times U)_\Gamma$ of this structure with U_Θ is just the structure making H a CAT imbedding into $E(\eta) \subset I \times E(\zeta)_\Theta \times R^s$. ∎

Remarks: This brief section contains the one essential intervention of the stable existence and uniqueness theorems for normal microbundles given in Appendix A. The reader may be convinced that this is the right tool to use here by the

Observation: Proposition 5.1 can conversely be used to prove the stable relative existence theorem A.1 for normal TOP microbundles.

Indeed as A.1 is suitably relative we need only prove A.1 when the ambient manifold is covered by one co-ordinate chart; then 5.1 further reduces A.1 to the DIFF relative existence theorem for DIFF normal microbundles, which is trivially proved (see end of proof of 1.2 in [III]).

In the same vein, one can observe that the classification theorem 4.1 lets one quickly prove 5.1. Indeed a simple pursuit of definitions suffices, if, in 5.1, N is an open subset of euclidean space and ζ is trivial. But as 5.1 is relative the general version of 5.1 follows.

It is reasonable to conclude that the cunning geometry in Appendix

A is the keystone of the classification theorem 4.1, indeed of this essay; it should not be neglected.

§6. BIJECTIVITY OF THE SMOOTHING RULE; ITS INVERSE.

Proof of the classification theorem 4.1.

(i) *σ is onto*†: *$TOP/CAT(r^*\tau(M)$ rel $C, \xi_o) \to \mathcal{S}_{CAT}(M$ rel $C, \Sigma_o)$*.

Let a CAT structure Σ representing a given $x \in \mathcal{S}_{CAT}(M$ rel $C, \Sigma_o)$. Then $\Sigma \times R^n$ gives a CAT manifold structure Θ on $E(r^*\tau(M))$, using the natural open embedding

$$h : E(r^*\tau(M)) \hookrightarrow M \times R^n$$

sending $i(M)$ to $M \times 0$. Then 5.1 finds an integer s and a concordance of $\Theta \times R^s$ to Θ' making $r^*\tau(M) \oplus \epsilon^s$ a CAT microbundle over N representing (say) $y \in TOP/CAT(r^*\tau(M)$ rel $C, \xi_o)$. Transferring the concordance by $h \times (id | R^s)$ into $M \times R^{n+s}$, we see that the smoothing rule σ sends y onto x. ■

(ii) *σ is injective.*

Let ξ_1, ξ_2 be two CAT structures on $r^*\tau(M) \oplus \epsilon^s$, $s \geq 0$, agreeing near C in the total space with $\Sigma_o \times R^{n+s}$. Supposing $\sigma(\xi_1) = \sigma(\xi_2)$, we are required to show that ξ_1 and ξ_2 are stably concordant rel $i(C)$. Supposing $\sigma(\xi_1) = \sigma(\xi_2)$ we do clearly get, using h, a conditioned concordance Γ rel C, from an open neighborhood U of M in $E(\xi_1)$, to U as open CAT submanifold of $E(\xi_2)$.

Proposition 5.1 applied over $I \times N'$ (where $N' = N \cap U$) says that, after U is cut down if necessary but not $N' = N \cap U$, there is a stable concordance of $(I \times U)_\Gamma$ rel $(\partial I) \times C \cup I \times C$, to a CAT structure Γ' on $I \times U \times R^t$ making $I \times \{r^*\tau(M) \oplus \epsilon^{s+t}\} | I \times N'$ a CAT microbundle over $I \times N'$. This Γ' gives a stable concordance ζ rel C from $\xi_1 | N'$ to $\xi_2 | N'$.

Using a CAT homotopy of $id | N$ fixing a neighborhood of C to a CAT map $f : N \to N'$, one easily derives, via the microbundle

† This also follows from 4.2 (proved below), but this argument will help the reader to understand the proof of *injectivity* to follow.

homotopy theorem, a stable concordance rel C of ξ_1 to ξ_2. Indeed, pulling back by f we get a concordance $f^*\zeta : f^*\xi_1 \doteq f^*\xi_2$ rel C, but pulling back by the homotopy $\mathrm{id}|N \simeq f$ rel C we get concordances $\xi_i \doteq f^*\xi_i$ rel C, $i = 1, 2$. Here the micro-bundles involved are all topologically micro-identified to $r^*\tau(M)$ or $I \times r^*\tau(M)$ via a single application of the relative topological micro-bundle homotopy theorem to the homotopy $(\mathrm{id}|N) \simeq f$ rel C. ■

This completes Theorem 4.1. ■

Remark 6.1. Without the Product Structure Theorem, the classification theorem 4.1 and its proof would remain valid if $\mathcal{S}_{CAT}(M \text{ rel } C, \Sigma_0)$ were replaced by the set of equivalence classes of CAT structures Σ on $M \times R^s$, $s \geqslant 0$, coinciding with $\Sigma_0 \times R^s$ near $C \times R^s$, under the equivalence $\doteq$ defined by agreeing that $\Sigma \doteq \Sigma \times R$, and $\Sigma \doteq \Sigma'$ if Σ and Σ' are concordant rel $C \times R^s$.

Proof of Theorem 4.2 : $\sigma\pi$ = *identity*.

This does not require §5; it is for the most part a patient pursuit of definitions. In fact it could well have been presented at the end of §3!

Given ξ representing $[\xi] \in TOP/CAT(r^*\tau(M)$ rel $C, \xi_0)$ obtained from $x \in \mathcal{S}_{CAT}(M$ rel $C, \Sigma_0)$ by the pull-back rule, i.e. $\pi(x) = [\xi]$, we are required to show that $\sigma[\xi] = x$.

Recall that $Er^*(\tau M)$ is naturally homeomorphic to $M \times N$ so that $r^*\tau(M)$ becomes $N \xrightarrow{i} M \times N \xrightarrow{p_2} N$ where $i(y) = (r(y), y)$. Also the imbedding $h : Er^*\tau(M) \to M \times R^n$ is then $h(x, y) = (x, y - x)$.

To establish $\sigma[\xi] = x$ amounts to showing that, when ξ is regarded as a CAT manifold structure defined on a neighborhood of the zero-section $i(N)$ in $Er^*\tau(M)$, then there is a neighborhood U of $i(N)$ and a CAT concordance of $E(\xi|U)$ to $h^{-1}(\Sigma \times R^s)|U$.

Following the pull-back rule, let $H : I \times N \to M_\Sigma$ be a homotopy, fixing a neighborhood of C in M, from r to a CAT map $r' : N \to M_\Sigma$. Then

$$E(H^*\tau(M)) = \{(t, y, H(t, y), x) \in I \times N \times M \times M\}$$

is naturally identified to $I \times M_\Sigma \times N$ by $(t, y, H(t, y), x) \leftrightarrow (t, x, y)$. With this identification the zero-section j of $H^*\tau(M)$ is $j(t, y) = (t, H(t, y), y)$ and the projection is the natural one

$I \times M \times N \to I \times N$. Thus

$$H^*\tau(M) : I \times N \xrightarrow{j} I \times M_\Sigma \times N \xrightarrow{p} I \times N .$$

Note that $H^*\tau(M)|0 \times N$ is $0 \times r^*\tau(M)$ and $H^*\tau(M) | 1 \times N$ with total space $1 \times M_\Sigma \times N$ is the CAT bundle $1 \times r'^*\tau(M_\Sigma)$.

Using the microbundle homotopy theorem, find a micro-isomorphism

$$f : I \times r^*\tau(M) \to H^*\tau(M)$$

that is the micro-identity over $0 \times N$ and is the natural micro-isomorphism over a neighborhood of $I \times C$. Let Γ be a CAT manifold structure defined near the zero-section in $I \times Er^*\tau(M)$ so that f is represented by a CAT imbedding into $I \times M_\Sigma \times N = E(H^*\tau(M))$.

Certainly Γ agrees with $0 \times \Sigma \times N$ and is a product along I near $I \times i(C)$. The pull-back rule *defines* ξ so that $\Gamma | 1 \times M \times N$ gives to $1 \times r^*\tau(M)$ the CAT bundle structure $1 \times \xi$. Thus

$$\Gamma : \Sigma \times N \triangleq E(\xi) \text{ rel } i(C) .$$

To conclude it suffices to give a further concordance

$$\Gamma' : \Sigma \times N \triangleq h^{-1}(\Sigma \times R^n) \text{ rel } i(C) ,$$

which is in fact easy ; we set $\Gamma' = J^{-1}(I \times \Sigma \times R^n)$ where

$$J : I \times M \times N \to I \times M_\Sigma \times R^n$$

is defined by $J(t, x, y) = (t, x, y - g_t(x))$ with $g_t : M \to N$ a homotopy, fixing a neighborhood of C, from $M \hookrightarrow N$ to a CAT map $g_1 : M_\Sigma \to N$. This completes the proof. ■

§7. A PARALLEL CLASSIFICATION THEOREM

When the manifold M has a given CAT structure, an alternative classification of CAT structures is available. Its directness and simplicity are appealing; what is more, it provides an abelian group structure on $\mathcal{S}_{CAT}(M)$ since Whitney sum makes TOP/CAT(ϵ(M)) an abelian group, ϵ(M) being the trivial bundle over M.

THEOREM 7.1.

Let M be a CAT manifold of dimension $\geqslant 5$, without boundary. Let $C \subset M$ be a closed subset and let $\epsilon(M)$ be the trivial (zero-dimensional) bundle over M. Then there is a natural bijection

$$\sigma_1 : \text{TOP/CAT}(\epsilon(M) \text{ rel } C) \rightarrow \mathcal{S}_{CAT}(M \text{ rel } C).$$

The mapping σ_1, parallel to the smoothing rule of §3, is defined as follows. If Θ is a CAT structure on a neighborhood of $M \times 0$ in $M \times R^s$ giving $\epsilon^s \doteq \epsilon(M)$ a CAT microbundle structure representing $x \in$ TOP/CAT(M rel C), then $\sigma_1(x)$ is represented by any CAT manifold structure Σ on M such that, for some neighborhood U of $M \times 0$, the structure $(\Sigma \times R^s)|U$ is concordant rel $C \times 0$ to $\Theta|U$. It is, to be sure, the Product Structure Theorem that assures us such a Σ exists and has a well-defined class $\sigma_1(x)$ in $\mathcal{S}_{CAT}(M \text{ rel } C)$, cf. [III, §5.1.1].

Proof of Theorem 7.1. A direct application of Proposition 5.1! ■

In the remainder of this section we relate 7.1 to the classification theorem 4.1 and then discuss an exact sequence of abelian groups entrapping TOP/CAT(ϵ(M)).

LEMMA 7.2.

Consider: M a CAT object; η a CAT microbundle over M; ξ a TOP microbundle over M; ξ_0 a CAT structure on the restriction of ξ to a neighborhood of a closed subset C of M.

In this situation, Whitney sum induces a bijection

$$\text{“}\eta\oplus\text{”} : \mathrm{TOP/CAT}(\xi \text{ rel } C, \xi_0) \to \mathrm{TOP/CAT}(\eta \oplus \xi \text{ rel } C, \eta \oplus \xi_0).$$

Proof of 7.2 : Choosing a CAT microbundle inverse η' to η (using [Mi_5], cf. Appendix A), we get $\eta' \oplus \eta \cong \epsilon^k$ where ϵ^k is the trivial bundle of dimension k say, over M. An inverse to “$\eta\oplus$” is provided by the composition of “$\eta'\oplus$” with the natural isomorphism $\mathrm{TOP/CAT}(\eta' \oplus \eta \oplus \xi \text{ rel } C) \to \mathrm{TOP/CAT}(\xi \text{ rel } C)$ coming from the evident micro-isomorphisms $\eta' \oplus \eta \oplus \xi \cong \epsilon^k \oplus \xi \cong \xi \oplus \epsilon^k$. Since we are dealing with *stable* CAT structures on ξ, the verification that this is the inverse of “$\eta\oplus$” requires a CAT isotopy of $\mathrm{id}\,|\,R^{k+s}$ to the permutation map $(x_1, \ldots, x_{k+s}) \leftrightarrow (x_{k+1}, \ldots, x_{k+s}, x_1, \ldots, x_k)$ for s even. ■

We now have a triangle (for the data of 7.1)

$$\mathrm{TOP/CAT}(\epsilon(M) \text{ rel } C) \xrightarrow{\ \sigma_1 \text{ (from 7.1)}\ } \mathcal{S}_{\mathrm{CAT}}(M \text{ rel } C)$$

$$\text{“}\tau(M)\oplus\text{”} \big\downarrow \text{ (from 7.2)}$$

$$\mathrm{TOP/CAT}(\tau(M) \text{ rel } C) \xrightarrow{\ \sigma_0 \text{ (from 4.1)}\ } \mathcal{S}_{\mathrm{CAT}}(M \text{ rel } C).$$

whose three sides are bijections.

Assertion 7.3. *This triangle is commutative.*

Proof of 7.3 : Let $\nu(M)$ be a CAT normal microbundle to M appropriately imbedded in euclidean space R^{m+n}, n large; denote the projection $r : E(\nu) \equiv N \to M$ and choose an isomorphism $\nu(M) \oplus \tau(M) \cong \epsilon^{m+n} \doteq \epsilon(M)$, to get as in the proof of 7.2 an inverse “$\nu\oplus$” to “$\tau(M)\oplus$”.

Evidently it suffices to check that $\sigma_1 \circ$ “$\nu\oplus$” $= \sigma_0$. For this we observe natural CAT isomorphisms of manifolds

$$E(\nu(M) \oplus \tau(M)) \cong E(r^*\tau(M)) \cong M \times N$$

and identify M with the common zero-section ΔM. Now $\sigma_1 \circ$ “$\nu\oplus$” is determined by the rule that assigns to a CAT manifold structure Θ defined near $M = \Delta M \times 0$ in $M \times N \times R^s$, $s \geqslant 0$, and standard near $C = \Delta C \times 0$, a new CAT structure on M obtained via the Product Structure Theorem applied rel C using the trivialized normal microbundle $\nu \oplus \tau \oplus \epsilon^s \cong \epsilon^{m+n+s}$ to M. On the other hand σ_0 is determined similarly, but using the distinct normal microbundle to M having fibers $x \times N \times R^s$, $x \in M$. Fortunately these two CAT normal microbundles to M in $M \times N \times R^s$ are CAT isotopic, if s is large, by Appendix A. It follows immediately that these two rules give the same map, i.e. $\sigma_1 \circ$ "$\nu\oplus$" $= \sigma$ as required. ∎

To conclude this section, we briefly discuss an exact sequence of abelian groups involving TOP/CAT(ϵ(M)). Here M denotes a DIFF manifold if CAT = DIFF or polyhedron if CAT = PL.

PROPOSITION 7.4. *There is a natural exact sequence of abelian groups as described below:*

$$\mathrm{CAT}(M) \xrightarrow{\alpha} \mathrm{TOP}(M) \xrightarrow{\beta} \mathrm{TOP/CAT}(\epsilon(M)) \xrightarrow{\gamma} \mathrm{B\,CAT}(M) \xrightarrow{\delta} \mathrm{B\,TOP}(M).$$

(A version relative to a closed subset $C \subset M$ could be described similarly.)

B CAT(M) is the abelian group (see [Mi_5]) of stable isomorphism classes of CAT microbundles over M, the addition coming from Whitney sum of microbundles (now CAT = TOP is allowed).

TOP/CAT(ϵ(M)) also has its addition from Whitney sum. It is an exercise (using CAT isotopics of even permutations on R^n to the identity, as for 7.2) to verify that this addition is well-defined and abelian.

CAT(M) is the set of which a typical element is represented by a CAT micro-isomorphism $f : \epsilon^m \to \epsilon^m$ of a trivial bundle $\epsilon^m : M \xrightarrow{\times 0} M \times R^m \xrightarrow{p_1} M$. Another such $g : \epsilon^n \to \epsilon^n$ represents the same element of CAT(M) if there exists a CAT micro-isomorphism $H : I \times \epsilon^s \to I \times \epsilon^s$) coincides with $0 \times f \times (\mathrm{id}|R^{s-m})$ and $H|(1 \times \epsilon^s)$ coincides with $1 \times g \times (\mathrm{id}|R^{s-n})$. We say that h is *stably isotopic* to h′ and write $h \sim h'$. An abelian group structure in CAT(M) is induced by composition of micro-isomorphisms: thus, given x, x′ in CAT(M), represent them by $h, h' : \epsilon^s \to \epsilon^s$; then $h'h$ represents $x+x'$. To verify that this addition is abelian, i.e. $h'h \sim hh'$, we use

Lemma. $h \oplus (\mathrm{id}|\epsilon^n) \sim (\mathrm{id}|\epsilon^n) \oplus h$ (n even).

To prove this lemma one can conjugate using a CAT isotopy of $\mathrm{id}|R^{s+n}$ fixing 0 to the permutation $(x_1, x_2, \ldots, x_{s+n}) \mapsto (x_{s+1}, \ldots, x_{s+n}, x_1, \ldots, x_s)$. ∎

From this lemma it follows that $hh' \sim h \oplus h' \sim h' \oplus h \sim h'h$. Thus CAT(M) is indeed an abelian group.

The maps α and δ are obvious forgetting maps. At the level of representatives, β assigns to a TOP micro-isomorphism $h : \epsilon^n \to \epsilon^n$ the CAT structure ξ^n on ϵ^n that makes $h : \xi^n \to \epsilon^n$ a CAT micro-isomorphism; and γ assigns to a CAT structure ξ^n on ϵ^n the CAT microbundle ξ^n.

This completes our description of the sequence of Proposition 7.4; it is clearly an exact sequence of group homomorphisms. ∎

§8. CLASSIFYING SPACES OBTAINED BY E. H. BROWN'S METHOD

In this section and the next we aim to review the classification of reductions of micro-bundles by vertical homotopy classes of sections of suitable fiber spaces. Combined with the results of earlier sections this will yield in §10 a 'homotopy' classification of CAT manifold structures on any topological manifold of high dimension.

The situation is much as for principal bundles with fiber a topological group (see [Hus]). For any topological group G, the contractible infinite join $E = G*G*G*\ldots$ with the natural right G-action is a universal numerable principal G-bundle with base E/G called B_G. For any subgroup $H \subset G$ the restricted action of H on E is again principal, and since E is contractible we dare to write $E/H = B_H$ †. Now B_G is naturally a quotient of B_H, indeed we have a bundle

$$\mathcal{B} : G/H \to B_H \xrightarrow{\text{quot.}} B_G$$

which is none other than the numerable bundle with fiber G/H associated to the universal principal bundle $E \to B_G$ (read [Hus, p. 70]). Hence reductions (= restrictions) to group H of any numerable principal G-bundle ξ over a space X correspond naturally and bijectively to liftings to B_H of any fixed classifying map $c : X \to B_G$ for ξ. (Recall that such a lifting of c is equivalent to a section of the

† This notation suggests $E \to B_H$ is universal. Here are two cases where it is:

(i) If $G \to G/H$ has local sections (equivalently, is a locally trivial H-bundle) then $E = G*G*\cdots \to B_H$ is evidently locally trivial and hence universal (by [Hus, p.57]), at least for locally trivial principal H-bundles over arbitrary CW complexes. A. Gleason showed that $G \to G/H$ does have local sections if H is a Lie group and is closed in G while G is Hausdorff, see [Ser].

(ii) If G is a complete metric space, and $H \subset G$ is closed and LC^∞, then Michael's selection theorems [Mic] show that every pull-back of the principal bundle $E \to B_H$ over a CW complex is locally trivial. And it follows (as on p.57 of [Hus]) that $E \to B_H$ is again universal for locally trivial principal H-bundles over CW complexes.

associated bundle $\xi[G/H]$.)

If H is a normal closed subgroup of G, then G/H is a topological group and $\mathcal{B}$ is a principal bundle with group G/H. In this case, when one lifting $c' : X \to B_H$ of c is fixed, all other liftings $c'' : X \to B_H$ of c are in bijective correspondence $c'' \leftrightarrow \Delta(c'', c')$ with maps $\Delta(c'', c') : X \to G/H$ by the rule defining $\Delta(c'', c')(x) \in G/H$ by

$$c''(x) = c'(x) \cdot \Delta(c'', c')$$

dot denoting the right action of G/H. This can be further explained by observing that c' provides a section of the induced bundle $c^*\mathcal{B}$ over X, hence a trivialization of it as G/H bundle; for sections of $c^*\mathcal{B}$ are precisely liftings of c.

This is the behavior we encounter with reductions of microbundles. In fact, reductions of locally trivial topological R^n bundles to vector bundles can be discussed in this framework; it is really only PL bundles that force us to adopt less concrete methods below.

PROPOSITION 8.1 . (CAT = PL or TOP) .

There exists a simplicial complex $B_{CAT(n)}$ (locally finite if CAT = PL) and a CAT n-microbundle γ^n_{CAT} over $B_{CAT(n)}$ that is universal in the following relative sense.

(*) Let ξ be a CAT n-microbundle over an euclidean neighborhood retract X (polyhedron if CAT = PL); let U be an open neighborhood of a closed set $C \subset X$; and let $\mathbf{f}_o : \xi | U \to \gamma^n_{CAT}$ be a CAT microbundle map. Then there exists a microbundle map $\mathbf{f} : \xi \to \gamma^n_{CAT}$ so that $\mathbf{f} = \mathbf{f}_o$ near C.

Complement to 8.1 . In case CAT = DIFF Proposition 8.1 is valid except that, to allow $B_{DIFF(n)}$ to be a DIFF manifold, we define it to be the Grassmannian $G_n(R^{n+k})$ of n-planes in R^{n+k}, k large, rather than $G_n(R^\infty)$ which is not a DIFF manifold, and let γ^n_{DIFF} be the standard vector bundle that enjoys a weakened universality, namely (*) for DIFF manifolds X of dimension $< k$. For convenience in what follows we choose $B_{DIFF(n)}$ inductively so that $k = k_n > \dim(B_{DIFF(n-1)}) + n$. ■

PROOF OF PROPOSITION 8.1 (cf. [Wes] [Whi]).

This takes up the rest of the section.

There are semi-simplicial proofs [Mi_3] [RS_0]; however we will establish 8.1 using E. H. Brown's theory of representability, which has earned a reputation for being widely useful although its results are not always the most precise.

Let $\mathcal{C}$ be the category of connected simplicial complexes with base point and base point preserving piecewise linear maps. The simplicial complexes are *not* supposed locally finite†; the compactly generated (CW complex) topology is used. A *piecewise-linear* map $f: X \to Y$ of such complexes is a continuous map such that for each simplex (or finite subcomplex) Δ in X there is some linear subdivision Δ' of Δ such that f maps each simplex of Δ' linearly into some simplex of Y.

If $X \in \mathcal{C}$ is not locally finite a PL n-microbundle ξ over X is a TOP n-microbundle η over X together with a PL structure $\xi|\Delta$ on $\eta|\Delta$ for each simplex (or finite subcomplex) Δ of X; the PL structures $\xi|\Delta$ are supposed to agree wherever they overlap. This certainly extends the definition valid for locally finite complexes given in §1.

A PL n-microbundle ξ over $X \in \mathcal{C}$ with basepoint x_o will be said to be *rooted* [Mi_5, §7] if it is equipped with a microbundle map $b: \xi|x_o \to \epsilon^n$ to the trivial bundle $\epsilon^n: 0 \to R^n \to 0$. Two rooted n-microbundles (ξ, b) and (ξ', b') over X are called *equivalent* if there exists a PL micro-isomorphism $h: \xi \to \xi'$ so that $b'h = b$ near x_o in $\xi|x_o$.

Let $H(X)$ be the set of equivalence classes of rooted PL n-microbundles over X. The standard rule of pull-back makes of H a contravariant functor

$$H: \mathcal{C} \to \mathcal{S}$$

to the category $\mathcal{S}$ of pointed sets. It is straightforward to check that H satisfies the classical axioms of E. H. Brown. First some notation: if $f: X \to Y$ in $\mathcal{C}$, the induced map $H(f): H(X) \leftarrow H(Y)$ is written f^*, and if f is an inclusion we write $f^*(u) = u|X$, $u \in H(Y)$.

† The use of arbitrary simplicial complexes is a device of D. White [Whi] that serves to get a maximum of information from E. H. Brown's method, and with a minimum of effort.

(1) **Homotopy axiom**. *If* $f, g: X \to Y$ *are PL homotopic as pointed maps then* $f^* = g^*$ *as maps* $H(X) \leftarrow H(Y)$.

Proof. This follows from the microbundle homotopy theorem (§1), applied (sharply) relative to base point. This relative microbundle homotopy theorem is valid for arbitrary simplicial complexes, in the form sharply relative to subcomplexes, since one can use the case of a simplex relative to its boundary to obtain the general case via an induction over skeleta. ∎

(2) **Glueing (or Mayer-Vietoris) axiom**. *If* $X = X_1 \cup X_2$ *with common base point in* $X_0 = X_1 \cap X_2$ *and* $x_1 \in H(X_1)$ *and* x_2 *in* $H(X_2)$ *satisfy* $x_1 | X_0 = x_2 | X_0$, *then there exists* $x \in H(X)$ *such that* $x | X_1 = x_1$ *and* $x | X_2 = x_2$. *And* x *is unique if* X *is the wedge (sum with common base point)* $X = X_1 \vee X_2$. ∎

(3) **Wedge axiom**. *The natural map*

$$H(\vee_\alpha X_\alpha) \to \prod_\alpha H(X_\alpha)$$

is a bijection for the wedge $\vee_\alpha X_\alpha$ *of any collection* $\{X_\alpha\}$ *of pointed complexes*. ∎

E. H. Brown's celebrated representation theorem [Brn] [Sp] [Ad] asserts that any functor $H : \mathcal{C} \to \mathcal{E}$ verifying (1), (2) and (3) is representable in the sense that it admits a 'classifying' space B and 'universal' element u as follows:

> (**) *There is an object* B *in* $\mathcal{C}$ *and an element* u *in* $H(B)$ *so that* $T_u : [X, B] \to H(X)$, *defined by sending* $f: X \to B$ *to* $f^*u \equiv H(f)(u) \in H(X)$, *is bijective for every* X *in* $\mathcal{C}$. *(Here* $[X, B]$ *denotes the set of pointed PL homotopy classes of pointed PL maps* $X \to B$.*)*

It is more usual for $\mathcal{C}$ to be the category $\mathcal{C}_0$ of pointed CW complexes and pointed continuous maps. But, given the relative simplicial approximation theorem of [Ze_2], Brown's arguments apply to our PL category, cf. [Whi]; indeed Brown even axiomatizes the properties of $\mathcal{C}$ that are essential [Brn]. Alternatively, since each CW complex is homotopy equivalent to a simplicial complex and since continuous maps of simplicial complexes can be homotoped to PL maps in a relative way (for both facts use the relative simplicial approximation theorem in a skeletal induction), one can extend H in an essentially unique way to

$$H : \mathcal{C}_o \to \mathcal{E} .$$

This functor still verifies axioms (1), (2) and (3). (Indeed (1) and (3) are evident; and (2) follows from the fact that any triad $(X ; X_1 , X_2)$ of CW complexes is homotopy equivalent to a simplicial triad, cf. [III, §5.2].)

Clearly B can be replaced by any complex homotopy equivalent in $\mathcal{C}$ (or $\mathcal{C}_o$). Hence B can chosen to be countable because $\pi_k B = H(S^k)$ is countable for all k† ‡; and then in addition B can be made a locally finite simplicial complex, cf. [Mi_2]. Then $u \in H(B)$ is represented by a 'universal' microbundle γ. We write B and γ now as $B_{PL(n)}$ and γ^n_{PL}.

Similarly, in fact more easily, one defines $B_{TOP(n)}$ and γ^n_{TOP}.

To establish 8.1 it remains only to prove.

ASSERTION 8.3 (for CAT = PL or TOP). *Let γ^n_{CAT} be the CAT microbundle over $B_{CAT(n)}$ constructed above. The universality property (**) implies the relative universality property (*) of 8.1.*

The most obvious approach to the proof fails because one finds that a micro-automorphism of ξ near C intervenes. Instead we argue roughly that (*) holds for $(X , C) = (S^k , 0)$, hence also for $(B^k , \partial B^k)$, which implies the general case; the details are given in four steps, which apply to most bundle theories.

Step (1). *B and γ remain universal in the sense of (**) if a new base point $b' \in B$ is chosen and γ is suitably re-rooted.*

Proof of (1). There is a (pointed) homotopy equivalence $(B , b) \to (B , b')$ homotopic (freely) to $id | B$, hence covered by a micro-bundle

† The countability of $H(S^k)$ follows quickly from the fact that, up to isomorphism, there are only countably many simplicial maps $f : X \to Y$ of finite simplicial complexes. Incidentally, if we were working in the topological category, the Černavskii local contractibility theorem [$Če_{1,2}$] would assure that $H(S^k)$ is countable, cf. [Si_{12}, p.162].

‡ Proving that if $\pi_* B$ is countable then B is homotopy equivalent to a countable complex can be made an exercise with Brown's method for representing the functor $X \mapsto [X , B]$.

map $\gamma \to \gamma$. From this fact step (1) follows easily. ■

Lemma 8.4 . *Let* $f: \epsilon^n \to \epsilon^n$ *be a CAT micro-isomorphism of the trivial bundle* ϵ^n *over* R^k *to itself such that* f *is the identity on* $\epsilon^n | 0$, *where* $0 =$ *origin of* R^k. *There exists another CAT micro-isomorphism* $f' : \epsilon^n \to \epsilon^n$ *equal* f *near* $R^k - 2B^k$ *(where* $B^k =$ *k-ball in* R^k*) and equal the identity near* B^k.

Proof of lemma : Indeed f' can be the pullback of f by a CAT map $R^k \to R^k$ that maps $\frac{3}{2} B^k$ to the origin and fixes $R^k - \frac{5}{3} B^k$.
■

Step (2) . *Property (*) holds when* $X = S^k$, $k \geq 0$, *and* C *is a hemisphere of* S^k.

Proof of (2). We are given a CAT micro-bundle map $\mathbf{f}_o : \xi | U \to \gamma$, U being an open neighborhood of C. Choose a base point $x_o \in C$ for S^k, and using step (1) let $f_o(x_o)$ ($\mathbf{f}_o$ covering f_o) be the base point for B. By (**) and the homotopy extension property we can find a classifying map $f : S^k \to B$ with $f = f_o$ on a neighborhood V of C and a micro-isomorphism

$$\varphi : \xi \to f^*\gamma .$$

Note that, as $f^*\gamma | V \doteq f_o^*\gamma | V$, there is a natural micro-identification

$$\varphi_o : \xi | V \to f^*\gamma | V$$

which, composed with $f^*\gamma \to \gamma$, yields $\mathbf{f}_o$. And (**) tells us that φ can be chosen equal to φ_o on $\xi | x_o$. By Lemma 8.4 we can alter φ near C to

$$\varphi' : \xi \to f^*\gamma$$

so that $\varphi' = \varphi$ near C. Then the composed microbundle map

$$\mathbf{f} : \xi \xrightarrow{\varphi'} f^*\gamma^n \xrightarrow{\text{nat.}} \gamma^n$$

equals $\mathbf{f}_o$ near C. ■

Step (3) . *Property (*) holds when* X *is a k-disc* D^k, $k \geq 0$, *and* $C = \partial D^k$.

Proof of (3). We can assume D^k is a hemisphere of S^k. Extend

$f_o : S^{k-1} = \partial D^k \to B$ over the complementary hemisphere D^k_- to get $f_- : D^k_- \to B$. Identifying $f^*_-\gamma^n | S^{k-1}$ naturally with $\xi | S^{k-1}$ near S^{k-1} form a microbundle ξ' over S^k extending ξ and $f^*_-\gamma^n$, together with a microbundle map $\mathbf{f}'_o : \xi' | U' \to \gamma^n$, $U' = U \cup D^k_-$, extending $\mathbf{f}_o$. Now apply Step (2). ∎

Step (4). *Property (*) holds in general.*

Proof of (4). We can assume X is a simplicial complex. If CAT = TOP, arrange this by passing to a neighborhood in euclidean space having the original X as retract. Triangulate the new X finely and construct the required extension by induction applying step (3) simplex by simplex. ∎

This completes the proof of Assertion 8.3 and of Proposition 8.1.

∎

§9. CLASSIFYING REDUCTIONS

The universal microbundle γ^n_{CAT}, (CAT = DIFF or PL) can be regarded as a TOP n-microbundle. As such it admits a map $j_n : \gamma^n_{CAT} \to \gamma^n_{TOP}$ covering a map $j_n : B_{CAT(n)} \to B_{TOP(n)}$. And by 8.1 j_n is unique up to homotopy. Using the well-known device of Serre we now convert j_n into a Hurewicz fibration $p_n : B'_{CAT(n)} \to B_{TOP(n)}$. Thus we define $B'_{CAT(n)}$ to be the set of commutative squares of continuous maps

$$\begin{array}{ccc} 0 & \xrightarrow{\text{inclusion}} & I \\ \downarrow & & \downarrow \\ B_{CAT(n)} & \xrightarrow{j_n} & B_{TOP(n)} \end{array}$$

endowed with the compact-open topology. (It can be described as the space of paths in $B_{TOP(n)}$ each issueing from some specified point of $B_{CAT(n)}$.) There is an imbedding $i_n : B_{CAT(n)} \to B'_{TOP(n)}$ as constant paths and the initial point map is a deformation retraction $B'_{CAT(n)} \to B_{CAT(n)}$ while the final point map $p_n : B'_{CAT(n)} \to$ $\to B_{TOP(n)}$ is a Hurewicz fibration extending $j_n : B_{CAT(n)} \to$ $\to B_{TOP(n)}$, i.e. p_n has the homotopy lifting property.

PROPOSITION 9.1.

Let ξ^n be a TOP n-microbundle over a CAT object X (= DIFF manifold or polyhedron) and let $\mathbf{f} : \xi \to \gamma^n_{CAT}$ *over $f : X \to B_{TOP(n)}$ classify ξ.*

Then, there is a natural bijective correspondence between concordance classes of CAT microbundle structures on ξ^n and vertical homotopy classes of liftings of f to $B'_{CAT(n)}$. (Vertical means that composition with p_n gives the constant homotopy of f.)

This bijection is denoted

$$\varphi_n : \mathrm{TOP}_n/\mathrm{CAT}_n(\xi) \to \mathrm{Lift}(f).$$

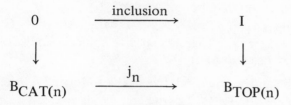

Proof of 9.1 (cf. Browder [Br_1 , §4]) .

To lighten notation here we delete CAT , replace TOP by + and usually suppress n .

Here is a rule that defines φ_n . Given a CAT structure η on ξ choose a CAT microbundle map $\mathbf{G} : \eta \to \gamma^n$ and consider the composed TOP microbundle map $\xi = \eta \to \gamma \to p^*\gamma_+$, which we also call $\mathbf{G}$. We will presently explain how to form a microbundle homotopy $\mathbf{H} : I \times \xi \to p^*\gamma_+$, $\mathbf{H} : \mathbf{G} \simeq \mathbf{F}$ so that $p_\# \mathbf{F}$ is the given map $\mathbf{f} : \xi \to \gamma_+$; then if $\mathbf{F} : \xi \to p^*\gamma_+$ is over say $F : X \to B'$ we have $pF = f$. The rule $\eta \leftrightarrow F$ defines φ_n .

To form $\mathbf{H}$ above , first use 8.1 to obtain a TOP microbundle homotopy $\mathbf{H}_+ : \dot{I} \times \xi \to \gamma_+$, $\mathbf{H}_+ : p_\# \mathbf{G} \simeq \mathbf{f}$, over $H_+ : I \times X \to B_+$, $H_+ : pG \simeq f$, say . Then apply the homotopy lifting property of $p : B' \to B_+$ to obtain a homotopy $H : I \times X \to B'$, from G to say F , and lifting H_+ (i.e. $pH = H_+$) . Finally note that $H : I \times X \to B'$ and $\mathbf{H}_+ : I \times \xi \to \gamma_+$ together specify a microbundle homotopy $\mathbf{H} : I \times \xi \to p^*\gamma_+$ as required .

Observation : The above rule is relative in the following sense : Let η_o be a CAT structure on $\xi_0 = \xi | U$ where U is an open neighborhood in X of a given closed set $C \subset X$. Suppose the above construction has been carried out for ξ_0 , η_o and $\mathbf{f}_o = \mathbf{f} | U$, yielding say $\mathbf{H}_o : \mathbf{G}_o \simeq \mathbf{F}_0$. *Then the construction can be carried out for* ξ , η *and* $\mathbf{f}$ *yielding say* $\mathbf{H} : \mathbf{G} \simeq \mathbf{F}$ *in such a way that* $\mathbf{H}$ *equals* $\mathbf{H}_o$ *near* $I \times C$. ■

Applied to the bundle $I \times \xi$ (in place of ξ) with $\{0 , 1\} \times X$ as C this observation shows that φ_n is *well-defined* .

Here is a rule defining a mapping ψ_n that will be the inverse of φ_n . Given a lifting $F : X \to B'$ of f , note that F and $\mathbf{f}$ together specify a unique $\mathbf{F} : \xi \to p^*\gamma_+$ lifting $\mathbf{f}$. *Since* B *is a strong deformation retract of* B' , we can find $\mathbf{H} : I \times \xi \to p^*\gamma_+$, $\mathbf{H} : \mathbf{G} \simeq \mathbf{F}$ with $\mathbf{G}$ over a CAT map $G : X \to B \subset B'$. (Use 8.1 and CAT approximation of continuous maps here .) Now ξ has a canonical CAT structure η such that $\mathbf{G} : \eta = \xi \to p^*\gamma_+ | B \doteq \gamma$ becomes a CAT microbundle map $\eta \to \gamma$. The rule $F \to \eta$ defines ψ_n .

The construction for ψ_n has a relative form like that for φ_n . Hence ψ_n is well-defined . It is clear that φ_n and ψ_n are inverse to one another . Thus 9.1 is proved . ■

Warning. The bijective correspondence of 9.1 is *not* well-determined without a specific TOP microbundle map $\mathbf{f} : \tau(M) \to \gamma^n_{TOP}$ being given (even with ξ and $f : X \to B_{TOP(n)}$ well specified). Without $\mathbf{f}$ we have instead a classification of CAT structures on $f^*\gamma^n_{TOP}$. The map $\mathbf{f}$ amounts to giving in addition a micro-isomorphism $\xi \cong f^*\gamma^n_{TOP}$, and there may be many essentially different ones. See end of §5 in [Si_{10}] for an example of the influence of the choice of $\mathbf{f}$ when $X = S^3$ and ξ is trivial, and compare §7.4 above.

We now formulate a relative version of this proposition. Let ξ_0 be a CAT structure on $\xi | U$ where U is an open neighborhood in X of a given closed subset $C \subset X$. We wish to analyse the set

$$TOP_n/CAT_n(\xi \text{ rel } C, \xi_0)$$

of concordance classes rel C of CAT structures on ξ that coincide near C with ξ_0.

For simplicity suppose now that for a CAT classifying map $\mathbf{F}_o : \xi_o \to \gamma^n_{CAT}$ one has $\mathbf{j}_n \mathbf{F}_0 = \mathbf{f} | (\xi | U)$ as maps to γ^n_{TOP}. Using 8.1 this can always be arranged, *provided* that we are free to rechoose $\mathbf{f}$. (If not see 9.2.1 below.)

Define

$$\text{Lift}_n(f \text{ rel } C, F_0)$$

to be the set of liftings of f to $B'_{CAT(n)}$ that equal F_0 near C, divided by the equivalence relation of vertical homotopy rel C.

PROPOSITION 9.2 (relative version of 9.1).

With the above data there exists a natural bijection

$$\varphi_n : TOP_n/CAT_n(\xi \text{ rel } C, \xi_0) \to \text{Lift}_n(f \text{ rel } C, F_0).$$

Remark 9.2.1. This bijection can also be set up in case we can not or will not make a special choice of $\mathbf{f}$ as above. Instead we must specify (a) a CAT classifying map $\mathbf{G}_o : \xi_o \to \gamma^n_{CAT}$ and (b) a TOP microbundle homotopy $\mathbf{H}_o : I \times \xi_0 \to p_n^*\gamma^n_{TOP}$, $\mathbf{H}_o : \mathbf{G}_o \simeq \mathbf{F}_0$ from $\mathbf{G}_o$ to say $\mathbf{F}_0 : \xi_0 \to p_n\gamma^n_{TOP}$ over (say) $F_0 : U \to B'_{CAT(n)}$ such that $(p_n)_\# \mathbf{F}_0 = \mathbf{f} | (\xi | U)$ and hence $p_n F_0 = f | U$. ∎

The proof of 9.2 is a straightforward generalization of 9.1 and we leave it aside. (Cf. [V, §2].) ∎

A word about the *naturality* in 9.2. If $\mathbf{g} : \xi' \to \xi$ is a TOP microbundle map covering a CAT map $g : X' \to X$ it means we have a commutative square

$$\begin{array}{ccc} TOP_n/CAT_n(\xi \text{ rel } C, \xi_0) & \xrightarrow{\varphi_n} & Lift_n(f \text{ rel } C, F_0) \\ \alpha \downarrow & & \beta \downarrow \\ TOP_n/CAT_n(\xi' \text{ rel } C', \xi'_0) & \xrightarrow{\varphi_n} & Lift_n(f' \text{ rel } C', F'_0) \end{array}$$

where $C' \subset g^{-1} C$, $\xi'_0 = (g | g^{-1} U)^* \xi_0$, $\mathbf{f'} = \mathbf{fg}$, $F'_0 = (F_0 g) | (g^{-1} U)$, and where α arises from pull-back while β is defined at the level of representatives by sending a lifting F of f to Fg. ■

The *space* TOP_n/CAT_n is defined to be the fiber (over a base point) of $p_n : B'_{CAT(n)} \to B_{TOP(n)}$. It is related to the *function* $\xi \mapsto TOP_n/CAT_n(\xi)$ above by 9.1, 9.2; the homotopy type of TOP_n/CAT_n puts up the obstructions to finding liftings of $f : X \to B_{TOP(n)}$ to $B'_{CAT(n)}$, or to homotoping two such liftings, etc.

Stable classification.

There is a homotopy commutative ladder

$$(L) \qquad \begin{array}{ccc} \cdots \to B_{CAT(n)} & \xrightarrow{s_n} & B_{CAT(n+1)} \to \cdots \\ \downarrow j_n & & \downarrow j_{n+1} \\ \cdots \to B_{TOP(n)} & \xrightarrow{s'_n} & B_{TOP(n+1)} \to \cdots \end{array}$$

where the horizontal CAT map s_n is defined up to homotopy by the condition that it be covered by a CAT microbundle map $\mathbf{s}_n : \gamma^n_{CAT} \oplus \epsilon^1 \to \gamma^{n+1}_{CAT}$. It represents stabilization. Similarly for $\mathbf{s}'_n$. Recall that j_n represents the forgetful functor and is covered by a TOP microbundle map $\gamma^n_{CAT} \to \gamma^n_{TOP}$.

With the help of mapping cylinder constructions arrange inductively that all the stabilizations s_n and s'_n are inclusions of subcomplexes. Then, using inductively the homotopy extension property for these inclusions, make the ladder (L) strictly commutative. Now we have "stable" classifying spaces (complexes !)

$$B_{CAT} = \bigcup_n B_{CAT(n)}, \quad B_{TOP} = \bigcup_n B_{TOP(n)}$$

and a "forgetful" map $j = \cup j_n : B_{CAT} \to B_{TOP}$. Let B'_{CAT} be the space of paths in B_{TOP} issueing each from a specified point in B_{PL}; and adopt the compact-open topology for B'_{CAT}. One has

$$B'_{CAT} = \bigcup_n B'_{CAT(n)} .$$

In fact the following classical lemma applied to $B_{TOP} = \cup B_{TOP(n)}$ shows that any such path in B'_{CAT} lies in some $B'_{CAT(n)}$. More generally, using the exponential law [Sp, p.6], the lemma shows that any compactum in B'_{CAT} lies in some $B'_{CAT(n)}$.

Lemma. *Let X be a topological space having its topology coinduced ([Sp, p.4]) from a nested sequence of subspaces $X_1 \subset X_2 \subset X_3 \subset \cdots$ with $X = \bigcup_n X_n$. If $K \subset X$ is compact Hausdorff, then K lies in some X_n.*

Proof of lemma (by contradiction). If $x_n \in K \cap (X - X_n)$ is a sequence of distinct points, then the set $\{x_n \mid n=1,2,3,\ldots\} \subset K$ has discrete topology coinduced since it meets each X_n in a finite set. ∎

Choosing base points to make the maps in (L) pointed, we next define the space TOP/CAT to be the fiber of the natural Hurewicz fibration $p : B'_{CAT} \to B_{TOP}$ and observe that

$$\mathrm{TOP/CAT} = \bigcup_n \mathrm{TOP}_n/\mathrm{CAT}_n .$$

Our definitions assure commutativity of the square

$$\begin{array}{ccc} \mathrm{TOP}_n/\mathrm{CAT}_n(\xi \text{ rel } C, \xi_o) & \xrightarrow[\cong]{\varphi_n} & \mathrm{Lift}_n(f \text{ rel } C, F_o) \\ \downarrow & & \downarrow \\ \mathrm{TOP}_{n+k}/\mathrm{CAT}_{n+k}(\xi \oplus \epsilon^k \text{ rel } C, \xi_o \oplus \epsilon^k) & \xrightarrow[\cong]{\varphi_{n+k}} & \mathrm{Lift}_{n+k}(f \text{ rel } C, F_o) \end{array}$$

where the vertical arrows come from stabilization. Taking the limit on k we get a bijection

$$(9.3) \quad \mathrm{TOP/CAT}(\xi \text{ rel } C, \xi_o) \xrightarrow[\cong]{\varphi} \varinjlim_k \mathrm{Lift}_{n+k}(f \text{ rel } C, F_o) .$$

The right hand side maps naturally to $\mathrm{Lift}(f \text{ rel } C, F_o)$ defined for $p : B'_{CAT} \to B_{TOP}$. This mapping is bijective whenever X is compact, because any map of a compactum into B'_{CAT} must lie in $B'_{CAT(n)}$ for some n. Thus we have verified the following theorem in case X is compact .

THEOREM 9.4. *The natural mapping (from 9.3)*

$$\theta : \mathrm{TOP/CAT}(\xi \text{ rel } C, \xi_o) \to \mathrm{Lift}(f \text{ rel } C, F_o)$$

is bijective provided the CAT object X is of finite dimension .

To prove 9.4 , it remains to establish that the natural map

$$\lim_{\vec{k}} \mathrm{Lift}_{n+k}(f \text{ rel } C, F_o) \to \mathrm{Lift}(f \text{ rel } C, F_o)$$

is also bijective in case X is noncompact .

A suitable stability lemma for the sequence $\mathrm{TOP}_n/\mathrm{CAT}_n$ will permit one to establish this by building vertical deformations into some $B'_{\mathrm{CAT}(n)}$ of lifts to B'_{CAT} using a cell by cell skeletal induction in X . Clearly the following result will suffice :

Stability Theorem (proved in [KS_1] [LR_2] [Si_{10} , §5] [V , §5.2]) .
If $k \leqslant n$ *and* $n \geqslant 5$, *then* $\pi_k(\mathrm{TOP/CAT}, \mathrm{TOP}_n/\mathrm{CAT}_n) = 0$.

We , however , prefer to use a weaker stability result whose proof we shall give here using only methods familiar in this essay . The above stability result requires (in all proofs given so far) some ideas from the rival immersion theoretic methods of classification .

Weak Stability Lemma 9.5 . *Given integers k and n , each* $\geqslant 0$, *there exists an integer* $s \geqslant 0$ *such that the stabilisation map*

$$\pi_k(\mathrm{TOP}_{n+s}/\mathrm{CAT}_{n+s}, \mathrm{TOP}_n/\mathrm{CAT}_n) \to \pi_k(\mathrm{TOP/CAT}, \mathrm{TOP}_n/\mathrm{CAT}_n)$$

is surjective .

This is clearly also strong enough to prove 9.4 (but just !) . Beware however that 9.4 is presumably† false for infinite dimensional objects such as $X = \coprod_k S^k$. At any rate , it is easy to prove using Pontrjagin characteristic classes that $[\amalg S^k, B_{\mathrm{CAT}}] \neq \lim_{\vec{n}} [\amalg S^k, B_{\mathrm{CAT}(n)}]$.

Proof of 9.5. We can suppose with no loss of generality that $k+n \geqslant 6$; for the special case formally implies the general case . Using the method of proof of 9.1 the reader will verify that 9.5 is equivalent to the following geometric assertion.

Assertion : Let ξ^{n+x} be a CAT structure on the trivial bundle ϵ^{n+x} over B^k such that near ∂B^k the structure ξ is a product with R^x. Then there exists an integer s depending on n and k (*but not on* ξ) such that, for some $y \geq 0$ depending on ξ, there is a concordance $\zeta^{n+x+y} : \xi \oplus \epsilon^y \doteq \eta^{n+s} \oplus \epsilon^{x+y-s}$ rel ∂B^k, where η is a CAT structure on ϵ^{n+s}, and ζ is a CAT structure on $I \times \epsilon^{n+x+y}$.

Proof of assertion : To simplify notation we can assume ξ^{n+x} represented by a CAT manifold structure on all of $B^k \times R^{n+x}$. We remark that, in the proof below, B^k could be any CAT manifold. Regarding ξ as a mere manifold structure, apply the Product Structure Theorem rel boundary to obtain a conditioned concordance $\Gamma : \xi \doteq \sigma \times R^x$ rel $\partial B^k \times R^{n+x}$, where σ is a CAT manifold structure on $B^k \times R^n$. Now apply the main lemma 5.1 to Γ and the bundle $I \times \epsilon^{k+n+x}$ over $I \times B^k$, always working rel $(I \times \partial B^k) \cup (0 \times B^k)$: first near the end $1 \times B^k$, where Γ is conditioned, to produce η^{n+s}; and second rel $1 \times B^k$ to produce ζ^{n+x+y} : $\xi \oplus \epsilon_y \doteq \eta^{n+s} \oplus \epsilon^{x+y-s}$. Observe that η^{n+s} depends only on $(B^k \times R^n)_\sigma$ via 5.1, whence s depends only on k (and not on ξ^n; nor even on n, an unnecessary bonus). ■

Remark : The early (but rigorous) treatment of smoothing theory by Lashof and Rothenberg [LR_1] modestly restricted attention essentially to compact manifolds ; we have just faced the one technical problem that presumably discouraged treating arbitrary noncompact manifolds.

§10. HOMOTOPY CLASSIFICATION OF MANIFOLD STRUCTURES

Combining the isomorphism of 9.4 relating reductions to liftings, with the isomorphisms of the classification theorems 3.1, 4.1, 4.2, 4.3, 7.2 relating manifold structures to reductions and with a theorem to absorb the boundary 2.1 we immediately arrive at a theorem relating CAT manifold structures to liftings of classifying maps, from B_{TOP} to B'_{CAT} in the Hurewicz fibration (of §9) $p : B'_{CAT} \to B_{TOP}$ with fiber TOP/CAT.

CLASSIFICATION THEOREM 10.1.

Consider the data: M^m *a topological m-manifold (metrizable, and possibly with boundary);* Σ_0 *a CAT manifold structure defined on an open neighborhood* U *of a closed* $C \subset M$; $\mathbf{F}_0 : \tau(M)|U \to \gamma^n_{CAT}$ *a microbundle map over* $F_0 : U \to B_{CAT}$ *that is CAT for the CAT microbundle structure on* $\tau(M)|U$ *given by* Σ_0; $\mathbf{f} : \tau(M) \to \gamma^m_{TOP}$ *a TOP microbundle map over* $f : M \to B_{TOP(m)}$ *such that* $\mathbf{j}_m \mathbf{F}_0 = \mathbf{f}$ *near* C.†

Suppose $m = \dim M \geqslant 6$, *(or* $m = 5$ *and* $\partial M \subset C$*). Then* M *admits a CAT manifold structure* Σ *equal* Σ_0 *near* C *if and only if* f *has a lifting* $M \to B'_{CAT}$ *equal to* F_0 *near* C. *In fact there is a bijection* θ

$$\mathcal{S}_{CAT}(M \text{ rel } C, \Sigma_0) \xrightarrow{\theta} \text{Lift}(f \text{ rel } C, F_0)$$

mapping from concordance classes rel C *of such structures* Σ *and mapping to vertical homotopy classes rel* C *of liftings of* f *to* B'_{CAT} *equal* F_0 *near* C. *Further, when one such* Σ *is singled out, there is a bijection* θ_1

$$\mathcal{S}_{CAT}(M \text{ rel } C, \Sigma_0) \xrightarrow{\theta_1} [M \text{ rel } C, \text{TOP/CAT}]$$

to the homotopy classes rel C *of maps* $M \to TOP/CAT$ *that send a neighborhood of* C *to the base point.*

† One can relax this restriction on **f** by giving more data, see 9.2.1.

The bijections θ *and* θ_1 *above are natural for restriction to open subsets* $\hat{M}$ *of* M. *Thus there is a commutative square*

$$\begin{array}{ccc} \mathcal{S}_{CAT}(M \text{ rel } C,\Sigma_o) & \xrightarrow{\theta} & \text{Lift}(f \text{ rel } C,F_o) \\ \text{restriction} \downarrow & & \downarrow \text{restriction} \\ \mathcal{S}_{CAT}(\hat{M} \text{ rel } \hat{C},\hat{\Sigma}_o) & \xrightarrow{\theta} & \text{Lift}(f|\hat{M} \text{ rel } \hat{C},\hat{F}_o) \end{array}$$

in which $\hat{\Sigma}_o = \Sigma_o|(U \cap \hat{M})$, $\hat{C} = C \cap M$, $\hat{F}_o|(U \cap \hat{M})$. *And there is a similar square with* θ_1 *in place of* θ.

10.2. Concerning boundary. When $\partial M \neq \phi$, one defines $\tau(M)$ to be $\tau(M')|M$ where M' is M with the collar $\partial M \times [0,\infty)$ attached by identifying $\partial M = \partial M \times 0$ as in 2.1. The case with $\partial M \neq \phi$ of 10.1 follows from 2.1 after an application to M' of the case without boundary, if we note that restriction gives an isomorphism

$$\text{Lift}(f' \text{ rel } C,F'_o) \to \text{Lift}(f \text{ rel } C,F_o)$$

whenever $f' : M' \to B_{TOP}$ extends f, while $F'_o : U' \to B'_{CAT}$ lifts $f'|U'$ and extends F_o.

10.3. Description of θ. If Σ is a CAT structure representing $x \in \mathcal{S}_{CAT}(M)$ then $\theta(x) \in \text{Lift}(f)$ can be quickly constructed as follows. (See §4 and §9 for proof.) Let $\mathbf{G} : \tau(M_\Sigma) \to \gamma^n_{CAT}$ over $G : M \to B_{CAT}$ be a CAT microbundle map. Then 8.1 lets us form a TOP microbundle homotopy $\mathbf{H} : I \times \tau(M) \to \gamma^n_{TOP}$ over $H : I \times M \to B_{TOP}$ from the composition $\mathbf{j}_n\mathbf{G} : \tau(M) \to \gamma^n_{CAT} \to \gamma^n_{TOP}$ to be *given* $\mathbf{f} : \tau(M) \to \gamma^n_{TOP}$. With our specific definition† of B'_{CAT}, the maps G and $H : I \times M \to B_{TOP}$ (with $0 \times G$ lifting $H|(0 \times M)$) already constitute a lifting $F : M \to B'_{CAT}$ of f; this represents $\theta(x)$.

Now the relative case $(C \neq \phi)$. If Σ represents $x \in \mathcal{S}_{CAT}(M \text{ rel } C,\Sigma_o)$ we can assure successively that $\mathbf{G}$ equals $\mathbf{F}_o$ near C and that $\mathbf{H}$ is the constant homotopy of $\mathbf{G}$ near $I \times C$. Then the lifting F equals F_o near C in X, and F once more represents $\theta(x)$.

10.4. Description of θ^{-1}. If $y \in \text{Lift}(f)$ is represented by

$F : M \to B'_{CAT}$, then $\theta^{-1}(y) \in \mathcal{S}_{CAT}(M)$ can be defined as follows when $\partial M = \phi = C$. As in §3, let M be imbedded in R^n, n large, and let $r : N \to M$ be a retraction to M of an open neighborhood of M in R^n. We arrange that $F(M) \subset B'_{CAT(t)}$, $m \leqslant t < \infty$ (see 9.4 and related discussion). Then F and $\mathbf{f} : \tau(M) \oplus \epsilon^{t-m} \to \gamma^m_{TOP} \oplus \epsilon^{t-m}$ together provide a map $\mathbf{F} : \tau(M) \oplus \epsilon^{t-m} \to p_t^* \gamma^t_{TOP}$ over F. Find a microbundle homotopy of its composition with $r^*\tau(M) \oplus \epsilon^{t-m} \to \tau(M) \oplus \epsilon^{t-m}$ to a microbundle map $\varphi : r^*\tau(M) \oplus \epsilon^{t-m} \to \gamma^t_{CAT}$ over a CAT map $N \to B_{CAT(t)} \subset B'_{CAT(t)}$. Then $r^*\tau(M) \oplus \epsilon^{t-m}$ gets a canonical CAT microbundle structure making φ CAT. But a neighborhood of M in $E(r^*\tau(M) \oplus \epsilon^{t-m}) \approx M \times N \times R^{t-n}$ imbeds naturally onto a neighborhood of $M \times 0$ in $M \times R^t$, as explained in §3. Then the Product Structure Theorem deduces from the last mentioned CAT structure a CAT manifold structure Σ on $M \times 0 = M$. This Σ represents $\theta^{-1}(y)$. The map θ^{-1} can be described similarly in the relative case.

10.5. Either surgery or STABLE structures. Theorem 10.1 unfortunately depends on surgery through the Product Structure Theorem (which contains the Annulus Theorem of [Ki_1], which in turn uses surgery). However, when 10.1 and the Product Structure Theorem are reformulated with the Brown-Gluck notion of STABLE topological manifold and STABLE homeomorphism in place of the usual TOP notions, then the dependence on surgery and the Annulus Theorem vanishes. Compare [I, §5]. Of course, the reformulated result is equivalent by [Ki_1] but we don't know this without surgery.

10.6. Sections can replace liftings. Observe that liftings of f correspond naturally and bijectively to sections of that fibration over M which is the pull-back of the fibration $B'_{CAT} \to B_{TOP}$ by the classifying map $f : M \to B_{TOP}$ for $\tau(M)$ (cf. the introduction §0).

10.7. TOP/CAT as an H-space. The set [M rel C, TOP/CAT] is by 9.2 naturally isomorphic to TOP/CAT(ϵ(M) rel C), which is an abelian group under Whitney sum. Indeed TOP/CAT like B_{CAT} and B_{TOP} can readily be given H-space structures representing the Whitney sum operation, cf. [Ad]. For stronger results along this line, see [Bo] [BoV] [Seg].

10.8. Obstructions appear with untwisted coefficients . The stable homeomorphism [Ki_1] theorem shows that $Z_2 = \pi_0(TOP_m) = \pi_1 B_{TOP(m)} = \pi_1 B_{CAT(m)}$, for $m \geq 5$. Hence $\pi_0(TOP/CAT) = 0$. Further , $Z_2 = \pi_1(B_{TOP})$ acts trivially on $\pi_*(TOP/CAT)$ in the fibration $TOP/CAT \to B'_{CAT} \xrightarrow{p} B_{TOP}$. To see this , observe that on $TOP/CAT(\epsilon(S^k)\mathrm{rel} *) = \pi_k(TOP/CAT)$ it acts by reflection in any factor of the trivial bundle over S^k . Alternatively , as p is an H-map of H-spaces by 10.7 , it follows , cf. [Sp , p.385] , that $\mathrm{Im}(\pi_1 B'_{CAT}) \subset \pi_1 B_{TOP}$ acts trivially on the fiber TOP/CAT . Thus one has a classical obstruction theory with untwisted coefficients to calculate $\mathrm{Lift}(f \text{ rel } C,F_o)$ and $[M \text{ rel } C,TOP/CAT]$.

10.9. The obstructions appear in Čech cohomology .

The obstructions to finding an element of $\mathrm{Lift}(f \text{ rel } C,F_o)$ lie in the Čech cohomology groups $\check{H}^i(M , C ; \pi_{i-1})$ (coefficients in $\pi_{i-1} = \pi_{i-1}(TOP/CAT)$ understood) . More generally , if D is closed in M , $D \supset C$, and we seek a neighborhood V of D and a lift of $f|V$ to B'_{CAT} equal F_o near C , then the obstructions lie in the Čech groups $\check{H}^i(D , C ; \pi_{i-1})$.

The proof that Čech groups appear requires only the taughtness (continuity) property of Čech theory [Sp , p.316 , p.334] , naturality of obstructions , and the fact that we are dealing with sections of a bundle over an ENR (an ANR should do) .

Here are the details . Making M a retract of an open neighborhood in euclidean space , we see that it suffices to give a proof when M is a simplicial complex . Consider closed subpolyhedra $L \supset K$ of M , neighborhoods of D and C respectively , so that $K \subset U$ (= domain of F_o) . A smaller such pair will be denoted $\hat{L} \supset \hat{K}$. Suppose now that F is a given (i–1)-lifting of $f|L$ equal F_o on K . There is classically a unique obstruction $\mathfrak{G}(F) \in H^i(L , K ; \pi_{i-1})$ to extending F to an i-lifting equal to F_o on K . When k-liftings are appropriately defined (as below) one observes that F restricts canonically to an (i–1)-lifting $\hat{F}$ of $f|\hat{L}$ equal F_o on $\hat{K}$, and $\mathfrak{G}(F)$ restricts to $\mathfrak{G}(\hat{F}) \in H^i(\hat{L} , \hat{K} ; \pi_{i-1})$. It follows immediately that the obstruction to finding an i-lifting G of f , defined on some neighborhood of D in M , equal F_o on some neighborhood of C , and extending F , is none other than the image of $\mathfrak{G}(F)$ in

$$\varinjlim H^i(\hat{L} , \hat{K} ; \pi_{i-1}) = \check{H}^i(D , C ; \pi_{i-1})$$

the limit being taken over all small neighborhoods $\hat{L}$, $\hat{K}$ of D, C as described.

It remains to define a *k-lifting* ($k \geq 0$) of $f|L$ equal F_o on K. There are two valid alternatives that are related but not identical:

Classical definition. F is represented by a true lifting G of f extending $F_o|K$ and defined on $K \cup L_\tau^{(k)}$ where $L_\tau^{(k)}$ is the k-skeleton of some PL triangulation τ of L making K a subcomplex. It is essential to agree that another such lifting G', defined say on $K \cup L_{\tau'}^{(k)}$, represents the same k-lifting F if there exists a lifting of $f|\{(I \times K) \cup (I \times L)_\sigma^{(k)}\}$ extending $0 \times G$, $1 \times G'$ and $I \times (F_o|K)$, where σ is a PL triangulation of $I \times L$ making $I \times K$ a subcomplex and extending $0 \times \tau$ and $1 \times \tau'$.

Modern definition: F is precisely a true lifting $F : L \to B_k$ of $f|L$ into the k-th stage fibration $p_k : B_k \to B_{TOP}$ of the Postnikov decomposition of the fibration $p : B_{CAT} \to B_{TOP}$, such that $F|K = q_k F_o|K$, where

$$p : B'_{CAT} \xrightarrow{q_k} B_k \xrightarrow{p_k} B_{TOP}$$

is the k-th stage factorization of p. Recall that $p_{k+1} = p_k \alpha_{k+1}$, where $\alpha_{k+1} : B_{k+1} \to B_k$ is a fibration with fiber an Eilenberg-MacLane space $K(\pi_k(TOP/CAT), k)$.

10.10. Interpreting the Postnikov decomposition of $p : B'_{CAT} \to B_{TOP}$.

The terminology has just been established above. In case $CAT = PL$ it turns out that $TOP/PL \simeq K(Z_2, 3)$; thus one chooses $B_k = B_{TOP}$ for $k \leq 3$ and $B'_k = B'_{PL}$ for $k \geq 4$! In case $CAT = DIFF$ it turns out that B_k is itself a stable classifying space of a manifold theory; vertical homotopy classes of liftings of f to B'_k correspond (for $\partial M = \phi$ and $\dim M \geq 5$) to conditioned concordance classes of **k-smooth** manifold structures on M. The pseudo-group of k-smooth homeomorphisms of open subsets of R^n, $n \geq 5$, is defined to contain $h : U \to V$ if h is C^∞ nonsingular except perhaps on a closed subset of U having Štanko homotopy dimension $\leq (n-k-1)$ (such as the dual of the k-skeleton of some triangulation), cf. [III, Appendix C]. Thus one has an infinite sequence of manifold theories† strung out between the smooth and the topological, corresponding to $B'_{CAT} \to \cdots \to B_k \to B_{k-1} \to \cdots \to B_0 =$

† See also N. Levitt's article [Lt] where essentially equivalent theories, independently constructed, are discussed. (The above pseudo-groups are not used).

$= B_{TOP}$. Finally we note that PL is essentially one of these theories ; indeed $B_k \simeq B_{PL}$ for $4 \leqslant k \leqslant 7$ simply because the canonical forgetful map $B_{DIFF} \to B_{PL}$ (from Whitehead's triangulation theorem) is 7-connected . We shall not attempt to prove these things here ; instead see [Si_{15}] .

10.11. Classification relative to an open subset .

For the data of 10.1 , consider the problem of discovering and classifying CAT structures on M that coincide with Σ_0 on *all* of the open subset U (rather than just near C) . Agree that two such , Σ' and Σ'', are equivalent if there exists a CAT concordance $\Gamma : \Sigma' \mathrel{\hat{=}} \Sigma''$ so that $\Gamma | (U \times I) = \Sigma_0 \times I$; and denote the set of equivalence classes

$$\mathcal{S}_{CAT}(M \text{ rel } U, \Sigma_0) .$$

Fortunately this problem is isomorphic to an associated problem in dimension $m+1$ which is solved by 10.1 . Consider

$$Z = M \times I - (M - U) \times 0 .$$

Assertion : $\mathcal{S}_{CAT}(M \text{ rel } U, \Sigma_0) \cong \mathcal{S}_{CAT}(Z \text{ rel } U \times 0 , \Sigma_0 \times I)$.

The asserted isomorphism is defined by sending Σ representing $x \in \mathcal{S}_{CAT}(M \text{ rel } U, \Sigma_0)$ to $(\Sigma \times I)|Z$. To show that this map is onto , consider Θ representing $y \in \mathcal{S}_{CAT}(Z \text{ rel } U \times 0 , \Sigma_0 \times I)$. Apply the Concordance Implies (small !) Isotopy Theorem to $\Theta | (U \times I)$ to get an isotopy h_t of $(\mathrm{id}|Z)$ rel $U \times 0$ fixing points outside $U \times I$ so that $h_1(\Theta)|(U \times I) = \Sigma_0 \times I$. Extend $h_1\Theta$ to a CAT structure on $Z \cup M \times [1,\infty)$, that on $M \times [1,\infty)$ is a product $\Sigma \times [1,\infty)$ and on $U \times [0,\infty)$ is a product with $[0,\infty)$. (For DIFF one needs local collaring theorems here , from [I , Appendix A] .) Translating uniformly to ∞ in the R factor we clearly obtain a concordance (on Z) of $h_1\Theta$ rel $U \times \dot{I}$ to the structure $(\Sigma \times I)|Z$.

A similar (but relative) use of the Concordance Implies Isotopy Theorem shows this map is injective , which proves the assertion .

Finally , note that $H^*(Z , U \times 0) \cong H^*(M , U)$ for any theory H^* — by the projection $Z \to M$, (use the 5-lemma) . *Thus , for purposes of calculation the problem behaves as if U were a closed ENR in M*. However , beware that F_0 obviously may not extend beyond U even if Σ_0 does ; in general it will only extend up to homotopy .

10.12 The Homotopy Groups of TOP/CAT .

We indicate how this vital determination can best be carried out using our present classification result and surgery.

For readers interested in understanding why TOP/PL is *not* contractible rather than in fully calculating $\pi_*(\mathrm{TOP/CAT})$, we have written Appendix B that reduces the prerequisites to mere handlebody theory and V.A. Rohlin's signature theorem.

For $m \geqslant 5$, there is a group isomorphism

$$\varphi : \pi_m(\mathrm{TOP/CAT}) \xrightarrow{\cong} \Theta_m^{\mathrm{CAT}}$$

It is the forgetting map from $\pi_m(\mathrm{TOP/CAT}) = \mathcal{S}_{\mathrm{CAT}}(S^m)$, cf. 10.8 , to the abelian group Θ_m^{CAT} of oriented isomorphism classes of oriented CAT m-manifolds homotopy equivalent to S^m , addition being connected sum [KeM] . Indeed *surjectivity* of φ is the (weak) CAT Poincaré theorem, which is proved readily by engulfing for $m \geqslant 5$, cf. [St_2] [Lu] [Hu_2] [RS_3] . *Additivity* is easy to check. To check *injectivity* suppose S^m_Σ goes to zero in Θ_m^{CAT} ; then S^m_Σ is the boundary of a contractible CAT manifold W^{m+1}, and as W^{m+1} is homeomorphic to B^{m+1} by the same engulfing or Smale's h-cobordism theorem [Mi_8] [Hu_2] , we see by coning that Σ extends over B^{m+1} , whence $[\Sigma] = 0$ in $\pi_m(\mathrm{TOP/CAT})$.

The group Θ_m^{PL} is zero for $m \geqslant 5$. The PL h-cobordism theorem suffices to prove this for $m \geqslant 6$, via coning. For $m = 5$, one first shows using classical PL-DIFF smoothing theory † and Cerf's $\Gamma_4 = 0$ [Ce_5] , that any PL homotopy 5-sphere M^5 admits a Whitehead compatible DIFF structure σ . Then the equation $\Theta_5^{\mathrm{DIFF}} = 0$ of [KeM] implies that M_σ is diffeomorphic to S^5 , whence, by Whitehead's smooth triangulation uniqueness theorem [Wh_1] [Mu_1] , M^5 is PL homeomorphic to S^5 .

The group Θ_m^{DIFF} of smooth homotopy spheres was made famous by Milnor and Kervaire [KeM] who proved that it is *finite* for $m \geqslant 5$, and indeed largely determined it. For $m = 5, 6, 7, 8, 9, 10, 11$, its values are $0, 0, Z_{28}, Z_2, Z_2 \oplus Z_2 \oplus Z_2, Z_6, Z_{992}$. See also [Bru] [Br_1] .

Whether $\pi_m(\mathrm{TOP/CAT}) = \Theta_m^{\mathrm{CAT}}$ for all m , is an intriguing

† See comment in Appendix B (below B.2).

question (especially since an affirmative answer contradicts the classical Poincaré conjecture). Nothing is known about Θ_m^{CAT} for $m = 3, 4$, (except that $\Theta_m^{PL} = \Theta_m^{DIFF}$ by classical smoothing theory [Mu3] [Hi3]).

For $k \leqslant 6$, one exploits homotopy tori to prove that

$$\pi_k(TOP/CAT) \cong \pi_k(K(Z_2,3)) \quad ,$$

i.e. that it is nonzero only if $k = 3$, when it is Z_2. See [KS1]. The proof we recommend to readers of this essay is very close to the one presented in §5.3 of Essay V. It is already accessible to readers of this essay, and we shall merely add a few comments here to let the reader adapt that argument efficiently: Work with the 6-manifold $I^k \times T^{6-k}$ to calculate $\pi_k(TOP/CAT)$. Replace TOP_m/CAT_m by TOP/CAT and when a classification theorem of Essay V is called upon use 10.1 instead. Thereby a map

$$(\#) \qquad \pi_k(TOP/CAT) \to \mathcal{S}_o^*(I^k \times T^{6-k} \text{ rel } \partial)$$

is constructed to the set of those 'homotopy-CAT' structures on $I^k \times T^{6-k}$ rel ∂ that are invariant under passage to standard coverings. This set is $\pi_k(K(Z_2,3))$ by a surgical calculation that is described in [V, Appendix B]. Then the device of passage to a standard covering of T^{6-k} plus the s-cobordism theorem and the *local contractibility principle* for homeomorphisms [Če1,2] [EK] are combined to show that (#) is bijective; the details are in [V, §5.3].

We now indicate how to dispense with the local contractibility principle in the above argument (this would be its first intervention in Essays I and IV). It is used in [V, §5.3] to prove injectivity of (#) and again to prove surjectivity. In each case, it reveals that a certain small self-homeomorphism, α say, of $I^k \times T^n$ rel ∂ is isotopic rel ∂ to the identity.

1) For *injectivity* of (#) one can instead observe that it suffices to prove that a universal cover $\widetilde{\alpha} : I^k \times R^n \to I^k \times R^n$ of α is isotopic rel ∂ to the identity. Now $\widetilde{\alpha}$ is rel ∂ (by a suitable choice) and we can extend it by the identity to a self-homeomorphism of $R^k \times R^n = R^{k+n}$. As $\widetilde{\alpha}$ is of bounded distance to $id \mid R^{k+n}$ the formula $\widetilde{\alpha}_t(\vec{x}) = t\widetilde{\alpha}(\vec{x}/t)$, $0 < t \leqslant 1$, gives an isotopy with support in $\mathring{I}^k \times R^n$ from $\widetilde{\alpha} = \widetilde{\alpha}_1$ to the identity, cf. [Ki1].

2) For the proof of *surjectivity* of (#) , α becomes a CAT automorphism of $I^2 \times T^3$ rel ∂ , and this time we observe that it suffices instead to obtain a topological *pseudo-isotopy* (= concordance) rel ∂ from α to the identity. It is easy to deduce from the definition of $\mathcal{S}_o^*$ that α is CAT concordant rel ∂ to its standard λ^3-fold covering (λ any integer > 1). Then a pleasant infinite iteration process pictured in [Si_{10}, §3] provides the wanted topological concordance from α to $\mathrm{id}\,|(I^2 \times T^3)$ rel ∂ .

We have established a homotopy equivalence $\mathrm{TOP/PL} \simeq K(Z_2,3)$. This means that the primary obstructions met in calculating PL structure sets are the only ones. More precisely, using 10.8 , 10.9 , 10.11 , we see that in the situation of 10.1 the TOP manifold M admits a PL manifold structure equal to Σ_0 near C , if and only if a well-defined obstruction $\Delta(M)$ in $\check{H}^4(M,C;Z_2)$ vanishes. When one such structure is singled out, the pointed set $\mathcal{S}_{PL}(M \text{ rel } C,\Sigma_0)$ is naturally isomorphic to $\check{H}^3(M,C;Z_2)$. Also C may be open in M instead of closed.

Knowledge of $\pi_*(\mathrm{TOP/DIFF})$ on the other hand, is merely a good first step towards evaluating $\mathcal{S}_{DIFF}(M \text{ rel } C,\Sigma_0)$ in 10.1 . The next big step is perhaps to describe the homotopy type and H-space structure of TOP/DIFF ; this is a subject of current research requiring methods little related to the present essays, see [Sull] [MorgS] [BruM] [MadM] .

Appendix A. NORMAL BUNDLES

We shall explain M. Hirsch's proof [Hi_5] of the relative stable existence and uniqueness theorems for normal microbundles. (If our exposition is at all more accessible than in [Hi_5] it is only because we have abandoned some of the generality there.) †

We adopt the PL or TOP category consistently throughout, and we use the conventions of § 1 concerning microbundles.

A normal microbundle to M in $W \supset M$ is a microbundle ξ over M whose total space $E = E(\xi)$ is a neighborhood of M in W. Two such, ξ_0 and ξ_1, are (sliced) **concordant** if there is a normal microbundle η to $M \times I$ in $W \times I$, $I = [0, 1]$ with $\xi_k \times k = \eta | M \times k$ for $k = 0$ and 1, such that for each $t \in I$, the restriction $\eta | M \times t$ is a normal microbundle to $M \times t$ in $W \times t$. We write $\eta : \xi_0 \Leftrightarrow \xi_1$. Observe that, by the microbundle homotopy theorem, one has $\xi_0 \Leftrightarrow \xi_1$ if and only if ξ_0 is **isotopic** to ξ_1 in the sense that there is a neighborhood V of M in $E(\xi_0)$ and an isotopy $F : V \times I \to W \times I$ (an open embedding fixing $M \times I$ with $p_2 = p_2 F$) from $F_0 = \text{inclusion} : V \hookrightarrow W$ to F_1 a micro-isomorphism $\xi_0 \to \xi_1$. We write $F : \xi_0 \approx \xi_1$. As in [I, § 2] we can equally well discuss concordance and isotopy *relative* to a closed subset $C \subset M$, i.e. constant on a neighborhood of C in W.

We shall encounter **composed microbundles**. If $\xi : X \hookrightarrow Y \xrightarrow{p} X$ and $\eta : Y \hookrightarrow Z \xrightarrow{q} Y$ are microbundles, then the composition $\xi \circ \eta : X \hookrightarrow Z \xrightarrow{p \circ q} X$, is again a microbundle as a simple pursuit of definitions shows.

Data. $M^m \subset W^w$ denote henceforth manifolds without boundary with M imbedded as a closed subset of W. A closed subset $C \subset M$ is given, with an open neighborhood U of C in M and a normal microbundle γ to U in W.

† For stronger results consult R. Stern [Ste]; in particular, provided $w - m \geqslant 5 + i$, (a) existence holds for $n = 0$ if $w \geqslant 2m - i - 1$, $i=0,1,2$, and (b) uniqueness holds for $n = 0$ if $w \geqslant 2m - i$, $i=0,1,2$. For examples of imbeddings for which normal bundles fail to exist or fail to be unique, see [$\text{RS}_{0,4}$].

STABLE EXISTENCE THEOREM A.1.

For some integer† $n \geq 0$, $M \times 0$ *admits a normal microbundle* ξ *in* $W \times R^n$ *so that* ξ *equals* $\gamma \oplus \epsilon^n$ *near* $C \times 0$ *in* $W \times R^n$.

STABLE UNIQUENESS THEOREM A.2.

Suppose ξ_0 *and* ξ_1 *are normal microbundles to* M *in* W. *Then there exists an integer*† $n \geq 0$ *such that* $\xi_0 \oplus \epsilon^n$ *and* $\xi_1 \oplus \epsilon^n$ *are isotopic normal microbundles to* $M \times 0$ *in* $W \times R^n$.

The following strongly relative form holds. Suppose ξ_0 *coincides with* ξ_1 *near* C *in* W; *let* $D \subset M$ *be a given closed set and* V *an open neighborhood of* D *in* M. *Then there exists an integer* $n \geq 0$, *an open neighborhood* E' *of* $M \times 0$ *in* $W \times R^n$ *and an isotopy* $f_t : E' \to W \times R^n$, $0 \leq t \leq 1$, *rel* $C \cup (M-V)$ *and fixing* M *from the inclusion* f_0 *to* f_1 *giving (by restriction) a micro-isomorphism* $(\xi_0 \oplus \epsilon^n)|(C \cup D) \to (\xi_1 \oplus \epsilon^n)|(C \cup D)$.

As usual results for manifolds with boundary can be deduced.

Uniqueness implies existence. We recall how the existence theorem A.1 follows from the uniqueness theorem A.2. We can assume M is locally flat in W since a construction given in §3 shows M is locally flat in $W \times R^m$. Then we can use a partition of unity to find a covering of M by $\leq m+1$ open sets $M_o, \ldots, M_m$ in M each with a trivial normal microbundle in W. An evident m+1 step application of the strongly relative uniqueness result then constructs ξ.

A key case of uniqueness: $D = M^m \subset R^m$.

Consider the two natural tangent microbundles

$$\tau_k(M) : M \xrightarrow{\Delta} M \times M \xrightarrow{p_k} M, \quad k = 1, 2,$$

where $p_k(x_1, x_2) = x_k$. Identifying M to ΔM these are two normal microbundles to ΔM in $M \times M$.

Fact 1. $\tau_1 M \cong \tau_2 M$

Proof: This is at least clear for $M = R^1$, where it is a trivial case of the (bi)collaring uniqueness theorem; it follows for $M = R^m$ by producting; then for M (open!) in R^m it follows simply by restriction. ■

† The proof of Hirsch presented here does show that the stabilising integer n can be chosen as a function of $m = \dim M$ only. *This* fact *is* used in §9.4.

Cutting back W if necessary we choose a retraction $r : W \to M$.

Fact 2. *The homeomorphism* $h : W \times R^m \to W \times R^m$, $h(x, y) = (x, y+r(x))$ *carries* $M \times 0$ *onto* $\Delta M \subset W \times R^m$, *sending the normal microbundle* $\xi \oplus \epsilon^m$ *of* $M \times 0$ *in* $W \times R^m$ (ξ *being* ξ_0 *or* ξ_1) *to the following composed normal microbundle to* ΔM *in* $W \times R^m$:

$$(\tau_1 M) \circ (\xi \times M) : M \overset{\Delta}{\to} E(\xi) \times M \overset{\pi}{\to} M$$

where $\pi(x, y) = p_\xi(x)$.

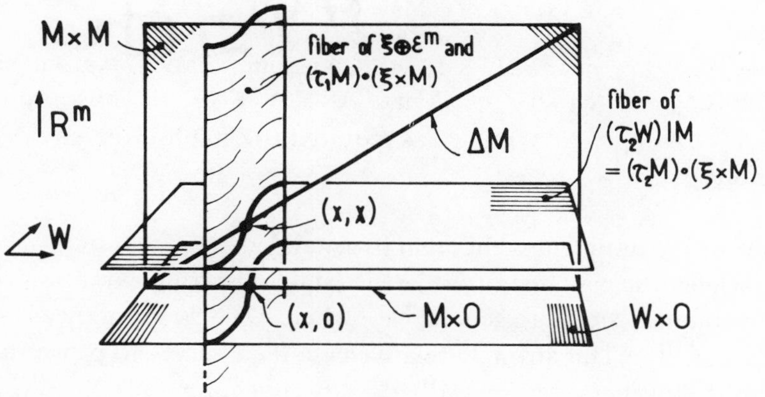

Figure A-1

Proof: The one total space is contained in the other $E(\xi) \times M \subset E(\xi) \times R^m$ and the projections agree : $(x, y) \mapsto p_\xi(x)$; but h respects this projection and carries the one zero section to the other : $\Delta(x) = h(x, 0)$. ■

Fact 3. *A sliced concordance* $\eta : \tau_1 M \Leftrightarrow \tau_2 M$ *(from Fact 1) yields canonically a sliced concordance*

$$\zeta_k : (\tau_1 M) \circ (\xi_k \times M) \Leftrightarrow (\tau_2 M) \circ (\xi_k \times M), \quad k = 0, 1.$$

Proof: Indeed the concordance is $\zeta_k = \eta \circ (\xi_k \times M \times I)$. ■

Fact 4. $(\tau_2 M) \circ (\xi \times M) \doteq \tau_2 W | M$.

Proof: The composed microbundle is $M \overset{\Delta}{\to} E(\xi) \times M \overset{p_2}{\to} M$. ■

Conclusion. We now have sliced concordances

$$\zeta_k : (\tau_1 M) \circ (\xi_k \times M) \;\text{⇳}\; \tau_2 W | M \,, \qquad k = 0, 1 \,.$$

Thus by Fact 2, $\xi_0 \oplus \epsilon^m \approx \xi_1 \oplus \epsilon^m$ which was the first assertion to prove. Unfortunately it is not immediately clear that $\xi_0 \oplus \epsilon^m \approx \xi_1 \oplus \epsilon^m$ rel C, since $\eta : \tau_1 M$ ⇳ $\tau_2 M$ is not rel C, which implies ζ_k is not rel C, $k = 0, 1$.

However $\xi_0 = \xi_1$ near C does imply that $\zeta_0 = \zeta_1$ near $\Delta C \times \times I$ in $W \times M \times I$. Applying the microbundle theorem to ζ_0 and ζ_1 respectively we obtain isotopies in $W \times M$

$$f_t : \tau_2 W | M \;\approx\; (\tau_1 M) \circ (\xi_0 \times M)$$

$$g_t : \tau_2 W | M \;\approx\; (\tau_1 M) \circ (\xi_1 \times M)$$

Since $\zeta_0 = \zeta_1$ near $\Delta C \times I$ we can assume that $f_t = g_t$ near ΔC in $W \times M$. Then $h^{-1} g_t \circ f_t^{-1} h$, $0 \leqslant t \leqslant 1$, is an isotopy $\xi_0 \oplus \epsilon^m \approx \xi_1 \oplus \epsilon^m$ rel C as required to establish the key case of uniqueness. ■

Proof of the uniqueness theorem in general.

Given the case just proved, the (stable) isotopy extension result below implies the more general case where $D \neq M$ is allowed, but $M^m \subset R^m$. This strongly relative case then serves to prove the completely general case ($M^m \subset R^m$ not supposed) by a routine (m+1)-step induction using a covering of M by m+1 coordinate charts. ■

It remains to prove

ISOTOPY EXTENSION LEMMA A.3. (for usual data)†

Let E be an open neighborhood of C in W, and let $f_t : E \to W$ be an isotopy of $E \hookrightarrow W$ fixing $M \cap E \supset C$ There exists a neighborhood E' of $M \times 0$ in $W \times R$ and an isotopy $H_t : E' \to W \times R$ of $E' \hookrightarrow W \times R$ fixing all of $M \times 0$ and equal $f_t \times R$ near $C \times 0$.

Proof of A.3 : Consider the isotopy

$$F_t : E \times R \;\to\; W \times R \,, \qquad F_t(x, s) = (f_{\tau(s,t)}(x), s)$$

† M and W need not be manifolds here (we just use normality of M); but if they are manifolds and M is locally flat in W then the stabilisation is known to be unnecessary and H_t could be an ambient isotopy, cf. [EK].

where $\tau: R \times I \to I$ satisfies :

$$\tau(s,t) = t \text{ for } s \geqslant -1 ; \quad \tau(s,t) = 0 \text{ for } s \leqslant -2 \text{ or } t = 0 .$$

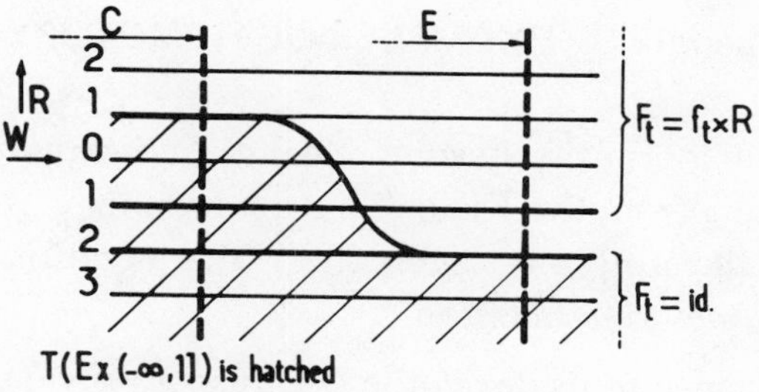

Figure A-2

Choose a function $\varphi : W \to R$ equal to 0 near C and equal to 3 near W – E (using normality of W), and form the homeomorphism

$$T : W \times R \to W \times R , \quad T(x,s) = (x, s - \varphi(x)) .$$

We can set

$$H_t = T^{-1} F_t T : E \times (-1,1) \to W \times R$$

and extend H_t by the identity to

$$H_t : W \times (-1,1) \to W \times R .$$

We set $E' = W \times (-1,1)$. The reader will easily verify , cf. Figure A–2 , that this is an isotopy as asserted by A.3 . ■

Appendix B. EXAMPLES OF HOMOTOPY TORI

We explain a beautifully simple 'handle' construction of A. Casson† starting from the Klein-Poincaré homology 3-sphere $P^3 = SO(3)/A_5$, the group A_5 being the 60 orientation preserving symmetries of the dodecahedron.

Its importance to us lies in the fact that it provides just the limited information about homotopy tori that is required to carry out the very succinct disproof of the Hauptvermutung in [Si_{10}, §2]. (A less efficient ad hoc construction was given in [Si_8, §5]; for the full classification, see Essay V, Appendix B).

As usual, we work in a CAT (= DIFF or PL) setting. The set $\mathcal{S}(I^3 \times T^n \text{ rel } \partial)$ has typical element represented by a CAT manifold M^{3+n} homotopy equivalent to T^n together with a CAT identification of a neighborhood of ∂M to a neighborhood of $\partial I^3 \times T^n$. And another such manifold N^{k+n} represents the same element $[M] = [N]$ in $\mathcal{S}(I^3 \times T^n \text{ rel } \partial)$ when there is a CAT isomorphism $M^{k+n} \to N^{k+n}$ that is the identity near ∂. (It is easy to show using a little obstruction theory that $\mathcal{S}(I^3 \times T^n \text{ rel } \partial)$ coincides with the set of homotopy-CAT structures on $I^3 \times T^n$ rel ∂ as defined in [V, §5.4 & App. B].)

We shall use the fact that P^3 is (CAT isomorphic to) the boundary of the Milnor plumbing P^4 of 8 copies of the unit tangent disc bundle of S^2 according to the E_8 graph, cf. [Mi_6].

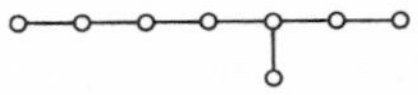

For this identification see [KiK], [Mi_{10}], and perhaps simplest of all [Rol]. The intersection matrix for $H_2(P^4)$ has determinant 1 and signature 8. From the determinant 1, it is readily deduced that $\partial P^4 \cong P^3$ has indeed the same integral homology as S^3.

† Casson has exploited this construction to study 4-manifolds; compare [Sch_0].

ASSERTION B.1.

If $3+n \geqslant 5$ *, there exists an element* $[M^{3+n}]$ *in* $S(I^3 \times T^n \text{ rel } \partial)$ *that is nontrivial, but is nevertheless invariant under passage to standard coverings of* $I^3 \times T^n$.

By matching M^{3+n} *with a copy of* $I^3 \times T^n$ *along the common boundary* $\partial I^3 \times T^n$ *one obtains* † *a CAT manifold homotopy equivalent to* $S^3 \times T^n$ *but not CAT isomorphic to* $S^3 \times T^n$ *, i.e. one obtains a CAT exotic homotopy* $S^3 \times T^n$.

Again, by indentifying the opposite ends of the three interval factors of I^3 *one derives from* M^{3+n} *a CAT exotic homotopy* T^{3+n} .

The Construction .

The whole construction of Casson is determined (up to various isomorphisms) by one arbitrary choice of an element γ that is in the perfect group $\pi = \pi_1 P^3$ but not in the subgroup $Z_2 = \pi_1 SO(3)$. As A_5 is simple, γ kills π . ‡ ‡‡

Let P^3_o be the interior connected sum $I^3 \# P^3$, cf. [KeM] ; this is homeomorphic to P^3 minus a standard open ball.

Attach a compact 2-handle to $[0,1] \times P^3_o \times I^n$ along $1 \times \text{int}(P^3_o \times I^n)$, choosing the attaching tube to represent γ and so as to keep the result parallelizable. Since $4+n \geqslant 5$, the result is a simply connected CAT cobordism rel ∂ from $0 \times P_o \times I^n$ to a simply connected CAT manifold Q^{3+n}_o having the homology of $I^{3+n} \# (S^2 \times S^{n+1})$, just as if P^3_o were I^3 .

† In the DIFF construction some routine rounding of corners is required here and later on (see [III, §4.2]) ; we shall leave this up to the reader.

‡ This means that the least normal subgroup (γ) containing γ is the whole group π . Indeed, A_5 is the full even permutation group on five objects (inscribed tetrahedra) hence is simple; this implies (γ) has index $\leqslant 2$. But index 2 contradicts π being perfect, as follows: The exact sequence $0 \to Z_2 \to \pi \to A_5 \to 0$ would then be *split* exact, while Z_2 is central, being the kernel of the (universal) covering homomorphism of connected Lie groups $(\pi \subset) S^3 \to SO(3)$.

‡‡ The reader who does not understand why $P^3 \cong \partial P^4$ may wish to replace P^3 by ∂P^4 , and, without bothering to calculate $\pi_1 \partial P^4$, carry out Casson's process using not one element γ of $\pi_1 \partial P^4$ but rather any finite collection $\gamma_1, \dots, \gamma_k$ sufficient to kill $\pi_1 \partial P^4$. (Obviously $\leqslant 8$ will do).

Attach a 3-handle to $\operatorname{int} Q_o^{3+n}$ to kill the 2-dimensional homotopy class. This operation is possible (and even unique up to isotopy), since $3+n \geqslant 5$ and $\pi_1 Q_o^{3+n} = 0$.

It is easy to see that the manifold

$$[0,1] \times P_o \times I^n \cup \{\text{the handles}\}$$

gives a contractible cobordism, rel boundary, from $P_o \times I^n = 0 \times P_o \times I^n$, to a contractible manifold that we denote $(P_o \times I^n)^{\boxtimes}$.

Then, enlarging from n-cube to n-torus,

$$X^{4+n} = [0,1] \times P_o \times T^n \cup \{\text{the handles}\}$$

is a cobordism rel ∂ from $P_o \times T^n$ to

$$(P_o \times T^n)^{\boxtimes} = (P_o \times I^n)^{\boxtimes} \cup P_o \times (T^n - \operatorname{int} I^n)$$

This manifold $(P_o \times T^n)^{\boxtimes}$ is the asserted M^{3+n} . ∎

The invariance under 'transfer' .

We now verify that, for any CAT covering map $c: T^n \to T^n$, of degree d say, the corresponding cover $\widetilde{M}$ of $M = (P_o^3 \times T^n)^{\boxtimes}$ satisfies $[\widetilde{M}] = [M] \in \mathcal{S}(I^3 \times T^n \text{ rel } \partial)$. To establish this consider the covering $\widetilde{X}^{4+n} \supset \widetilde{M}^{3+n}$ of X^{4+n} corresponding to c . It is a copy of $[0,1] \times P_o^3 \times T^n$ with 2d handles attatched Match $\widetilde{X}^{4+n}$ to X^{4+n} along $0 \times P_o^3 \times T^n$ (only), forming Y^{4+n} . Inspecting the handle decomposition of Y^{4+n} on M^{3+n} , one sees that Y^{4+n} is an h-cobordism *rel boundary* with zero torsion. Then the s-cobordism theorem provides the required isomorphism $\widetilde{M} \to M$. ∎

Two theorems, neither dependent on surgical techniques, will play a vital role in establishing the asserted non-triviality of M^{3+n} .

B.2. V. A. Rohlin's Signature Theorem .

Every closed oriented smooth 4-manifold M^4 *of which the second Stiefel-Whitney class* $w_2(M)$ *is zero has signature* $\sigma(M) \in Z$ *divisible by* 16 .

We suggest the reader study, first of all, the elementary reduction of Milnor and Kervaire [MiK] to the isomorphism $Z_{24} \cong \pi_{3+n} S^n$, n large, a result of Serre [Se_1] , also proved in [MiK] [Toda] , and in textbooks. A proof along the lines of Rohlin's original geometric arguments [Roh_1] using surfaces in M^4, and exploiting an Arf invariant, has been pointed out by A. Casson, see [FrK] (and also [Roh_2]).

If M is merely a (simplicial) homology manifold, $\sigma(M)$ is still divisible by 8 since the quadratic form is even (see [MiH]). But divisibility by 16 fails; indeed the Milnor plumbing P^4 produces, on smashing ∂P^4, a key example $P^4_* = P^4/\partial P^4$ with signature exactly 8. Rohlin's theorem does hold for PL manifolds by classical smoothing theory [Mu_3] [Hi_3] [HiM_2] ; it is undecided for TOP manifolds.

Comment on classical smoothing theory. For low dimensions where the obstruction groups Γ_i vanish, we like best the simple handle by handle approach of M. Hirsch's article [Hi_3]. It relies on the 'classical' PL-DIFF analogues of the results of our Essay I, as proved in Hirsch's article [HiM_2,I]. Granting these, Hirsch argues as follows. Uniqueness questions reduce to existence questions by the Concordance Implies Isotopy Theorem. To answer existence questions it is enough to show how to extend (the germ of) a smoothing defined near the boundary sphere ∂B^k in an open k-handle $R^k \times R^n$ to a (germ of) a smoothing near $B^k = B^k \times 0$. The (local) PL-DIFF Product Structure Theorem reduces this to finding a concordance from a smoothing of ∂B^k to the standard one. Finally the PL-DIFF Concordance Implies Isotopy Theorem together with Whitehead's DIFF triangulation theorems and a little PL topology quickly show that concordance classes of smoothings of ∂B^k correspond bijectively to oriented diffeomorphism classes of smooth twisted (k−1)-spheres [Mi_8, p.110], which correspond bijectively to the elements of Γ_{k-1}.

The second theorem comes from handlebody theory.

B.3. F. T. Farrell's Fibration Theorem

Consider a map $f: M^m \to T^1$ of a compact CAT manifold to the circle such that $f|\partial M^m$ is a CAT locally trivial fibration. The following assumptions assure that f is homotopic fixing $f|\partial M$ to a CAT locally trivial fibration $M^m \to T^1$.

(a) $\dim M \geqslant 6$.

(b) The infinite cyclic covering $\widetilde{M}$ of M, pulled back from $R \to R/Z \cong T^1$ by f, has finite homotopy type.

(c) $\pi_1(M)$ *is free abelian (or free or poly-Z).*

In [Fa] , Farrell gives a brief and elegant reduction to the s-cobordism theorem, the thesis [Ke_2] , and the nullity $Wh(\pi_1 M) = 0 = \widetilde{K}_0(\pi_1 \widetilde{M})$ proved in [BHS] [FH_1] . For more general 'splitting' theorems of this sort are known, see [FH_2] [Cap] .

Note that condition (b) is assured if f is the T^1-component of a homotopy equivalence $M \to X \times T^1$ where X is homotopy equivalent to a finite complex .

Proofs of nontriviality

If $[M^{3+n}]$ were trivial in $\mathcal{S}(I^3 \times T^n \text{ rel } \partial)$, it would follow that, on matching M^{3+n} to $I^3 \times T^n$ along the common boundary, we get a manifold M_2 that is CAT isomorphic to $S^3 \times T^n$.

But supposing $M_2 \cong S^3 \times T^n$ we shall now derive a contradiction to Rohlin's theorem. Note that M_2 is the boundary of $W_o = P^4 \times T^n \cup \{\text{two handles}\}$. Using the supposed isomorphism $M_2 \cong S^3 \times T^n$, attach to W_o a copy of $B^4 \times T^n$ to get a closed CAT manifold W^{4+n} . One has $W^{4+n} \simeq P^4_* \times T^n$; indeed, a copy of $P^4 \times T^n$ is in a standard way a subset of each, the remainders being homotopy equivalent to T^n ; and there is no obstruction to extending this identification of subsets to a homotopy equivalence.

Applying Farell's theorem $(n-1)$ times, find a sequence of CAT manifolds $W^{4+n} \supset W^{3+n} \supset \ldots \supset W^5 \cong P^4_* \times T^1$. Then, making the composed map $W^5 \to T^1$ at least transverse to 0 , we get an orientable CAT manifold $W^4 \subset W^5$ which, we assert, contradicts Rohlin's theorem.

The fact that $w_2(W^{4+n}) = 0$, is seen by restricting to $P^4 \times T^n$, which is parallelizable. Since each W^k is bicollared in W^{k+1}, one can check that $w_2(W^k) = 0$, $k \geqslant 4$.

Finally $\sigma(W^4) = \sigma(CP_2 \times W^4)$, and $CP_2 \times W^4$ is cobordant to a manifold $V^8 \simeq CP_2 \times P^4_*$ that comes from applying Farrell's theorem to the equivalence $CP_2 \times W^5 \to CP_2 \times P^4_* \times T^1$. (The cobordism lies in the ∞-cyclic covering of $CP_2 \times W^5$). Thus $\sigma(W^4) = \sigma(V^8) = \sigma(CP_2 \times P^4_*) = \sigma(P^4_*) = 8$, which brings the required contradiction to $M_2 \cong S^3 \times T^n$. ∎

To check the last nontriviality statement of B.1 let $M_3 \simeq T^{3+n}$ be the manifold obtained from $M^{3+k} = (I^3 \times P_0^3)^{\sharp}$ by identifying opposite ends of the three interval factors of I^3 . If M_3 were PL isomorphic to T^{3+n} , we could construct a PL manifold $W'^{4+n} \simeq \simeq (T^4 \# P_*^4) \times T^n$, which would contradict Rohlin's theorem, much as above. For details see [Si_{10}, §2] . ■

Application to Hauptvermutung and Triangulation Conjecture .

For the disproof in [Si_{10}, §2] , the reader will need the PL automorphism rel ∂

$$\alpha : I^2 \times T^n \to I^2 \times T^n \ , \ 2+n \geqslant 5 \ ,$$

unique up to PL pseudo-isotopy (= concordance) that one obtains by trivializing $(P_0^3 \times T^n)^{\sharp}$ regarded as an s-cobordism rel ∂ from $0 \times I^2 \times T^n$ to $1 \times I^2 \times T^n$. In addition he needs a PL pseudo-isotopy H rel ∂ from α to α_1 , the standard 2^n-fold covering of α . It exists because we have constructed a CAT identification rel ∂ from $(P_0^3 \times T^n)^{\sharp}$ to its standard 2^n-fold covering. (The reader may find it amusing to make the construction of α and H more concrete still.)

At this point the catastrophe pictured in [Si_{10}, Fig.2a] intervenes to provide a TOP pseudo-isotopy rel ∂ from α to $\mathrm{id}\,|(I^2 \times T^n)$.

From this, one rapidly deduces, for example, that M_2 is (topologically!) homeomorphic to $S^3 \times T^n$ and that a TOP manifold $W^{4+n} \cong P_*^4 \times T^n$ exists although we have observed that there exists no PL manifold of this homotopy type.

Essay V

CLASSIFICATION OF SLICED FAMILIES OF MANIFOLD STRUCTURES

by

L. Siebenmann

— *dedicated to Claude Morlet* —

Section headings in Essay V

§0. INTRODUCTION

The classification we shall establish asserts†, in its simplest form 2.3, a homotopy equivalence $\mathrm{CAT}(M^m) \simeq \mathrm{Lift}(f \text{ to } B_{\mathrm{CAT}(m)})$ from a space $\mathrm{CAT}(M^m)$ of sliced families of CAT (= smooth C^∞ or piecewise linear) manifold structures on a topological m-manifold M^m without boundary ($m \neq 4$) to a space of liftings to $B_{\mathrm{CAT}(m)}$ of any fixed classifying map $f : M \to B_{\mathrm{TOP}(m)}$ for the tangent microbundle of M.

The sliced concordance implies isotopy theorem, proved by some careful engulfing in Essay II, brings this classification theorem within easy reach. The simplest version of it sufficient for our needs below is the

BUNDLE THEOREM 0.1. *Let M be a metrizable topological manifold with $\dim M \neq 4 \neq \dim \partial M$, and let Σ be a CAT manifold structure on $\Delta^k \times M$ such that projection $p_2 : (\Delta^k \times M)_\Sigma \to \Delta^k$ is a CAT submersion onto the standard k-simplex ($k \geqslant 0$). Then $p_2 : (\Delta^k \times M)_\Sigma \to \Delta^k$ is a trivial CAT bundle.*‡

This result was established as Theorem 1.8 in [II, §1]. A structure Σ as described is none other than a typical k-simplex of the (semi-simplicial) space $\mathrm{CAT}(M^m)$. This bundle theorem reveals that if M^m is a CAT manifold, and $\mathrm{Aut}_{\mathrm{CAT}}(M)$ denotes a similarly defined semi-simplicial space of CAT automorphisms of M, then

$$\pi_k(\mathrm{Aut}_{\mathrm{TOP}}(M), \mathrm{Aut}_{\mathrm{CAT}}(M)) = \pi_k(\mathrm{CAT}(M)), \quad k > 1,$$

(see 1.5). Thus the classification theorem is a key tool for comparing the spaces of TOP and CAT automorphisms of M.

† The present formulation is Morlet's [Mor_3] [Mor_4], it improves on earlier ideas of ours [KS_1].

‡ This modest result may be our largest contribution to the final classification theorem; we worked it out in 1969 in the face of a widespread belief that it was irrelevant and/or obvious and/or provable for all dimensions, (cf. [Mor_3] [Ro_2] and the 1969 version of [Mor_4]). Such a belief was not so unreasonable since 0.1 *is* obvious in case M is compact: every *proper* CAT submersion *is* a locally trivial bundle.

Our dedication to Claude Morlet is made in recognition of his important and precocious contributions to this classification theory in [Mor_2] [Mor_3] [Mor_4] . Morlet showed us that a nonstable classification theorem for sliced families was a much better goal for the handle-by-handle classification proceedure we sketched in [KS_1] . He observed that the remarkable equations $\pi_{m+k+1}(TOP_m/CAT_m) = \pi_k(Aut_{CAT}(B^m \text{ rel } \partial B^m))$ (see §3) follow naturally from the sliced theory and the topological Alexander isotopy .

Credit for first discovering such an equation goes to Jean Cerf . In fact , in 1962 [Ce_0] , Cerf derived the case CAT = DIFF from certain technical conjectures [Ce_0 ; p.18 , p.23] , all of which were then known for $m = 3$ through work of Moise and can now be verified readily for $m \neq 4$ by use of the version with epsilons of the Sliced Concordance Implies Isotopy Theorem of [II , §2.1] . Although Cerf's equation looks different, it is exactly equivalent - see §3.4 .

M. W. Hirsch too had done some early work on sliced classification theorems ; unfortunately , his results were never announced correctly [Hi_9 , p.75] , cf. [Ro_2] . R. K. Lashof's work on classification also fits into this context .

All approaches to the sliced classification theory seem to exploit ideas that first arose in the Smale-Hirsch classification of immersions around 1957-9 [Sm] [Hi_1] . Even Cerf uses them in [Ce_0] .

Our exposition will follow closely the outline we gave in [KS_4] , cf. [Si_6 , §4] , and will exploit the ideas of immersion theory in a way formalized and popularized by M. L. Gromov [Gr] [Ha_2] . (This includes a well-oiled 'handle-induction' mechanism replacing that in [KS_1]) . This technique is eminenty worth learning as it has a great many other applications far and near ; for example the classification of [Si_6 , §4] is one of the many results of Haefliger (cf. [Ha_0] [Ha_1] [RS_2]) that can be pleasantly reproved and strengthened by this means.

Morlet's original line of argument , cf. Mor_2] , seems to have been rather less direct than our present one, as it involved spaces of immersions; it is perhaps now fully explained in [BuL] . C. Rourke was working out a direct proof of the sliced classification theorem at the same time we were (see [Ro_2]), and we would like to acknowledge a mutually helpful exchange of ideas (through R. Kirby) in summer 1970. Probably all direct proofs will of necessity have much in common.

We round out our exposition with a review in §5 of known results (including our own and Morlet's) concerning the fiber TOP_m/CAT_m of the forgetful map $B_{CAT(m)} \to B_{TOP(m)}$. This is the fiber whose homotopy obstructs the liftings mentioned in the classification theorem.

We have largely neglected the classical problem of classification of DIFF (= smooth C^∞) structures on M^m compatible (in the sense of Whitehead) with a fixed PL (= piecewise linear) structure. Fortunately C. Rourke [Ro_2] has announced an exposition emphasizing this problem. Our discussion does apply with no essential change to this PL-DIFF version up to the end of the first classification theorem 1.4. (A notable feature is that the restrictions concerning dimension 4 vanish).

We adopt the terminology of [I, §2], all of it we hope self-explanatory. When an object of the form $I \times \Delta \times M$ is considered where $I = [0,1]$ and Δ is a manifold, it is convenient to write $\sqsupset(I \times \Delta \times M)$ or simply the heiroglyph $\sqsupset$ for $(I \times \partial\Delta \times M) \cup (I \times \Delta \times M)$. The expression *rel C* means 'leaving fixed the restriction to some neighborhood of C'.

§1 CLASSIFYING CAT STRUCTURES ON MANIFOLDS BY CAT STRUCTURES ON MICROBUNDLES

1.1 For a TOP manifold without boundary M^m we wish to acquire an understanding of the space CAT(M) of CAT manifold structures on M. It is defined as a complete semi-simplicial set (= css set = css space = css complex)† as follows. A typical k-simplex $\Sigma \in \mathrm{CAT}(M)^{(k)}$ of CAT(M) is a CAT (= DIFF or PL) manifold structure on $\Delta^k \times M$ such that projection $(\Delta^k \times M)_\Sigma \to \Delta^k$ is a CAT submersion. One often writes $\Sigma : \Delta^k \to \mathrm{CAT}(M)$. For any CAT map $\lambda : \Delta^\ell \to \Delta^k$, there is an induced map $\lambda^\# : \mathrm{CAT}(M)^{(\ell)} \leftarrow \mathrm{CAT}(M)^{(k)}$ sending Σ to the pullback $\lambda^\#\Sigma$ of Σ by λ. Here $(\Delta^k \times M)_\Sigma$ pulls back just like a CAT bundle over Δ^k. For the injective and surjective ordered simplicial maps $\lambda : \Delta^\ell \to \Delta^k$ we get the face and degeneracy operators in CAT(M).

1.2 Since the question of existence of a CAT structure on M has been extensively discussed elsewhere [Las$_2$] [IV] [Si$_{10}$, §4], it would not be unreasonable to suppose that M is CAT (so that CAT(M) is nonempty). This makes it easy to define a space CAT(τ(M)) of CAT structures on the tangent microbundle τ(M) or indeed on any microbundle ξ over M. Recall that τ(M) has total space $E(\tau M) = M \times M$, projection $p_2 : M \times M \to M$ (to second factor), and zero-section $\delta : M \to M \times M$, $\delta(x) = (x,x)$.

A CAT **structure** Σ on a TOP m-microbundle $\xi : X \xrightarrow{i} E(\xi) \xrightarrow{p} X$ over a CAT manifold X is by definition a CAT manifold structure Σ on an open neighborhood U in $E(\xi)$ of the zero section i(X) such that $p : U_\Sigma \to X$ is a CAT submersion. If in addition $i : X \to U_\Sigma$ is CAT we call Σ a CAT **microbundle**

† For instruction in semi-simplicial topology see [Lam] [May] [RS$_1$]. Let $\Delta^\cdot$ be the category whose objects are the standard simplices $\Delta^k \subset R^{k+1}$, $k = 0, 1, 2, \cdots$, and whose morphisms are the order preserving simplicial maps among these Δ^k. Then a css set X is by definition a contravariant functor from $\Delta^\cdot$ to the category of sets. We write $X : \Delta^k \mapsto X^{(k)}$, and call $X^{(k)}$ the set of k-simplices of X.

structure on ξ. Another such, Σ', has the same **germ** about the zero section if $\Sigma = \Sigma'$ on some open neighborhood of $i(X)$. We define a k-simplex of the space **CAT**(ξ) [respectively CAT(ξ)] to be the germ about the zero section, of Σ, a CAT structure [respectively a CAT microbundle structure] on the product bundle $\Delta^k \times \xi$ over $\Delta^k \times X$. (The inclusion **CAT**(ξ) $\hookrightarrow$ CAT(ξ) is usually and perhaps always a homotopy equivalence, see §2.1).

Analogous to the **differential** in immersion theory is a mapping of css complexes

$$d : \mathrm{CAT}(M) \to \mathbf{CAT}(\tau(M))$$

defined as follows when M is CAT. For $\Sigma \in \mathrm{CAT}(M)^{(k)}$ define $d\Sigma$ to be the germ about the zero-section of the CAT structure $\Sigma \times M$ on $E(\Delta^k \times \tau M) = \Delta^k \times M \times M$. We have expressly chosen as projection of $\tau(M)$ the *second* factor $p_2 : M \times M \to M$, $p_2(x,y) = y$, in order to make this most important rule optimally simple. We propose to prove in this section that d is a homotopy equivalence except perhaps in some cases when $m = 4$.

Beware that $d\Sigma$ does not lie in the subspace CAT($\tau(M)$) of **CAT**($\tau(M)$) unless the identity mapping $M \to M_\Sigma$ happens to be CAT. We will clear up this point in §2.1 below.

1.3 A slightly more general "differential" is wanted when the TOP manifold M without boundary has *no* given CAT manifold structure.† What one does is choose a topological embedding $M \to N$ into any CAT manifold N along with a continuous retraction $r : N \to M \subset N$. (As M is an ANR [Sp], r exists as soon as N is replaced by a small neighborhood of M). Consider the pullback $\hat{\tau} = r^*\tau(M)$ over N. It is

$$\hat{\tau} : M \times N \underset{\hat{\delta}}{\overset{p_2}{\rightleftarrows}} N, \text{ where } \hat{\delta}(y) = (r(y), y) \in M \times N.$$

As N is CAT, the space **CAT**($\hat{\tau}$) is well defined, and the rule $\Sigma \mapsto \Sigma \times N$ is a map CAT(M) $\to$ **CAT**($\hat{\tau}$); then passage to germs of structures about $\hat{\delta}(M)$ in $M \times N$ defines a complex **CAT**$\hat{\tau}$(M) = $= \varinjlim$ {**CAT**($\hat{\tau}$ | U); M $\subset$ U open in N} and a restriction map **CAT**($\hat{\tau}$) $\to$ **CAT** $\hat{\tau}$(M). The new **differential**

† The generality is also helpful in checking that the ultimate classifications of §§2, 3, 4 do not depend on certain choices, like the preferred structure on M in 1.2.

$$d : \mathrm{CAT}(M) \to \mathbf{CAT}\,\hat{\tau}\,(M)$$

is the composition of these two maps , and clearly coincides with the old differential if M = N .

1.4 FIRST CLASSIFICATION THEOREM .
For any TOP manifold M^m without boundary , $m \neq 4$,

$$d : \mathrm{CAT}(M) \to \mathbf{CAT}\,\hat{\tau}\,(M)$$

is a homotopy equivalence of css Kan complexes .

ADDENDUM 1.4.1 . *If $m = 4$ and CAT = PL , d is a homotopy equivalence provided no component of M is compact (cf. [Las_1] [Las_2][Si_{10} , §4])* † .

If CAT = PL , one verifies the Kan extension condition for these two complexes by applying the pullback rule to PL retractions $\Delta^k \to \Lambda_{k,i}$ (where $\Lambda_{k,i}$ is $\partial\Delta^k$ minus the interior of the face opposite the ith vertex) . The Kan condition is not so obvious if CAT = DIFF as there is no DIFF retraction $\Delta^k \to \Lambda_{k,i}$, and we shall postpone its verification (using 0.1) to the very end .

The homotopy theory of css Kan complexes is the classical homotopy theory of their geometric realizations (see especially [RS_1-I , §6]) . The reader is urged to exploit this fact constantly to avoid technical difficulties in verifying homotopy equivalences below . For a css complex X the geometric realization $|X|$ is the CW complex formed from the sum $\coprod_k \Delta^k \times X^{(k)}$ (where $X^{(k)}$ is regarded as an abstract *set* with discrete topology) by identifying each point $(x, \sigma) \in \Delta^k \times \sigma$, $\sigma \in X^{(k)}$ to $(\lambda x, \tau) \in \Delta^\ell \times \tau$, $\tau \in X^\ell$, whenever $\lambda : \Delta^k \to \Delta^\ell$ is an order preserving simplicial map and $\sigma = \lambda^\# \tau$. Geometrical realization is clearly a functor from css complexes to CW complexes . It is even faithful (injective) if we retain, in the CW complex structure of $|X|$, the natural characteristic maps $\Delta^k \to |X|$, one for each nondegenerate simplex ‡ of X . *This permits us to identify css complexes with their geometrical realizations* . Any ordered simplicial complex K is thus identified with (the geometrical realization of) the css complex of all order preserving

† For *closed* M^4 see the partial results of [LS] ; for CAT = DIFF , cf. 1.6(A).

‡ A *degenerate* simplex is one of the form $\lambda^\# \tau$ for some $\lambda : \Delta^k \to \Delta^\ell$ with $k > \ell$.

simplicial maps $\Delta^k \to K$, $k = 0,1,2, \ldots$. A map between css complexes is always understood to be a css map, until the contrary is stated.

Here is a useful semi-simplicial approximation lemma from [RS_1 - I, §5.3, §6.9]. Any merely continuous map $f : K \to X$ from an ordered simplicial complex K to a css Kan complex X, can be continuously homotoped to a css map $f' : K \to X$; further if f is css on a subcomplex $L \subset K$, the homotopy can leave $f|L$ fixed. This lemma will give us the liberty to choose convenient representatives for elements of homotopy groups.

A css map $\pi : X \to Y$ is called a *Kan fibration* if to any commutative diagram

$$\begin{array}{ccc} \Lambda_{k,i} & \to & X \\ \downarrow & & \downarrow \pi \\ \Delta^k & \to & Y \end{array}$$

one can add a map $\Delta^k \to X$ preserving commutativity. Observe that X is a Kan complex (i.e., satisfies the Kan extension condition) precisely if the constant map $X \to \Delta^0$ is a Kan fibration. It is a very convenient fact that the geometrical realization of a Kan fibration $\pi : X \to Y$ of Kan complexes is a Serre fibration [Q]; thus it has the homotopy lifting property for simplices--and hence for CW complexes.

PROOF OF THE CLASSIFICATION 1.4 FOR CAT = PL.

Given the Kan complex conditions, arguments conceived for immersion theory, (see especially [HaP], also [Gr] [Ha_2]) permit a proof of the homotopy equivalence as soon as we verify a few simple facts.

(1) *The rules* $U \mapsto CAT(U)$ *and* $U \mapsto$ **CAT**$\hat{\tau}(U)$ *are contravariantly functorial on inclusions of open subsets of* M. *They convert*

(i) *union of two into fiber product*

(ii) *disjoint discrete sum (possibly infinite) into cartesian product.* ■

Definition: For any subset $A \subset M$ we use injective limits to define

† These rules in fact constitute what are called sheaves of css sets, (see Appendix A.

$$\mathrm{CAT}_M(A) = \varinjlim \{\mathrm{CAT}(U)\ ;\ A \subset U \text{ open in } M\}$$

$$\mathbf{CAT}\hat{\tau}(A) = \varinjlim \{\mathbf{CAT}(\hat{\tau}|V)\ ;\ A \subset V \text{ open in } N\}$$

Note that we have a natural ' differential ' $d_A : \mathrm{CAT}_M(A) \to \mathbf{CAT}\hat{\tau}(A)$ still defined via the rule $\Sigma \leftrightarrow \Sigma \times N$.
Note that the extended rules $A \leftrightarrow \mathrm{CAT}_M(A)$ and $A \leftrightarrow \mathbf{CAT}\hat{\tau}(A)$ will also enjoy properties (i) & (ii) of (1), when restricted to *closed* sets A.

(2) *If B is a simplex (linear in some chart of M) and $A \subset B$ is a point then the restriction maps*

(i) $\mathrm{CAT}_M(B) \to \mathrm{CAT}_M(A)$, (ii) $\mathbf{CAT}\hat{\tau}(B) \to \mathbf{CAT}\hat{\tau}(A)$

are homotopy equivalences.

Proof of (2)(i). There exists a homotopy h_t, $0 \leqslant t \leqslant 1$, of $\mathrm{id}\,|\,M = h_0$ such that h_t, $0 \leqslant t < 1$, is an isotopy fixing A with $h_t(B) \subset B$, while h_1 crushes B onto A and induces a homeomorphism $M - B \approx M - A$. Using this isotopy h_t, $0 \leqslant t < 1$, one readily shows that the restriction (i) induces an isomorphism of semi-simplicially defined homotopy groups.

Proof of (2)(ii). The proof is similar but requires instead CAT homotopies in N. More precisely, using relative CAT approximation of continuous functions [Ze_2][Mu_1, §4], one obtains, for prescribed open neighborhoods $U \subset V$ of A and B respectively in N, a CAT homotopy f_t, $0 \leqslant t \leqslant 1$, of $\mathrm{id}\,|\,V = f_0$ fixing A so that $f_1(B) \subset U$. Such homotopies f_t let one show that restriction (ii) likewise induces an isomorphism of semi-simplicially defined homotopy groups. ■

(3) *If $A \subset M$ is a point, then $d : \mathrm{CAT}_M(A) \to \mathbf{CAT}\hat{\tau}(A)$ is a homotopy equivalence.*

Proof of (3). Indeed one shows that restriction $r : \mathbf{CAT}\hat{\tau}(A) \to \mathbf{CAT}(\hat{\tau}|A)$ is a homotopy equivalence by using a CAT deformation of an neighborhood of A to a retraction onto A. Now, (pleasant surprise!) the composition $rd : \mathrm{CAT}_M(A) \to \mathbf{CAT}(\hat{\tau}|A)$ is a true isomorphism of css sets. ■

(4) *For any compact pair $A \subset B$ in M, the two natural restriction mappings induced*

(i) $\mathrm{CAT}_M(B) \to \mathrm{CAT}_M(A)$, (ii) $\mathbf{CAT}\hat{\tau}(B) \to \mathbf{CAT}\hat{\tau}(A)$

are both Kan fibrations.

One can deduce the same for any *closed* pair $A \subset B$, by using (1).

Proof of (4) .

For A and B open (not compact) the sliced concordance extension theorem [II , §2.1] says precisely that (i) is a Kan fibration since $CAT_M(B)$ is a Kan complex . The same follows for quite arbitrary pairs $A \subset B$ in M by taking injective limits . The reader should also deduce fibration (i) for compact A and B from the less refined Bundle Theorem 0.1 with the help of the CAT and TOP isotopy extension theorems . We have here a very essential use of [II] .

As for fibration (ii) , given open neighborhoods $U \subset V$ in N of the closed sets $A \subset B$ respectively , and a CAT structure Σ on $\Lambda_{k,i} \times \tau(V) \cup \Delta^k \times \tau(U)$ we can find (as CAT = PL) a CAT map

$$r : \Delta^k \times V \to \Lambda_{k,i} \times V \cup \Delta^k \times U$$

respecting projection to V and fixing $\Lambda_{k,i} \times V$ union a neighborhood of $\Delta^k \times A$. Then existence of the pullback $r^*\Sigma$ as a well-defined CAT structure on $\Delta^k \times \tau(V)$ establishes the fibration (ii) . ■

Given these four properties , the proof of 1.4 , is reduced to homotopy theory by the famous 'handle induction machine' of immersion theory [HaP] [Gr] . Here is an outline .† For $A \subset M$, $\mathcal{E}(A)$ is the statement *that* $d_A : CAT_M(A) \to$ **CAT**$\hat{\tau}(A)$ *is a homotopy equivalence* . It is understood that the simplicial complexes mentioned below are simplicially imbedded in some co-ordinate chart of M .

(α) *$\mathcal{E}(A)$ holds for any simplex A* . Proof : Combine (3) with (2) .

(β) *If $\mathcal{E}$ holds for the compacta A , B , $A \cap B$, it also holds for $A \cup B$* . Proof : Apply (1) and (4) to the commutative square of four inclusions . (Cf. proof of A.2 in Appendix A .)

(γ) *$\mathcal{E}(A)$ holds for any finite simplicial complex A* . Proof : Use (α) , (β) and induction on the number of open simplices .

(δ) *$\mathcal{E}(A)$ holds for any compactum A in a co-ordinate chart* . Proof : A is an intersection of finite simplicial complexes .

† For the student it is eminently worthwhile to fill in the details , and as elegantly as possible .

(ε) *&(A) holds for an arbitrary compactum* $A \subset M$. Proof : A is a finite union of compacta in co-ordinate charts, a union to which one applies (β) inductively.

(ζ) *&(M) holds*. Proof : One can assume M connected, so that M is a union of nested compacta $A_1 \subset A_2 \subset \cdots$ with $A_i \subset \mathring{A}_{i+1}$. Define B_o [respectively B_e] to be the disjoint union of the compacta $A_{2i+1} - \mathring{A}_{2i}$ [respectively $A_{2i} - \mathring{A}_{2i-1}$] for i=0,1,2,... . Now apply the proof of (β) to the union $M = B_o \cup B_e$ using (4), (ε) and (1).

This concludes the proof of 1.4 for CAT = PL. ■

PROOF OF ADDENDUM 1.4.1 for CAT = PL.

The reader may omit this proof, as the result is of marginal importance. It does however apply to all dimensions.

The appearance of a weak homotopy lifting property and a homotopy micro-lifting property (= 'micro-gibki' property) in the relevant immersion theoretic method make it convenient to enrich our semi-simplicial complexes to become polyhedral quasi-spaces (as in [Si_{10}, §4], cf. Appendix A). Thus a map $P \to CAT^*(M)$ to the enriched space $CAT^*(M)$ is by definition a CAT structure Σ on $P \times M$ such that first factor projection $p_1 : (P \times M)_\Sigma \to P$ is a PL submersion with each fiber a PL manifold. And if ξ is a TOP microbundle over a locally compact polyhedron B, a map $P \to \mathbf{CAT}^*(\xi)$ is a PL structure on the microbundle $P \times \xi$. Thus our css complexes become contravariant functors from the category of compact polyhedra to the category of sets ; this change is indicated by an added star. We must show that the differential

$$d : CAT^*(M) \to \mathbf{CAT}^*\hat{\tau}(M)$$

is a weak homotopy equivalence of polyhedral quasi-spaces, i.e. gives an isomorphism on homotopy groups (defined as for ordinary spaces).

All the basic properties (1)–(4) remain essentially valid in this quasi-space context *except* the fibration (4)(i) above, established using $m \neq 4$. Note that the Kan fibration property becomes the Serre fibration property (= homotopy lifting property for compact polyhedra). Also weak homotopy equivalence replaces homotopy equivalence.

This missing fibration property (4)(i) is replaced here by the following *micro-gibki* property or 'homotopy micro-lifting' property.

(μ) *Given any commutative square*

$$\begin{array}{ccc} 0 \times P & \xrightarrow{F_0} & CAT^*_M(B) \\ \downarrow & & \downarrow \pi = \text{(restriction)} \\ I \times P & \xrightarrow{f} & CAT^*_M(A) \end{array} ,$$

where P *is a compact polyhedron and* (B,A) *is a compact pair in* M , *there exists an* $\epsilon > 0$ *and a map*

$$F_\epsilon : [0,\epsilon] \times P \to CAT^*_M(B)$$

extending F_0 *so that* $\pi F_\epsilon = f \mid ([0,\epsilon] \times P)$.

Proof of (μ) : This property follows from the PL *and* TOP (many-parameter) isotopy extension theorems via the 'union lemma for submersion product charts' [Si_{12} ; 6.9 , 6.10 , 6.15] , applied to charts around parts of the fiber $0 \times P \times M$ of the submersion $p_1 : I \times P \times M \to I$, in a form respecting all fibers of the projection p_2 to P .

More precisely, one obtains such a product chart $\varphi : [0,\delta] \times P \times V \to I \times P \times M$, V an open neighborhood of B in M , such that, representing f by $(I \times P \times U)_\Sigma$, one can restrict the chart φ to a PL product chart $\varphi \mid$ for the PL submersion $p_1 : (I \times P \times U)_\Sigma \to I$, with the image of $\varphi \mid$ a neighborhood of $0 \times P \times A$. Then if $(P \times V)_{\Sigma_0}$ represents F_0 , the image structure $\varphi(I \times \Sigma_0)$ yields F_ϵ (for all small ϵ) . ∎

By a remarkable geometric argument [Gr, §3.3.1] [Ha_1] that we will not repeat here, (μ) lets one prove that *restriction* $CAT^*(B) \to CAT^*(A)$ *is a Serre fibration after all in case* (B , A) *is a closed polyhedral pair and* $\dim(B-A) < m$. (B is supposed a subpolyhedron of some open subset of M that carries a PL manifold structure.)

Then the classical immersion theoretic method , cf. [HaP] or [Gr] goes on to prove that d_M is a homotopy equivalence as required in case M posesses by hypothesis some PL manifold structure. An infinite handle decomposition of M with no handle of index m is used . (Beware that the restrictions from thickening in this handle filtration have only the *weak* homotopy lifting property, cf. Appendix A .)

It remains then to show that M does have a PL manifold structure whenever **CAT** $\hat{\tau}$(M) is nonempty. The following fairly quick proof of this is due to R. Lashof [La_2] , cf. Appendix A .

Suppose 1.4.1 proved for manifolds covered by $< s$ co-ordinate charts, and let M be a manifold covered by s charts, of which U say is one, and V say is the union of the rest. Using the normality of M find a closed subset B of M in V such that $U \cup \mathring{B} = M$ and $B \cap U$ is a closed subcomplex of the chart U (suitably linearly triangulated). Let $C \subset U$ be the $(m-1)$-skeleton of $U - \mathring{B}$.

Recall that $\mathcal{E}(X)$ is the statement that d_A is a (weak) homotopy equivalence. Now $\mathcal{E}(V)$ holds by inductive hypothesis, and likewise $\mathcal{E}(B)$, $\mathcal{E}(C)$ and $\mathcal{E}(B \cap C)$ – by a limit argument. Since restriction from C to $B \cap C$ is a Serre fibration, one deduces that $\mathcal{E}$ holds for $B \cup C = B \cup U^{(m-1)}$, cf. ($\beta$) above. We can suppose $\mathbf{CAT}\,\hat{\tau}(M) \neq \phi$ (otherwise 1.4.1 is vacuously true). Then a neighborhood M_o of $B \cup U^{(m-1)}$ admits a PL manifold structure. If S is the discrete set in M consisting of the centers of all the m-simplices of U outside B, we easily construct open embeddings $M \to (M-S) \to M_o$. Hence M like M_o admits a PL structure, in which case $\mathcal{E}(M)$ follows, closing up the induction.

This completes our limited discussion of 1.4.1. ∎

PROOF OF THE CLASSIFICATION 1.4 FOR CAT = DIFF.

The above PL proof applies except for

(a) *verification of the Kan extension conditions for* $DIFF(M)$ *and* $DIFF\,\hat{\tau}(M)$, *for any* M *(as given)*,

(b) *verification that* $DIFF\,\hat{\tau}(B) \to DIFF\,\hat{\tau}(A)$ *is a Kan fibration, for compacta* $A \subset B$ *in* M.

In each case, the difficulty arises because simplices have corners. Both difficulties will be overcome by extensive use of the bundle theorem 0.1.

(a) The Kan complex conditions.

Lemma (see [May, §17.1, §18.2]). *Consider any css group* Y *and* $X \subset Y$ *a css subgroup. Then the coset complex* Y/X *is a Kan css complex and the quotient map* $Y \to Y/X$ *is a Kan fibration.* ∎

To apply this, let σ be a structure on M^m ; let $X = \mathrm{Aut}_{CAT}(M_\sigma)$ by the css group of CAT automorphisms of M_σ, and let $Y = \mathrm{Aut}_{TOP}(M_\sigma)$ be the css group of TOP automorphisms of M. A typical k-simplex $H \in Y^{(k)}$ is a TOP automorphism $H: \Delta^k \times M \to \Delta^k \times M$ sliced over Δ^k. There is a natural map $Y \to CAT(M)$ defined by the rule $H \mapsto H(\Delta^k \times \sigma)$. It clearly induces an injective css map

$$\sigma_\# : Y/X \to CAT(M)$$

But the Bundle Theorem 0.1 shows that, for $\dim M \neq 4 \neq \dim \partial M$, the image of $\sigma_\#$ is a union of connected components of CAT(M) one of which is certainly the component $CAT(M)_\sigma$ of $\sigma = \Delta^0 \times \sigma$. Varying σ now establishes the Kan complex condition for all of CAT(M). ■

The Kan fibrations (for $\dim M \neq 4 \neq \dim\partial M$ and CAT = PL or DIFF)

$$\mathrm{Aut}_{CAT}(M_\sigma) \to \mathrm{Aut}_{TOP}(M) \xrightarrow{\sigma_\#} CAT(M) \tag{1.5}$$

are themselves very worthwhile ; they show that, for $k \geqslant 1$,

$\pi_k(\mathrm{Aut}_{TOP}(M), \mathrm{Aut}_{CAT}(M_\sigma)) = \pi_k CAT(M)_\sigma$ (but not for $k = 0$). Here the base points for the groups are taken at the identity. We observe that (1.5) admits a version relative to a closed subset $C \subset M$. One uses (simplices of) automorphisms fixing some neighborhood of C in M and (simplices of) structures that coincide near C with the given structure σ ; the same proof applies using a relative version of the Bundle Theorem from [II, §1.1].

A similar argument will now show that **DIFF** $\hat{\tau}(M)$ is Kan.

Assertion. *For any locally topologically trivial bundle $\eta: E \to N$ with fiber R^m, $m \neq 4$, over a DIFF manifold N, the following css complex $DIFF_o(\eta)$ is Kan.*

A k-simplex of $DIFF_o(\eta)$ is a DIFF manifold structure Γ on $\Delta^k \times E$ such that projection $(\Delta^k \times E)_\Gamma \to \Delta^k \times N$ is a DIFF submersion. This is a DIFF bundle with fiber R^m by 0.1 since each point of $\Delta^k \times N$ has a neighborhood that is a smooth simplex. So by the DIFF bundle homotopy theorem† we have

(**) $$(\Delta^k \times E)_\Gamma \cong \Delta^k \times (E_\gamma)$$

for some γ, by an isomorphism sliced over $\Delta^k \times N$.

Now it follows by the argument above that for any 0-simplex $\Delta^0 \times \eta'$ of $\mathrm{DIFF}_0(\eta)$ we have a Kan fibration

$$\mathrm{Aut}_{\mathrm{DIFF}}(\eta') \to \mathrm{Aut}_{\mathrm{TOP}}(\eta) \to \mathrm{DIFF}_0(\eta)$$

identifying the quotient of the automorphism groups to certain components of $\mathrm{DIFF}_0(\eta)$, of which one certainly contains $\Delta^0 \times \eta'$. This establishes the assertion. ■

We now know that $\mathrm{DIFF}_0(\eta)$ is Kan for any R^m-bundle η contained in any restriction of τ. Recall that by the Kister-Mazur theorem [KuL1], every microbundle (over a triangulable space or retract of one) contains a $(R^m,0)$ bundle η with the same zero-section. It follows immediately that $\mathrm{DIFF}\,\hat{\tau}(M)$ is Kan. ■

(b) The fibration $\mathrm{DIFF}\hat{\tau}(B) \to \mathrm{DIFF}\hat{\tau}(A)$.

Using the notation established above and the property (**) one easily proves that for any compact (or closed) K in X the restriction $\mathrm{DIFF}_0(\eta) \to \mathrm{DIFF}_0(\eta \text{ near } K)$ is a Kan fibration, where by definition $\mathrm{DIFF}_0(\eta \text{ near } K)^{\bullet} = \varinjlim \mathrm{DIFF}_0(\eta | U)$, the injective limit being taken over open neighborhoods U of K in X. The desired fibration clearly follows. ■

1.6 REMARKS COMPLEMENTARY TO THE CLASSIFICATIONS 1.4 AND 1.4.1.

A) DIFF structures on open 4-manifolds. The main obstacle to proving the DIFF version of the PL classification theorem 1.4.1 for open 4-manifolds is to prove that the complexes involved are Kan complexes. We would not underestimate the difficulty of giving a proof, although we expect it is in no sense to be compared with the problem of proving

† To avoid any possible difficulty here involving corners, observe that the conditioning procedure below (see ρ^* in proof 2.4) produces a DIFF concordance $\Theta : \Gamma \sim \Gamma'$, sliced over $I \times \Delta^k \times N$, to a structure Γ' that extends to Γ'' on $R^{k+1} \times E$ sliced over $R^{k+1} \times N$ (recall $\Delta^k \subset R^{k+1}$). By one application of the bundle homotopy theorem we have $(\Delta^k \times E)_\Gamma \cong (\Delta^k \times E)_{\Gamma'}$ sliced over $\Delta^k \times N$. By a second $(k+1)$-fold application to $(I^{k+1} \times E)_{\Gamma''}$ we get a further isomorphism $(\Delta^k \times E)_{\Gamma'} \cong \Delta^k \times (E_\gamma)$ sliced over $\Delta^k \times M$.

the stronger Bundle Theorem 0.1 for open 4-manifolds. What one *can* prove without difficulty is the quasi-space version of 1.4.1 stating that the rule $d : (P \times M)_\Sigma \mapsto \Sigma \times N$ gives a weak homotopy equivalence

$$d : CAT^*(M) \to CAT^*\hat{\tau}(M)$$

of DIFF quasi-spaces for CAT = DIFF, defined just as for CAT = PL above but using arbitrary DIFF manifolds and maps in place of compact polyhedra and PL maps. Homotopy groups are understood to be defined using the standard smooth disks and spheres. The familiar DIFF devices for conditioning homotopies etc., using an DIFF homotopy of $id | [0,\infty)$, fixing 0 and $[2,\infty)$, to a map sending $[0, 1]$ to 0 serve to make the (weak) homotopy theory of these quasi-spaces function well. Difficulties with kinks and corners are thus avoided. It seems to us that every result in this essay which involves PL 4-manifolds, can be proved for DIFF with little more effort, on the condition that it be formulated and proved for DIFF quasi-spaces. The Kan property for the corresponding css complexes would let one deduce the css versions using the device of conditioning simplices (as in proving assertion 2.4 below).

B) Non-metrizable manifolds.

Such manifolds are usually regarded as remote curiosities and excluded from consideration. This many not be a wise attitude since some turn up quite naturally. To illustrate, if a non-compact Lie group G, like (R, +), acts smoothly and freely on a smooth manifold W in such a way that every point in W is 'wandering' ‡, then the orbit space W/G is a smooth manifold which is, in general, not even Hausdorff (see [Pa_2]) ; interesting examples appear already in the plane $W = R^2$, (see [Kap]). Thus in any classical dynamical system the open set of 'wandering' points always yields a possibly non-metrizable manifold as orbit space.

Recall that we proved in [II, § 2.1] that the restriction mapping $CAT(V) \to CAT(U)$ is a Kan fibration for arbitrary open subsets of a CAT manifold M with $\dim M \neq 4 \neq \dim \partial M$. The classification theorems of this section did not need the full strength of this result, but on non-metrizable manifolds it becomes a powerful tool. The rule $U \mapsto CAT(U)$ always constitutes a sheaf of css sets on the space M,

‡ The point p in W is *wandering* if there exists a neighborhood U of p such that $g(U) \cap U = \phi$ for all $g \in G$ outside some compactum in G (depending on p).

(cf. Appendix A). If $\Phi : U \mapsto \Phi(U)$ is a sheaf of css sets on a topological space M, we will say Φ is *strongly flexible* if the restriction $\Phi(V) \to \Phi(U)$ is a Kan fibration for every pair (V, U) of open sets in M. Using Zorn's lemma one readily verifies the

Proposition. *A sheaf Φ of css sets on an arbitrary topological space M is strongly flexible if it is locally strongly flexible, i.e. if the restriction $\Phi | U_\alpha$ is strongly flexible for every set U_α of some open covering $\mathfrak{U} = \{U_\alpha\}$ of M. If $d : \Phi \to \Phi'$ is a morphism of strongly flexible sheaves on M, then $d_M : \Phi(M) \to \Phi'(M)$ is a homotopy equivalence whenever $d_U : \Phi(U) \to \Phi'(U)$ is a homotopy equivalence for every open set U such that $U \subset U_\alpha$ for some $U_\alpha \in \mathfrak{U}$.*

This implies that $CAT(M^m)$ is contractible for any possibly non-metrizable m-manifold, in case CAT(U) is known to be contractible for every open $U \subset R^m_+$ $(m \neq 4)$. (Just make Φ' the constant sheaf $U \mapsto \Phi'(U) = \Delta^0$.) In fact Cerf has proved $CAT(M^m)$ contractible for metrizable M^m, provided $m \leqslant 2$, or provided CAT = PL and $m = 3$, (see [Ce_1], and a proof in §§4.4, 5.8, 5.9 below); the same now follows for possibly non-metrizable manifolds. Similarly one can show $DIFF(M^3)$ is 1-connected for possibly non-metrizable 3-manifolds (see §5.8).

It should be possible to prove that the sheaf $U \mapsto \mathbf{CAT}\hat{\tau}(U)$ is strongly flexible $(m \neq 4)$ and so obtain the equivalence $d : CAT(M) \to \mathbf{CAT}\hat{\tau}(M)$ of 1.4 for possibly non-metrizable manifolds.

Two words of caution:

(i) CAT sliced concordance does *not* imply CAT isomorphism on non-metrizable manifolds. For example, although CAT(M) is contractible for the long line $M^1 = (0, \Omega)$ of Cantor and Alexandroff, it is known that the long line has infinitely (uncountably?) many non-isomorphic CAT structures; indeed Koch and Puppe give a DIFF construction [KP]. To give a similar PL construction is an amusing exercise – disproving the Hauptvermutung on the long line.

(ii) What we have said under this rubric B) does not apply to DIFF structures Whitehead compatible with a fixed PL structure, cf. [I, §4.2].

§2 CAT STRUCTURES AND CLASSIFYING SPACES

To render useful the equivalences $d : CAT(M) \to \mathbf{CAT}\,\hat{\tau}(M)$ of §1 one needs an analysis of $\mathbf{CAT}\,\hat{\tau}(M)$ in terms of microbundle classifying spaces.

2.0. This analysis applies to any TOP m-microbundle ξ over a (metrizable) CAT object X (CAT = DIFF or PL). We will need a universal CAT m-microbundle γ^m_{CAT} over $B_{CAT(m)}$. It is known, cf. [IV, §1.1], that γ^m_{DIFF} can be a smooth universal m-vector-bundle†, while $B_{DIFF(m)}$ can be built up as an open subset of ℓ_2 (a smooth infinite dimensional Hilbert manifold). Also $B_{PL(m)}$ can be a locally finite simplicial complex [IV, §8].

We need a notion of CAT microbundle map $\theta : \xi_1 \to \xi_2$ of CAT microbundles $\xi_\alpha : X_\alpha \xrightarrow{i_\alpha} E_\alpha \xrightarrow{p_\alpha} X_\alpha$. This consists of an open neighborhood U of $i_1(X_1)$ in E_1 and CAT maps $H : U \to E_2$, $h : X_1 \to X_2$ so that (a) $p_2 H = h p_1$ and $H i_1 = i_2 h$, and (b) H is an open CAT imbedding of each fiber of $p_1 | U$ into a fiber of p_2. These compose naturally.

We fix from the outset a classifying (m-microbundle) map $\varphi : \xi \to \gamma^m_{TOP}$ covering a continuous map $f : X \to B_{TOP(m)}$. Also we fix a classifying TOP microbundle map $\gamma^m_{CAT} \to \gamma^m_{TOP}$ over a map $j_m : B_{CAT(m)} \to B_{TOP(m)}$. By Serre's device [IV, §9] we extend j_m to a Hurewicz fibration $j_m : \overline{B}_{CAT(m)} \to B_{TOP(m)}$. The subspace $B_{CAT(m)} \subset \overline{B}_{CAT(m)}$ is a deformation retract.

A k-simplex σ of the Kan complex Lift(f to $B_{CAT(m)}$) is defined to be a continuous map $\sigma : \Delta^k \times X \to \overline{B}_{CAT(m)}$ so that $j_m \sigma = f p_2 : \Delta^k \times X \to B_{TOP(m)}$ where p_2 is projection to X.

† This justifies writing $B_{O(m)}$ for $B_{DIFF(m)}$ and later on TOP_m/O_m for $TOP_m/DIFF_m$ etc.

THEOREM 2.1 *(above data).*

Assuming CAT(ξ) *is Kan* †, CAT(ξ) → **CAT**(ξ) *is a homotopy equivalence of Kan complexes.*

THEOREM 2.2 *(same data)*

There are canonical and natural homotopy equivalences of Kan complexes

$$\mathrm{CAT}(\xi) \leftarrow \mathrm{CAT}^{\mathrm{cls}}(\xi) \rightarrow \mathrm{Lift}(f \text{ to } B_{\mathrm{CAT}(m)}) \quad .$$

$\mathrm{CAT}^{\mathrm{cls}}(\xi)$ is defined in the proof of 2.2, which is postponed in favor of corollaries.

For any subset $A \subset X$ we can take injective limit over open neighborhoods of A in X to obtain from 2.1 and 2.2 homotopy equivalences (for same data) denoted :

2.1.1. **CAT**(ξ near A) ← CAT(ξ near A) (for $m \neq 4$ or CAT ≠ DIFF),

2.2.1. CAT(ξ near A) ← $\mathrm{CAT}^{\mathrm{cls}}$(ξ near A) → Lift(f to $B_{\mathrm{CAT}(m)}$ near A)

all natural for restriction (i.e. behaving contraviariant functorially with respect to inclusions). *The notation introduced here will be used repeatedly.*

To apply this result in the setting of § 1.1, we choose a microbundle map $\varphi : \tau(M) \rightarrow \gamma^m_{\mathrm{TOP}}$ over a map $f : M \rightarrow B_{\mathrm{TOP}(m)}$. This we extend to $\hat{\varphi} : \hat{\tau} \rightarrow \gamma^m_{\mathrm{TOP}}$ over $\hat{f} : N \rightarrow B_{\mathrm{TOP}(m)}$, by composition with the natural microbundle map $\hat{\tau} = r^* \tau(M) \rightarrow \tau(M)$ over r. We note that for any $A \subset M$, the restriction

$$\rho : \mathrm{Lift}(\hat{f} \text{ to } B_{\mathrm{CAT}(m)} \text{ near } A) \rightarrow \mathrm{Lift}(f \text{ to } B_{\mathrm{CAT}(m)} \text{ near } A)$$

is a homotopy equivalence; the inverse equivalence ρ' comes from using the retraction $r : N \rightarrow M$ to extend lifts. (Proof of equivalence: $\rho\rho' = \mathrm{id}$ is clear; to provide a homotopy $\rho'\rho \simeq \mathrm{id}$ use a homotopy $r_t : N \rightarrow N$, $0 \leq t \leq 1$, of $r_0 = \mathrm{id}|N$ fixing M so that $r_1 = r$ on a neighborhood of M in N. This homotopy exists because M and N are ANR's.)

† This is evident for CAT = PL. It is established for **DIFF**(ξ), for $m \neq 4$, below 1.5; the same argument applies to DIFF(ξ), *for all m*, in fact much more directly, using the DIFF microbundle homotopy theorem.

CLASSIFICATION THEOREM 2.3 .

Under the hypotheses of 1.4 or 1.4.1 , there is, given $\varphi : \tau(M) \to \gamma^m_{TOP}$ *, a homotopy equivalence* $\theta : CAT(M^m) \to Lift(f \text{ to } B_{CAT(m)})$ *well-defined up to homotopy . And for any subset* $A \subset M$ *there is a homotopy equivalence* $\theta_A : CAT_m(A) \to Lift(f \text{ to } B_{CAT(m)} \text{ near } A)$ *. These equivalences behave naturally under restriction, viz . when* $A \subset B$ *the square*

$$\theta_B \downarrow \overset{\rightarrow}{\underset{\rightarrow}{}} \downarrow \theta_A$$

in which the verticals are restrictions,is homotopy commutative .

Indeed θ_A arises from the sequence of four perfectly canonical and natural equivalences

2.3.1 . $\mathrm{CAT}_M(A) \overset{d}{\to} \mathbf{CAT}\hat{\tau}(A) \equiv \mathbf{CAT}(\hat{\tau} \text{ near } A) \leftarrow$

$$\leftarrow \mathrm{CAT}(\hat{\tau} \text{ near } A) \overset{\alpha}{\leftarrow} \mathrm{CAT}^{cls}(\hat{\tau} \text{ near } A) \overset{\rho\beta}{\to} \mathrm{Lift}(f \text{ to } B_{CAT(m)} \text{ near } A)$$

established by §1 and 2.1.1 and 2.2.1 . The existence of this canonical sequence is a good practical substitute for a canonical (prefered) choice of θ_A as css map . It will prevent any loss of precision in passing to relative versions and versions with boundary . ■

COMPLEMENT 2.3.2 . *The homotopy class of the equivalence* $\theta_A : CAT_M(A) \to Lift(f \text{ to } B_{CAT(m)} \text{ near } A)$ *does not depend on choice of the embedding* $M \hookrightarrow N$ *or the retraction* $r : N \to M$ *.*

A proof of 2.3.2 is given at the end of this section .

PROOF OF 2.1 : $\mathbf{CAT}(\xi) \simeq \mathrm{CAT}(\xi)$.

Suppose given

$$\Sigma : (\Delta^k, \partial\Delta^k) \to (\mathbf{CAT}(\xi), \mathrm{CAT}(\xi)) \quad .$$

In the microbundle

$$I \times \Delta^k \times \xi \; : \; I \times \Delta^k \times E(\xi) \underset{i}{\overset{p}{\rightleftarrows}} I \times \Delta^k \times X$$

we approximate i by a section i′ which is CAT on $0 \times \Delta^k \times X$ with respect to the CAT manifold structure† $I \times \Sigma$ and equals i on $\sqsupset = (I \times (\partial\Delta^k) \times X) \cup (1 \times \Delta^k \times X)$. This requires only chart by chart use of the relative approximation of continuous functions to the fiber R^m by CAT ones , as proved by Zeeman [Ze_2] for CAT = PL and by Whitney [Wh_1] [See] for DIFF . ‡ Replacing i by i′ we

† We are using the same symbol Σ for a CAT structure with germ Σ .

‡ The need of approximation *sharply* relative to ⊐ could be obviated by a conditioning procedure as in the proof of 2.4 below .

have a new TOP microbundle ζ over $I \times \Delta^k \times X$, to which we apply the microbundle homotopy theorem to get a TOP morphism $H : \zeta \to I \times \Delta^k \times \xi$ over $(\mathrm{id} | I \times \Delta^k \times \xi)$ and *equal* to the identity over $\sqsupset$. (Recall here that $(1 \times \Delta^k) \cup (I \times \partial\Delta^k) \approx 1 \times \Delta^k$ under a self-homeomorphism of all $I \times \Delta^k$). The image of $I \times \Sigma$ under H is a new CAT structure on $I \times \Delta^k \times \xi$, equal to $I \times \Sigma$ over $\sqsupset$ (near the zero section), and having a CAT zero section over $0 \times \Delta^k \times X$. Thus $H(I \times \Sigma)$ provides a homotopy of Σ fixing $\partial\Delta^k$ into $\mathrm{CAT}(\xi)$ as soon as $I \times \Delta^k$ is triangulated suitably. ■

THE EQUIVALENCES $\mathrm{CAT}(\xi) \overset{\alpha}{\leftarrow} \mathrm{CAT}^{\mathrm{cls}}(\xi) \overset{\beta}{\rightarrow} \mathrm{Lift}(f \text{ to } B_{\mathrm{CAT}(m)})$ OF 2.2.

We shall define natural 'forgetful' maps of css Kan complexes

$$\mathcal{B}_{\mathrm{CAT}}(X) \overset{\alpha}{\leftarrow} \mathcal{B}^{\mathrm{cls}}_{\mathrm{CAT}}(X) \overset{\beta}{\rightarrow} \{X, B_{\mathrm{CAT}(m)}\}$$

and pass to fibers.

A typical k-simplex of $\{X, Y\}$ is a continuous map $\Delta^k \times X \to Y$. And $\{X,Y\}_{\mathrm{CAT}}$ is the subcomplex of simplices that are CAT maps.

$\mathcal{B}^{\mathrm{cls}}_{\mathrm{CAT}}(X)$ has typical k-simplex represented by a CAT m-microbundle γ over $\Delta^k \times X$ equipped with a CAT morphism $g : \gamma \to \gamma^m_{\mathrm{CAT}}$. ($\mathcal{B}$ indicates m-micro*bundles* ; and cls *classifying* map). To avoid set-theoretical difficulties we can insist that $E(\gamma)$ be a subset of some $X \times R^n \subset X \times R^\infty$. Another such pair (γ', g') represents the same k-simplex iff (γ, g) and (γ', g') have the same germ, i.e. coincide (are identical) on a neighborhood of the respective zero-sections.

To define $\mathcal{B}_{\mathrm{CAT}}(X)$ we just omit mention of g.

For CAT = DIFF or PL we also define $\mathcal{B}^{\mathrm{cls}}_{\mathrm{CAT}}(X)$ containing $\mathcal{B}^{\mathrm{cls}}_{\mathrm{CAT}}(X)$ as deformation retract as follows. To the above definition of a typical k-simplex of $\mathcal{B}^{\mathrm{cls}}_{\mathrm{CAT}}(X)$ we add, to get all of $\mathcal{B}^{\mathrm{cls}}_{\mathrm{CAT}}(X)$, the germ of a morphism $h : I \times \gamma \to \gamma^m_{\mathrm{TOP}}$ such that $\gamma = 0 \times \gamma \overset{h}{\to} \gamma^m_{\mathrm{TOP}}$ coincides with $\gamma \overset{g}{\to} \gamma^m_{\mathrm{CAT}} \to \gamma^m_{\mathrm{TOP}}$. It is not difficult to show that the forgetting map:

$$\mathcal{B}^{\mathrm{cls}}_{\mathrm{CAT}}(X) \to \mathcal{B}^{\mathrm{cls}}_{\mathrm{TOP}}(X) \quad ,$$

which selects $\gamma = 1 \times \gamma \overset{h}{\to} \gamma^m_{\mathrm{TOP}}$, is a Kan fibration.

ASSERTION 2.4. $\mathcal{B}_{\mathrm{CAT}}(X) \overset{\alpha}{\leftarrow} \mathcal{B}^{\mathrm{cls}}_{\mathrm{CAT}}(X) \overset{\beta}{\rightarrow} \{X, B_{\mathrm{CAT}(m)}\}$ *are equivalences, where* α *and* β *merely forget.*

Proof of 2.4 .

β is a composition $\mathcal{B}^{cls}_{CAT}(X) \to \{X, B_{CAT(m)}\}_{CAT} \hookrightarrow \{X, B_{CAT(m)}\}$ of two easily verified natural equivalences of which the second requires relative CAT approximation of continuous maps [Ze$_2$] [Whn$_1$] .

It is the following universality of γ^m_{CAT} that establishes the equivalence α .

(*) *For any CAT microbundle η^m over Y (an ENR), any closed set $C \subset Y$, and any CAT microbundle map $g : (\eta | U) \to \gamma^m_{CAT}$ where U is an open neighborhood of C in Y, there exists a CAT morphism $G : \eta \to \gamma^m_{CAT}$ equal g over C.*

This is a standard version, cf [IV , §8.1]†.

It is not at first clear that this suffices, particularly for CAT = DIFF . So consider any CAT m-microbundle γ^m over $\Delta^k \times X$ and a morphism $g : \gamma^m | (\partial\Delta^k \times X) \to \gamma^m_{CAT}$ that is CAT (wherever this makes sense). Let $r_t : \Delta^k \to \Delta^k$, $0 \leq t \leq 1$, be a CAT homotopy of $id | \Delta^k$ respecting $\partial\Delta^k$ so that $r_1^{-1}(\partial\Delta^k)$ contains an open neighborhood U of $\partial\Delta^k$. (For CAT = DIFF , r_t *cannot* fix $\partial\Delta^k$ (pointwise); it can be constructed by composing $(k+1)$ homotopies that squash a collar of a face $\partial_i\Delta^k$ into $\partial_i\Delta^k$ following the lines radiating from v_i). Now define a self-map ρ of $I \times \Delta^k \times X$ by $\rho(t,u,x) = (t, r_t(u), x)$ and consider the pullback $\rho^*(I \times \gamma)$ over $I \times \Delta^k \times X$.‡

This bundle inherits from

$$\rho^*(I \times \gamma) \to I \times \gamma \to \gamma \quad \text{and} \quad g : \gamma \mid (\partial\Delta^k \times X) \to \gamma^m_{CAT}$$

a microbundle map $\rho^*(I \times \gamma) | A \to \gamma^m_{CAT}$, $A = (I \times \partial\Delta^k \cup 1 \times U) \times X$, that is CAT wherever this makes sense . By the standard CAT universality property cited this extends rel $I \times \partial\Delta^k \times X$ to all of $\rho^*(I \times \gamma) | \sqsupset$ to produce a map‡‡ $\sqsupset(I \times \Delta^k) \to \mathcal{B}^{cls}_{CAT}(X)$ which establishes the homotopy equivalence α . ∎

† The condition that Y be an ENR can be replaced by the condition that η^m be numerable if we use the more general form of the Kister-Mazur theorem in [Hol] together with Milnor's classifying space for R^m- bundles [Hus] .

‡ In terminology we have occasion to use elsewhere the CAT homotopy r_t of $id | \Delta^k$ yields a *conditioning* $\rho^*(I \times \gamma)$ of the simplex $\gamma : \Delta^k \to \mathcal{B}_{CAT}(X)$; the *conditioned* simplex is $\rho^*(I \times \gamma) | 1 \times \Delta^k \times X$.

‡‡ Determined by a convenient simplicial subdivision of the prism $I \times \Delta^k$.

Continuing the proof of 2.2 , consider now the natural commutative diagram of 'forgetful' maps

$$\begin{array}{ccccc} \mathcal{B}_{CAT}(X) & \overset{\bar\alpha}{\leftarrow} & \mathcal{B}^{cls}_{CAT}(X) & \overset{\bar\beta}{\rightarrow} & \{X, \bar{B}_{CAT(m)}\} \\ \downarrow & & \downarrow & & \downarrow \\ \mathcal{B}_{TOP}(X) & \overset{\alpha}{\leftarrow} & \mathcal{B}^{cls}_{TOP}(X) & \overset{\beta}{\rightarrow} & \{X, B_{TOP(m)}\} \end{array}$$

in which the verticals are known to be Kan fibrations while the horizontals are known to be homotopy equivalences . For the point (ξ, φ) of $\mathcal{B}^{cls}_{TOP}(X)$ we have $\alpha(\xi, \varphi) = \xi$ and $\beta(\xi, \varphi) = f$ on identifying $\Delta^0 \times X = X$. The resulting two equivalences between the three css fibers of the three vertical maps, over the points ξ , (ξ, φ) , and f (in this order) , are *by definition* the wanted natural equivalences of Kan complexes

$$CAT(\xi) \overset{\alpha}{\leftarrow} CAT^{cls}(\xi) \overset{\beta}{\rightarrow} \text{Lift}(f \text{ to } B_{CAT(m)}) .$$

This completes the proof of 2.2 and also of the classification theorem 2.3 . ∎

PROOF OF COMPLEMENT 2.3.2 .

Consider two choices of embedding $i : M \hookrightarrow N$ and retraction $r : N \to M$ distinguished by subscripts 0 and 1 , yielding equivalences θ^0_A , θ^1_A via 2.3 .

(a) *Suppose* $N_0 \subset N_1$ *and that there is a* CAT *retraction* $\rho : N_1 \to N_0$ *so that* $r_1 = r_0\rho$. *Then is is clear that* $\theta^0_A \simeq \theta^1_A$. ∎

Using (a) we easily reduce the proof of 2.3.2 to the case where N_0 , N_1 are open subsets of R^n , n large , and one has a topological embedding $\bar{i} : I \times M \hookrightarrow \bar{N} \subset I \times R^n$ and neighborhood retraction $\bar{r} : \bar{N} \to I \times M$ so that $\bar{i}$, $\bar{r}$ both are a product along I near 0 and 1 , and give concordances $\bar{i} : i_0 \sim i_1$ and $\bar{r} : r_0 \sim r_1$.

(b) *In this situation* $\theta^0_A \simeq \theta^1_A : CAT_M(A) \to \text{Lift}(f \text{ to } B_{CAT(m)} \text{ near } A)$.

To prove (b) we form a diagram

$$\mathrm{CAT}_M(A) \xrightarrow{\bar{d}} \mathrm{CAT}(\bar{\tau} \text{ near } \bar{\iota}(I \times A)) \xrightarrow{\Phi}$$

$$\xrightarrow{\Phi} \mathrm{Lift}(I \times f \text{ to } B_{\mathrm{CAT}(m)} \text{ near } I \times A) \xrightarrow{q_t} \mathrm{Lift}(f \text{ to } B_{\mathrm{CAT}(m)} \text{ near } A)$$

Here $\bar{\tau}$ is $\bar{N} \xrightarrow{j} \bar{M} \times \bar{N} \xrightarrow{p_2} \bar{N}$ with $j(y) = (p\bar{r}(y), y)$, p being the natural projection $\bar{\iota}(I \times M) \to M$; $\bar{d}$ is the differential $\Sigma \twoheadrightarrow \Sigma \times \bar{N}$; Φ is the equivalence of 2.2.1 composed with restriction to $I \times M$. (This equivalence of 2.2.1 being obtained using $\bar{\varphi} : \bar{\tau} \xrightarrow{\text{nat}} \tau(M) \xrightarrow{\varphi} \gamma^m_{\mathrm{TOP}}$ over $\bar{f} = fp\bar{r}$) . Finally q_t , $0 \leqslant t \leqslant 1$, comes from restriction to $t \times M$. It is easy to see that $\theta^0_A \simeq q_0 \Phi \bar{d}$ and $\theta^1_A \simeq q_1 \Phi \bar{d}$; hence the homotopy q_t , $0 \leqslant t \leqslant 1$, establishes $\theta^0_A \simeq \theta^1_A$ as required to complete the proof of 2.3.2 . ■

§3. THE RELATIVE CLASSIFICATION THEOREM

For the hypotheses of 1.4 or 1.4.1, the classification theorem 2.3 provides homotopy equivalences θ and θ_C in a homotopy commutative square

$$\begin{array}{ccc} \mathrm{CAT}(M) & \xrightarrow{\theta} & \mathrm{Lift}(f \text{ to } B_{\mathrm{CAT}(m)}) \\ r_1 \downarrow & & \downarrow r_2 \\ \mathrm{CAT}_M(C) & \xrightarrow{\theta_C} & \mathrm{Lift}(f \text{ to } B_{\mathrm{CAT}(m)} \text{ near } C) \end{array} \tag{3.1}$$

where the verticals are restrictions. We suppose $C \subset M$ is closed to assure that r_2 is a Kan fibration. The restriction r_1 is always a fibration for $m = \dim M \geqslant 5$, see property (4) in proof of 1.4 ; for $m = 4$ we secure this fibration by assuming $\check{H}^4(M, C; Z_2) = 0$ (for Čech cohomology), see proof of 1.4.1 and Appendix A. It follows now that there are equivalences of the fibers of r_1 and r_2 over components corresponding under θ_C. To fix notations let Σ_o be a CAT structure defined on an open neighborhood U_o of C in M, and let g_o be the lift corresponding to Σ_o under θ_{U_o}. Then we write

$$\mathrm{CAT}(M \text{ rel } C; \Sigma_o) \overset{\theta}{\to} \mathrm{Lift}(f \text{ to } B_{\mathrm{CAT}(m)} \text{ rel } C; g_o) \tag{3.2}$$

for the induced equivalence of the fibers over the germs of Σ_o and g_o respectively. Note that this θ again arises from a sequence of four perfectly canonical and natural equivalences.

We apply this now taking $M = N = R^m$, $m \geqslant 5$; $C = R^m - \mathring{B}^m$; Σ_o standard ; f and g_o constant. The left hand side of (3.2) becomes $\mathrm{CAT}(R^m \text{ rel } R^m - \mathring{B}^m) = \mathrm{CAT}(B^m \text{ rel } \partial B^m)$. The right hand side becomes the singular complex of the space of those maps $B^m/\partial B^m \to$ $\to \mathrm{TOP}_m/\mathrm{CAT}_m$ sending a neighborhood of the base point (quotient of ∂B^m) to the base point (image of g_o). Here $\mathrm{TOP}_m/\mathrm{CAT}_m$ is the fiber of $j : B_{\mathrm{CAT}(m)} \to B_{\mathrm{TOP}(m)}$ over the base point (image of f). Thus the right hand side is equivalent to the mth loop space $\Omega^m(\mathrm{TOP}_m/\mathrm{CAT}_m)$, and we have deduced an equivalence

$$\mathrm{CAT}(B^m \text{ rel } \partial B^m) \simeq \Omega^m(\mathrm{TOP}_m/\mathrm{CAT}_m)\,. \tag{3.3}$$

Now the loop space of $CAT(B^m \text{ rel } \partial B^m)$ pointed by the standard structure is (up to homotopy) the css group of CAT automorphisms $Aut_{CAT}(B^m \text{ rel } \partial B^m)$ because we have the natural Kan fibration†

$$Aut_{CAT}(B^m \text{ rel } \partial) \to Aut_{TOP}(B^m \text{ rel } \partial) \to CAT(B^m \text{ rel } \partial)$$

where the total space $Aut_{TOP}(B^m \text{ rel } \partial)$ is contractible by Alexander's device $[Al_1]$ ‡ . Thus we have verified

MORLET'S THEOREM 3.4 .

$Aut_{CAT}(B^m \text{ rel } \partial B^m) \simeq \Omega^{m+1}(TOP_m/CAT_m)$ *for* $m \neq 4$. ∎

Since $Aut_{PL}(B^m \text{ rel } \partial B^m)$ is contractible by the PL version of the Alexander device, one has, for $m \neq 4$, $\Omega^{m+1}(TOP_m/PL_m) \simeq$ (point) , that is, $\pi_{m+k}(TOP_m/PL_m) = 0$, $k \geqslant 1$.

Historical Remarks : Morlet first obtained and made famous this striking result for $PL_m/DIFF_m$ in his cours Peccot 1969 $[Mor_2]$. It was probably the raison d'être for his formulation of the sliced classification theorems $[Mor_3]$ $[Mor_4]$. Several mathematicians besides us have since taken the trouble to verify it themselves (A. Chenciner , C. Rourke, D. Burghelia, ...

Cerf's version of 3.4 for CAT = DIFF in $[Ce_4]$ ‡‡ is (with a slight shift of terminology)

$$\text{(3.4.1)} \quad \pi_i(Aut_{DIFF}(S^m),O(m+1)) = \pi_{i+m+1}(Aut_{TOP}(S^m),O(m+1)).$$

He regards $Aut_{CAT} S^m$ as a topological group, with C^∞ topology for CAT = DIFF , and with C^0 topology for CAT = TOP . Exploiting the natural fibrations $Aut_{CAT} S^m \to S^m$ and $O(m+1) \to S^m$, one shows without difficulty that 3.4.1 is equivalent to 3.4 in the equation form

$$\text{(3.4.2)} \quad \pi_i(Aut_{DIFF}(B^m \text{ rel } \partial B^m)) = \pi_{i+m+1}(TOP_m/DIFF_m) .$$

† Geometrically realized , this is a Serre fibration [Q] .

‡ One can homotop any map of the contractible spaces $Aut_{TOP}(B^m \text{ rel } \partial) \to$ $\to \Lambda\, CAT(B^m \text{ rel } \partial)$ to be fiber preserving over $CAT(B^m \text{ rel } \partial)$. There results a homotopy equivalence of the fibers $Aut_{CAT}(B^m \text{ rel } \partial) \to \Omega\, CAT(B^m \text{ rel } \partial)$.

‡‡ For interested readers unable to find $[Ce_4]$, Theorem 4 on p. 365 of $[Ce_3]$ closely approaches the case $m = 3$.

§4. CLASSIFICATIONS FOR MANIFOLDS WITH BOUNDARY

An interesting variant of Morlet's theorem ([Mor_2] [Mor_3]) comes from generalizing the classification theorem to manifolds with non-empty boundary M^m , $(m \neq 4 \neq m-1)$.

4.0. For technical convenience we initially single out an open neighborhood W of ∂M in M with an open embedding $\alpha : W \to \partial M \times [0,\infty)$ so that, for all x in ∂M , $\alpha(x) = (x,0)$. Also, we form M_+ by attaching $\partial M \times (-\infty,0]$ to M identifying $\partial M \times 0$ to ∂M . Write δM_+ for $\partial M \times (-\infty,0]$ viewed as a subset of M_+ ; write $W_+ = \delta M_+ \cup W$, and extend α trivially to $\alpha_+ : W_+ \to \partial M \times R$.

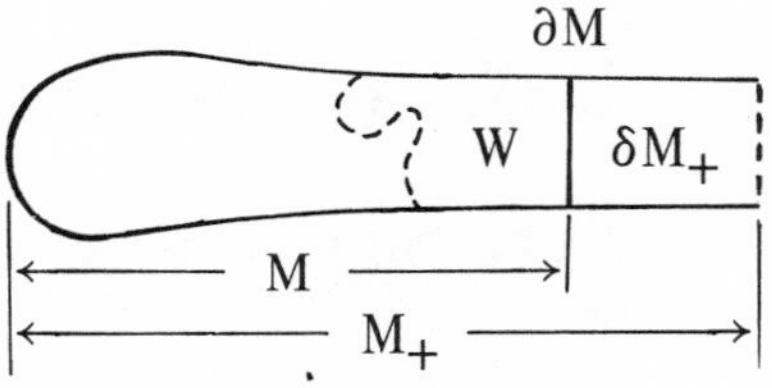

Consider the css complex $\mathrm{CAT}_\alpha(M)$ of which a typical k-simplex is a CAT structure $(\Delta^k \times M)_\Sigma$, with p_1 a CAT submersion to Δ^k , so that Σ is a product near $\Delta^k \times \partial M$ along the factor $[0,\infty)$ (offered by $\Delta^k \times \alpha$ near $\Delta^k \times \partial M$) . Then Σ extends canonically to $\Delta^k \times M_+$ as a product along $(-\infty,\infty)$ near $\Delta^k \times (-\infty,0] \times \partial M$, making $\mathrm{CAT}_\alpha(M)$ a subcomplex of $\mathrm{CAT}(M_+)$.

Observe that we have a fiber product square

$$\begin{array}{ccc} \mathrm{CAT}_\alpha(M) & \xrightarrow{\alpha_*} & \mathrm{CAT}(M_+) \\ \downarrow & & \downarrow \\ \mathrm{CAT}(\partial M) & \xrightarrow{\alpha_*} & \mathrm{CAT}_{M_+}(\delta M_+) \end{array} \tag{4.1}$$

in which α_* indicates maps naturally determined by α and the vertical maps are restrictions. The vertical maps are Kan fibrations (see §1); the horizontal maps are inclusions. As our classification theorem already applies to all but $CAT_\alpha(M)$ we will be able to get a result for $CAT_\alpha(M)$ too. We will prepare the usual equipment for this application with some extra care.

There is a commutative diagram relating stabilization to forgetting

$$\begin{array}{ccccccc} \gamma^{m-1}_{CAT}\oplus\epsilon^1 & \xrightarrow{s} & \gamma^m_{CAT} & \qquad & B_{CAT(m-1)} & \longrightarrow & B_{CAT(m)} \\ j'_{m-1}\oplus\epsilon^1 \downarrow & & j'_m \downarrow & & j_{m-1}\downarrow & & j_m \downarrow \\ \gamma^{m-1}_{TOP}\oplus\epsilon^1 & \longrightarrow & \gamma^m_{TOP} & & B_{TOP(m-1)} & \longrightarrow & B_{TOP(m)} \end{array}$$

in which s is CAT . (It is unique up to microbundle homotopy). We choose a microbundle map $\partial\varphi : \tau(\partial M) \to \gamma^{m-1}_{TOP}$, over say $\partial f : \partial M \to B_{TOP(m-1)}$. Note that $\tau(\partial M\times R) = (\tau(\partial M)\times R) \oplus \epsilon^1$ (canonically) and that α_+ gives a morphism $\tau(\alpha_+) : \tau(W_+) \to \tau(\partial M\times R)$ over α_+ . Thus we can make a special choice of morphism $\varphi_+ : \tau(M_+) \to \gamma^m_{TOP}$ over say $f_+ : M_+ \to B_{TOP(m)}$, as follows. Define φ_+ first near δM_+ as

$$\tau(W_+) \xrightarrow{\tau(\alpha_+)} \tau(\partial M\times R) \xrightarrow{nat} \tau(\partial M)\oplus\epsilon^1 \xrightarrow{\varphi\oplus\epsilon^1} \gamma^{m-1}_{TOP}\oplus\epsilon^1 \to \gamma^m_{TOP}$$

then extend rel δM_+ over all $\tau(M_+)$. Finally, choose a CAT manifold $N_+ \supset M_+$ and a retraction $r_+ : N_+ \to M_+$; for ∂M we insist on taking $N' = r_+^{-1} W_+ \supset \partial M$ with the composite retraction

$$N' \xrightarrow{r_+} W_+ \to \partial M\times R \to \partial M \ .$$

With all these choices made, one gets using 2.3 a well-defined homotopy equivalence, from the diagram 4.1 without $CAT_\alpha(M)$, to the *fiber product* diagram below without L .

(4.2)
$$\begin{array}{ccc} L & \longrightarrow & \mathrm{Lift}(f_+ \text{ to } B_{CAT(m)}) \\ \downarrow & & \downarrow \text{restriction} \\ \mathrm{Lift}(\partial f \text{ to } B_{CAT(m-1)}) & \xrightarrow{\alpha_*} & \mathrm{Lift}(f_+ \text{ to } B_{CAT(m)} \text{ near } \delta M_+) \end{array}$$

Here α_* is defined on a k-simplex $g : \Delta^k\times\partial M \to B_{CAT(m-1)}$ on the left by defining $\alpha_*(g)$ near $\Delta^k\times\delta M_+$ as

$$\Delta^k\times W_+ \xrightarrow{\Delta^k\times\alpha_+} \Delta^k\times\partial M\times R \xrightarrow{proj} \Delta^k\times\partial M \xrightarrow{g} B_{CAT(m-1)} \ .$$

To check that the classifying equivalences θ carry α_* in (4.1) to α_* in (4.2) is straightforward, given $\tau(\partial M \times R) = (\tau(\partial M)\times R)\oplus\epsilon^1$. Admittedly one has to check down through the stages of the construction of θ, (cf. 2.3.1). We then *deduce* an equivalence $CAT_\alpha(M) \simeq L$, since the vertical maps in both squares are Kan fibrations. By collaring theorems (existence for CAT † and uniqueness for TOP), the inclusion map is an equivalence $CAT_\alpha(M) \hookrightarrow CAT(M)$ to the space of all CAT structures on M (without corners if CAT = DIFF). And (4.2) is homotopy equivalent by restriction to the simpler fiber product square of Kan fibrations.

$$\text{(4.3)}\qquad \begin{array}{ccc} \mathrm{Lift}(f,\partial f \text{ to } B_{CAT(m)},B_{CAT(m-1)}) & \longrightarrow & \mathrm{Lift}(f \text{ to } B_{CAT(m)}) \\ \downarrow & & \downarrow \\ \mathrm{Lift}(\partial f \text{ to } B_{CAT(m-1)}) & \longrightarrow & \mathrm{Lift}(f\,|\,\partial M \text{ to } B_{CAT(m)}) . \end{array}$$

In all, we have an equivalence:

$$\text{(4.4)}\qquad CAT(M) \xrightarrow{\simeq} \mathrm{Lift}(f,\partial f \text{ to } B_{CAT(m)},B_{CAT(m-1)}) ,$$

valid ‡ for $\dim M \neq 4 \neq \dim \partial M$. This classification has a version relative to a closed subset $C \subset M$ for just the same reasons for which the absolute theorem 2.3 has the one formulated as (3.2). With notation generalizing that for (3.2) in the obvious way, it gives an equivalence

$$\text{(4.5)}\qquad CAT(M \text{ rel } C\ ;\ \Sigma_0) \xrightarrow{\simeq}$$
$$\xrightarrow{\simeq} \mathrm{Lift}(f,\partial f \text{ to } B_{CAT(m)},B_{CAT(m-1)} \text{ rel } C\ ;\ g_0,\partial g_0)$$

Applying this to $M = [0,\infty)\times B^{m-1}$ with $C = M - \{[0,1)\times B^{m-1}\}$ and with f, ∂f, g_0, ∂g_0 all constant maps we get (for $m \neq 4 \neq m-1$):

$$\text{(4.6)}\qquad \pi_i \mathrm{Aut}_{CAT}(I\times B^{m-1} \text{ rel} \sqsupset) = \pi_{i+m+1}(TOP_m/CAT_m, TOP_{m-1}/CAT_{m-1})$$

It says nothing new for CAT = PL. But, for CAT = DIFF, Cerf

†For CAT = DIFF, integration of vector fields may as well be used. For CAT = PL and for TOP exploit [I, Appendix A].

‡And also valid if (i) m = 4 and M connected non-compact, or (ii) m = 5 and ∂M has no nonempty compact component. The same proof applies, although for CAT = DIFF we would have to use DIFF quasi-spaces.

[Ce_3] has proved, by studying generic 1-parameter families of smooth real functions, that, for $m \geqslant 6$ and $i = 0$, the left side is zero. This adds one dimension to the stability of $TOP_m/DIFF_m$ implied by the s-cobordism theorem (see 5.2 below).

On the other hand, for $i = 1$, I. Volodin [Vol] has asserted that the left hand side is $Wh_3(0) \oplus Z_2$, provided m is large ($m \geqslant$). K. Igusa (thesis, Princeton Univ., 1976) has confirmed that it is non-zero, admitting a homomorphism onto the advertized Z_2. It is known that $K_3(Z) = Z_{48}$, (see [Kar] and joint articles of R. Lee and R.H. Szczarba to appear), and it is thought that $Wh_3(0)$ will be the order 2 quotient by Z_{24}.

§5. THE HOMOTOPY GROUPS $\pi_*(TOP_m/CAT_m)$.

5.0. The calculations we review can be summarized as follows: †

(1) $TOP_m/PL_m \cong K(Z_2,3)$ for $m \geqslant 5$. See 5.5 .

(2) TOP_m/PL_m is contractible for $m \leqslant 3$. See 5.8 , 5.9 .

(3) $\pi_{k+1+m}(TOP_m/DIFF_m) = \pi_k Aut_{DIFF}(B^m \text{ rel } \partial B^m)$, $m \neq 4$. See 3.4.

$\pi_{k+1+m}(TOP_m/DIFF_m, TOP_{m-1}/DIFF_{m-1}) = \pi_k Aut_{DIFF}(B^{m-1} \times I \text{ rel } \sqsupset)$ $m \neq 4 \neq m-1$. See 4.6 .

(4) $\pi_k(TOP/DIFF, TOP_m/DIFF_m) = 0$, $k \leqslant m+2$, $m \geqslant 5$, where $TOP/DIFF = \bigcup_m TOP_m/DIFF_m$. See 4.6 , 5.2 .

(5) $\pi_m(TOP/DIFF) = \Theta_m$, $m \geqslant 5$,

$= \pi_m K(Z_2,3)$, $m \leqslant 6$. See 5.5 .

(6) $\pi_k(TOP_3/DIFF_3) = 0$, $k \leqslant 4$. See 5.8 .

(7) $TOP_m/DIFF_m$ is contractible $m \leqslant 2$. See 5.9 .

5.1. Recall that TOP_m/CAT_m is by definition the standard homotopy fiber of the forgetful map $j_m : B_{CAT(m)} \to B_{TOP(m)}$ (see 2.0) over a chosen base point. One can arrange that all the squares

$$\begin{array}{ccc} B_{CAT(m)} & \xrightarrow{s} & B_{CAT(m+1)} \\ \downarrow & & \downarrow \\ B_{TOP(m)} & \xrightarrow{s} & B_{TOP(m+1)} \end{array}$$

relating stabilizations s to forgetting maps (verticals) are commutative and present each s as an inclusion. One need only enlarge the spaces by mapping cylinder devices and suitably choose the maps. This done, define

$$B_{CAT} = \bigcup_m B_{CAT(m)} \quad \text{and} \quad B_{TOP} = \bigcup_m B_{TOP(m)} \quad .$$

† Some further information can be found in [BuL] .

One notes that $s : TOP_m/CAT_m \to TOP_{m+1}/CAT_{m+1}$ is then also an inclusion and that, in analogy with TOP_m/CAT_m, the space $TOP/CAT = \bigcup_m (TOP_m/CAT_m)$ is precisely the fiber of the standard Serre path fibration $\bar{B}_{CAT} \to B_{TOP}$.

FIRST STABILITY THEOREM 5.2 . *(CAT = DIFF or PL ; $m \neq 4, 5$).*

For $0 \leqslant k \leqslant m$, $\pi_{k+1}(TOP_m/CAT_m, TOP_{m-1}/CAT_{m-1}) = 0$. Hence $\pi_i(TOP/CAT, TOP_m/CAT_m) = 0$ for $i \leqslant m+1$ and $m \geqslant 5$.

Proof of 5.2. By (4.5), this group is in bijective correspondence with $\pi_0 CAT(I \times B^k \times R^n \text{ rel } \sqsupset)$ where $1 + k + n = m$. But the latter is zero by the Concordance Implies Isotopy Theorem (already in its handle version [I, §3]).

Our present heavy machinery could eliminate parts of the proof of [I, §3], cf. proof of 5.3 below. But both the CAT s-cobordism theorem and some topological geometry (involving infinite processes or coverings, and the Alexander isotopy) do seem irreplaceable ingredients. The cases $m \leqslant 3$ will be covered again by the next result. ■

THEOREM 5.3 *(CAT = DIFF or PL)* [KS_1] .

Suppose $4 \neq m \leqslant 6$, and $k \leqslant m$. Then $\pi_k(TOP_m/CAT_m) = 0$ unless $3 = k < m$, in which case we have $\pi_3(TOP_m/CAT_m) = Z_2$.

Proof of 5.3 . ⟦ with arguments for $m = 3$ marked off ⟧ .

For suitable compact CAT manifolds $M^m = I^k \times T^n$, $m = k+n$, with boundary ∂M we will compare $\mathcal{S}(M \text{ rel } \partial) = \pi_0(CAT(M^m \text{ rel } \partial M))$ with the set $\mathcal{S}^*(M \text{ rel } \partial)$ of homotopy smoothings (or triangulations) of M rel boundary. An element of $\mathcal{S}^*(M \text{ rel } \partial)$ is represented by a homotopy equivalence $f_1 : (M_1, \partial M_1) \to (M, \partial M)$ from a compact CAT m-manifold M_1, such that $f_1 \mid \partial M_1$ is a CAT isomorphism $\partial M_1 \to \partial M$ and f_1 is a CAT embedding near ∂M_1. Another such equivalence $f_2 : (M_2, \partial M_2) \to (M, \partial M)$ represents the same element of $\mathcal{S}^*(M, \partial M)$ if and only if there is a CAT isomorphism $h : M_1 \to M_2$ so that $f_2 h$ is homotopic to $f_1 : (M_1, \partial M_1) \to (M, \partial M)$ fixing a neighborhood of the boundary. ⟦ In case $m = 3$, we supplement this definition by supposing that M_1 is Poincaré *i.e.,* contains *no fake* 3-disc (=compact CAT contractible 3-manifold not isomorphic to B^3). ⟧

The natural map $\mathcal{S}(M,\partial) \to \mathcal{S}^*(M,\partial)$ sends $[\Sigma]$ to the element represented by $id : (M_\Sigma, \partial M_\Sigma) \to (M, \partial M)$. ⟦ For $m = 3$, to see that this a valid definition one must know that $M_\Sigma = (I^k \times T^n)_\Sigma$ contains no fake CAT 3-disc. A proof is given by [Mac] [HeM] † in fact for the universal covering. We prefer not to appeal to Moise's proof of the Hauptvermutung in dimension 3 [Mo] . ⟧

Represent a given element x of $\pi_k(TOP_m/CAT_m)$ by a map $I^k \to TOP_m/CAT_m)$ sending ∂I^k to the base point (image of a classifying map for the standard structure). Then extend to $I^k \times T^{m-k}$ by projection to I^k . To this map there corresponds by 3.2 a well defined element $[\Sigma] \in \mathcal{S}(I^k \times T^{m-k}$ rel $\partial)$ represented by a structure Σ .

If $x \neq 0$ *, then the image* $[\Sigma]^*$ *of* $[\Sigma]$ *in* $\mathcal{S}^*(I^k \times T^{m-k}$ rel $\partial)$ *under the forgetful map* $\mathcal{S} \to \mathcal{S}^*$ *cannot be zero.* To see this, suppose it were, that is, suppose one had a CAT isomorphism $h : I^k \times T^n \to (I^k \times T^n)_\Sigma$, $n = m-k$, fixing a neighborhood of ∂ and homotopic to the identity rel ∂ . Then some standard finite λ^n-fold covering h_λ , $\lambda \geqslant 1$, of h will be seen to be topologically isotopic to the identity rel ∂ . Here h_λ fixes ∂ and makes commutative the square

$$\begin{array}{ccc} I^k \times T^n & \xrightarrow{h_\lambda} & (I^k \times T^n)_{\Sigma_\lambda} \\ {\scriptstyle id \times (\cdot\lambda)} \downarrow & & \downarrow {\scriptstyle id \times (\cdot\lambda)} \\ I^k \times T^n & \xrightarrow{h} & (I^k \times T^n)_\Sigma \end{array} ,$$

where $(\cdot\lambda) : T^n \to T^n$ indicates multiplication by λ and Σ_λ is defined to make the right hand map a CAT covering map (λ^n -fold) . For λ large one checks that the component of h_λ on T^n can approach that of the identity. Further, up to a standard isotopy, h_λ has component on I^k arbitrarily near the identity. This standard 'Alexander' type isotopy extends h_λ by the identity over all of $R^k \times T^n$ then conjugates h_λ by an isotopy that shrinks I^k in R^k radially to towards a point. From the local contractibility theorem [EK] for the

† An alternative proof of 5.3 for $m = 3$ (suggested in [Si_8 , §5]) neatly avoids this technicality by solving handle problems in coordinate charts where the Alexander 'Shoenflies theorem' [Al_2][Ce_1] precludes fake 3-discs. In following the technique of [KS_1] , one should use the Novikov imbedding $T^{n-1} \times R \hookrightarrow R^n$ in place of the Kirby immersion $(T^n\text{-point}) \to R^n$. This approach gives our favorite proof of the Hauptvermutung and the triangulation conjecture in dimension 3 . See also the recent article of A.J.S. Hamilton [Hal] , who has succeeded even with the Kirby immersion.

group of homeomorphisms of $I^k \times T^n$ fixed on the boundary ∂, we conclude that h_λ is isotopic to the identity rel ∂ for λ large, which shows $[\Sigma_\lambda] = 0$ in $\mathcal{S}(I^k \times T^n \text{ rel } \partial)$. Since a classifying map for Σ_λ is the covering $\text{id} \times (\cdot\lambda)$ followed by that for Σ, we can conclude that $x = 0$ by restricting the nul-homotopy of the classifying map of Σ_λ to $I^k \times (\text{point}) \subset I^k \times T^n$.

We have seen that $[\Sigma]^*$ lies in the subgroup

$$\mathcal{S}_o^*(I^k \times T^{m-k} \text{ rel } \partial)$$

of $\mathcal{S}^*$ consisting of those elements that are unchanged by passage to standard λ^{m-k}-fold coverings, $\lambda \geq 1$; i.e., those $[f] \in \mathcal{S}^*$, $f : M \to I^k \times T^{m-k}$, such that $[f] = [\tilde{f}]$ whenever one forms a fiber product square

$$\begin{array}{ccc} \widetilde{M} & \xrightarrow{\tilde{f}} & I^k \times T^{m-k} \\ \downarrow & & \downarrow \text{id} \times (\cdot\lambda) \\ M & \xrightarrow{f} & I^k \times T^{m-k} \end{array}$$

Thus the next step is to exploit the important surgical calculation of and $\mathcal{S}_o^*$ from [HS] and [Wa].

Theorem 5.4 . *(CAT = DIFF or PL)*

For $m \leq 3$, one has $\mathcal{S}^*(I^k \times T^{m-k} \text{ rel } \partial) = 0$.

For $m = 5$ or 6 (or $m \geq 5$ and CAT = PL), one has a bijection from $\mathcal{S}^*(I^k \times T^{m-k} \text{ rel } \partial)$ to $H^3(I^k \times T^{m-k}, \partial; Z_2) = H^{3-k}(T^{m-k}; Z_2)$. Under this bijection the self-maps of $\mathcal{S}^*$ and H^3 respectively, induced by the standard λ^{m-k}-fold covering, correspond to one another.

Since this λ^{m-k}-fold covering in fact induces multiplication by the integer λ^{3-k} in the cohomology group we have the

Corollary 5.4.1 . *For $4 \neq m \leq 6$, the set $\mathcal{S}_o^*(I^k \times T^{m-k} \text{ rel } \partial)$ of homotopy smoothings or triangulations invariant under passage to standard λ^{m-k}-fold coverings is zero unless $3 = k < m$, in which case $\mathcal{S}_o^*(I^k \times T^{m-k} \text{ rel } \partial) = Z_2$.* ∎

Comments on the proof of 5.4. (see also Appendices B and C).

The proof for $m > 4$ is given in [HS] and [Wa] (cf. [Si_8]) for

CAT = PL, and the DIFF proof is identical since $\pi_i(G/PL) = \pi_i(G/O)$ for $i \leqslant 6$. It requires the full strength of non-simply-connected surgery [Wa] and the periodicity in surgery related to producting with CP_2. The source of the Z_2 is Rohlin's theorem that closed smooth almost parallelizable 4-manifolds have signature divisible by 16 (not just 8). The reader should be reminded that the PL proof relies on the PL Poincaré theorem stating that $\mathcal{S}^*(S^5) = 0$. There is still no proof of the latter without Cerf's difficult proof that $\Gamma_4 = 0$ [Ce_1].

⟦ The proof of 5.4 for m = 3 is essentially due to Stallings [St_1]. His setting is PL, but there would be no obstacle to giving essentially the same proof for the DIFF category. To convert Stallings' construction of fiberings over the circle into the result 5.4 one uses the fact that any CAT automorphism of $I^k \times T^{2-k}$, $0 \leqslant k \leqslant 2$, rel boundary is CAT isotopic rel boundary to the identity. This easily is proved if CAT = PL [Sco], and the DIFF result follows using example 5.9 below. ⟧ ■

Returning to the proof of 5.3 recall that we have established an injection

$$(\#) \qquad \pi_k(TOP_m/CAT_m) \to \mathcal{S}_o^*(I^k \times T^{m-k} \text{ rel } \partial) \quad .$$

In view of the vanishing of $\mathcal{S}_o^*$ assured by 5.4.1 we know now that $\pi_k(TOP_m/CAT_m) = 0$ when $4 \neq m \leqslant 6$ and either $k \neq 3$ or $m \leqslant 3$.

In case k = 3 and m = 5 or 6, the target of the injection (#) is not zero but rather $Z_2 = \mathcal{S}^*(I^3 \times T^{m-k} \text{ rel } \partial)$. The source is naturally isomorphic to $\mathcal{S}(I^3 \times T^{m-k} \text{ rel } \partial)$; indeed, the relative classification theorem 3.2 yields $\mathcal{S}(I^3 \times T^{m-3} \text{ rel } \partial) = [I^3 \times T^{m-3}/\partial\, ; TOP_m/CAT_m] =$ $= H^3(I^3 \times T^{m-3}, \partial; \pi_3(TOP_m/CAT_m)) = H^0(T^{m-3}; \pi_3(TOP_m/CAT_m)) =$ $= \pi_3(TOP_m/CAT_m)$.

For m = 6, we now show that (#) is surjective. Thus $\pi_3(TOP_m/CAT_m) = Z_2$ for m = 6, and the same follows for all $m \geqslant 5$ by the stability 5.2.

Represent the nontrivial element of $\mathcal{S}^*(I^3 \times T^3 \text{ rel } \partial)$ by $f : (M, \partial M) \xrightarrow{\simeq} (I^3 \times T^3, \partial)$. Our aim is to show that f can be a homeomorphism. We can apply the CAT s-cobordism theorem to the CAT relative h-cobordism

$$c = (M\, ; f^{-1}(0 \times I^2 \times T^3), f^{-1}(1 \times I^2 \times T^3)) \quad ,$$

recalling that $Wh(\pi_1 T^n) = 0$. Thus we obtain a CAT isomorphism

$f' : M \to I^3 \times T^3$ equal f near $f^{-1}(\sqsupset) = f^{-1}((1\times I^2 \cup I\times \partial I^2)\times T^3)$. This provides a CAT automorphism $0\times g = f'f^{-1} : 0\times I^2 \times T^3 \to 0\times I^2 \times T^3$ equal the identity near $0\times \partial I^2 \times T^3$.

If g is topologically isotopic rel ∂ to the identity we can alter f' near $f^{-1}(0\times I^2 \times T^3)$ to obtain a homeomorphism $h : M \to I^3 \times T^3$ equal f near ∂M. Then h is necessarily homotopic to f rel ∂, the obstructions being in $H^*(I^3\times T^3, \partial; \pi_*(I^3 \times T^3)) = 0$. This would then complete the proof.

To establish this, find a topological isotopy rel ∂ to the identity from a standard λ^3-fold covering g_λ of g (λ a large integer), by precisely the argument used in establishing the injectivity of (#). Now it is clear that $g_\lambda = f'_\lambda f_\lambda^{-1} \mid 0\times I^2 \times T^3$, where $f'_\lambda, f_\lambda : M_\lambda \to I^3 \times T^3$ are the corresponding standard λ^3-fold coverings of f, f' respectively. Thus we now know that f'_λ can be replaced by a homeomorphism $h_\lambda : M_\lambda \to I^3 \times T^3$ homotopic rel ∂ to f_λ. Fortunately any standard covering $f_\lambda : M_\lambda \to I^3 \times T^3$ of f still represents the unique nontrivial element of $\mathcal{S}^* = \mathcal{S}^*_0$. Thus we have a CAT identification $M_\lambda = M$ so that h_λ is homotopic rel ∂ to f. The surjectivity of (#) is now established and with it Theorem 5.3. ■

Remarks:

(a) We have presented the basic elements of the above argument before [KS_1] [Ki_2] [Si_8] [Si_{10}]. The reader wishing to see variants or more details should consult these references.

(b) It is worth noting that, that if we accept the classification theorem of Essay IV rather the results of this essay, then the above argument does prove that $\pi_i(TOP/CAT) = \pi_i(K(Z_2,3))$ for $i \leqslant 6$. A similar remark applies to the next theorem.

We are now in a position to draw the threads together.

THEOREM 5.5 .

I) *For* $m \geqslant 5$, $TOP_m/PL_m \cong K(Z_2,3) \cong TOP/PL$.

II) *TOP/DIFF* $\equiv$ *TOP/O* † *has homotopy groups* $\pi_i \cong \pi_i K(Z_2,3)$ *for* $i \leqslant 6$ *and* $\pi_i = \Theta_i$ *for* $i \geqslant 5$. *Here* Θ_i *is the Kervaire-Milnor group [KeM] of oriented isomorphism classes DIFF homotopy spheres spheres under connected sum; and* $K(Z_2,3)$ *is, of course, the Eilenberg-Maclane space with* $\pi_3(K(Z_2,3) = Z_2$.

† See 2.0 .

Proof of II) : In view of 5.2, 5.3, and [Ki_1] † it remains only to prove $\pi_m(TOP_m/O_m) = \Theta_m$, $m \geq 5$. Borrowing notation from 5.3 we have a forgetful map $\mathcal{S}^*(S^m \text{ rel } B^m_-) \to \Theta_m$, where B^m_+ , B^m_- are the northern and southern hemispheres of S^m . It is bijective by the DIFF isotopy uniqueness of oriented imbeddings of B^m into S^m . By §3 , $\pi_m(TOP_m/O_m) = \mathcal{S}(S^m \text{ rel } B^m_-)$. But $\mathcal{S}(S^m \text{ rel } B^m_-) \xrightarrow{\cong} \mathcal{S}^*(S^m \text{ rel } B^m_-)$; indeed surjectivity results from the Poincaré theorem in dimension m (cf. [Mi_8]) and injectivity results from the Alexander isotopy. Thus $\pi_m(TOP_m/O_m) = \Theta_m$. (We let the reader verify that the bijection established is a group isomorphism.) ∎

Proof of I) : This result follows from 3.3 , 5.2 and the argument for II) above, in view of the PL Poincaré theorem that $\mathcal{S}^*(S^m) = 0$ for $m \geq 6$ [Sm_2] [Hu_2] . This completes the proof of 5.5 . ∎

5.6 . For a given topological manifold M^m , $m \geq 6$ (or $m \geq 5$ if $\partial M = \phi$) , we can now observe that the one obstruction to introducing a PL structure on M is a cohomology class $k(M) \in H^4(M;Z_2)$, namely the one obstruction to lifting the classifying map $f : M \to B_{TOP(m)} \subset B_{TOP}$ for $\tau(M)$ to $B_{PL(m)}$, or equivalently to B_{PL} . The obstruction to sectioning the fibration

$$(**) \qquad K(Z_2,3) \to B_{PL} \xrightarrow{j} B_{TOP}$$

is the *universal* triangulation obstruction $k \in H^4 B_{TOP}, Z_2)$ (also denoted Δ), and $k(M) = f^*k$ by naturality of obstructions. We represent k by a map $k : B_{TOP} \to K(Z_2,4)$ and observe that there is a commutative square unique up to homotopy:

$$\begin{array}{ccc} \bar{B}_{PL} & \xrightarrow{k'} & \Lambda K(Z_2,4) = (K(Z_2,4), \text{point})^{(I,0)} \\ j \downarrow & & \downarrow p \\ B_{TOP} & \xrightarrow{k} & K(Z_2,4) \end{array} .$$

(One can specify k' using any contraction of $kj|B_{PL}$ to the base point.) One observes easily that k' gives a homotopy equivalence on fibers:

$$TOP/PL \longrightarrow K(Z_2,3) = \Omega K(Z_2,4) \quad ,$$

and so a homotopy equivalence Lift(f to B_{PL}) → Lift(kf to $\Lambda K(Z_2,4)$). When we single out a fixed lift $g : M \to \Lambda K(Z_2,4)$ we deduce by

† Beware that 5.2 *failed* to prove: $\pi_0(TOP/CAT , TOP_m/CAT_m) = 0$, for $m \geq 5$. Cf. [III, Appendix A] .

matching common end points of paths, an equivalence
Lift(kf to $\Lambda K(Z_2,4)) \to K(Z_2,3)^M$ to the space † of maps of M to $K(Z_2,3) = \Omega K(Z_2 4)$ for which g also offers a natural inverse equivalence. We can conclude that when M^m is a PL manifold, with $m \geqslant 6$ (or $m \geqslant 5$ and $\partial M = \phi$), one has

(5.7) $$CAT(M) \;\cong\; K(Z_2,3)^M$$

Of course there is a relative version. It would be more usual to obtain this result by observing that a section of the pullback ξ of (**) by $f : M \to B_{TOP}$ amounts to a homotopy trivialization of ξ .

For more information concerning the triangulation obstruction, the reader should consult [Si_{10}] [HoM] [Mor] [BruM] .

To tie up the loose ends let us review the situation in low dimensions.

Nothing is known about the structure of TOP_4/CAT_4 , although something is known of its significance [LS] [CLS] .

For dimension 3 one has

THEOREM 5.8 . (J. Cerf)

I) *TOP_3/PL_3 is contractible.*

II) $\pi_i(TOP_3/O_3) = 0$ *for* $i \leqslant 4$.

Proof: $\pi_i(TOP_3/CAT_3) = 0$ for $i \leqslant 3$ by 5.3 . And by Morlet's theorem $\Omega^4(TOP_3/CAT_3) \cong Aut_{CAT}(B^3 \text{ rel } \partial B^3)$ which is contractible for CAT = PL (Alexander), and connected by Cerf [Ce_1] for CAT = DIFF . ■

Conjecture (Smale 1958) . $Aut_{DIFF}(B^3 \text{ rel } \partial B^3)$ *is contractible* .

This has often been announced (at least once in print [Ak] [Ro_1]); no correct proof has appeared.

The picture is complete for $m \leqslant 2$.

THEOREM 5.9 . TOP_m/CAT_m *is contractible for* $m \leqslant 2$.

Proof .

For CAT = PL , the proof of 5.8 applies.

† Recall that this space is naturally homotopy equivalent to its singular complex [Mi_2] .

For CAT = DIFF , observe that TOP_m/CAT_m contractible means that $j : B_{CAT(m)} \to B_{TOP(m)})$ is a homotopy equivalence. In view of the Kister-Mazur theorem 'microbundles are bundles' the latter is equivalent to *the inclusion* $O(m) \hookrightarrow \mathrm{Homeo}(R^m,0)$ *being a homotopy homotopy equivalence.* This last condition is what we establish.

m = 1 : The orientation preserving self-homeomorphisms of $(R^1,0)$ form a convex hence contractible subset of the continuous maps $R^1 \to R^1$.

m = 2 : Already in 1926 [Kn] H. Kneser published a pleasant proof based on conformal mapping theory. In 1958 [Sm_1] Smale proved that that $\mathrm{Aut}_{DIFF}(B^2 \text{ rel } \partial B^2)$ is contractible, cf. [Ce_1] ; this accomplishes an indirect second proof, via the argument presented for 5.8 . ∎

We wish to advertise an elementary proof that is a variant of the reasoning in [Ki_3] † . Bjorn Friberg gives a full discussion in [Fr] .

We denote by TOP_2^+ the space of orientation preserving homeomorphisms of the Gauss plane $C \approx R^2$ fixing the point 1 (not 0). Its topology is, of course, the compact-open topology. There is a locally trivial fibration

$$H^+ = p^{-1}(0) \hookrightarrow TOP_2^+ \xrightarrow{p} C - \{1\}$$

where p selects the image of $0 \in C$. Pulling back this fibration over the universal cover of $C - \{1\}$, one sees that $O(2) \hookrightarrow \mathrm{Homeo}(R^2,0)$ is a homotopy equivalence if and only if the inclusion $H^+ = p^{-1}(0) \hookrightarrow TOP_2^+$ is deformable to the constant map.

To establish this deformability, consider a typical $h \in H^+$. There is an isotopy $h \sim h_1$ fixing 0 and 1 and depending continuously on h to a homeomorphism h_1 such that $\log|h_1(z)| - \log|z| < \frac{1}{10}$. This involves shuffling circles with center 0 with their images under h .

Then there is a wrapping and unwrapping device (see [EK, Appendix] and [Si_{12}, §4.9]) , yielding $h_2 \in H^+$ such that $h_2(ez) = eh_2(z)$, where $e = \exp(1)$, and $h_2 = h_1$ on a neighborhood of 1 independent of h . The rule $h_1 \mapsto h_2$ is again continuous.

Lifting $h_2|C^*$ in the universal covering $\exp : C \to C-\{0\} = C^*$, there results a unique covering h_3 of $h_2|C^*$ with $\exp \circ h_3 = h_2 \circ \exp$ and $h_3(0) = 0$. This h_3 commutes with translation by the points

† Reasoning that is unfortunately fallacious in dimensions > 2 , but see [KS_5] .

$m + 2\pi i n$ in C with $m,n \in Z$. Thus $|h_3(z) - z|$ is bounded for $z \in C$, and so the formulae $h_{3,0} = id|C$, $h_{3,t}(z) = th_3(z/t)$, $0 < t \leqslant 1$, give an isotopy $id|C \sim h_3$.

Find a homeomorphism $\varphi : C \to C$ equal to exp near 0. Set $h_4 = \varphi h_3 \varphi^{-1}$, and observe that $\exp \circ h_3 = h_2 \circ \exp$ and $h_3(1) = 1$ imply that $h_4 = h_2 = h_1$ near 1. As $h_4 h_1^{-1} = id|C$ near 1, there is an Alexander isotopy $h_4 h_1^{-1} \sim id|C$ fixing 1. Also note that $id \sim h_3$ yields $id \sim h_4$ by conjugation.

One now has isotopies $h \sim h_1 \sim h_4 \sim id|C$ in TOP_2^+ depending continuously on h. ■

APPENDIX A. THE IMMERSION THEORETIC METHOD WITHOUT HANDLE DECOMPOSITIONS

This appendix was written to complete the proof of the 'foliated' classification theorem of [Si_{10}, §4]† for pseudo-group structures on a connected open topological manifold. It is needed in case that manifold admits no handle-decompositions; such manifolds unfortunately exist in dimension 4 and/or 5. The chart-by-chart procedure given here may appear to some to be a pleasantly uniform way to prove, without handlebodies Morse functions or global triangulations, those many other classification theorems which rely on the "micro-gibki" property (= homotopy micro-lifting property) singled out by Gromov [Gr]. One example is the sliced structure classification theorem for open 4-manifolds given in §1.4.1 (and §1.6) of this essay.

This mild extension of the immersion theoretic procedure is made possible by the following sequence of lemmas each of which is an exercise in elementary topology. After these lemmas, we shall make precise the extension we have in mind; however the reader motivated by any of the above applications should by himself be able to put these lemmas to work.

(a) Collapsing lemma. *Let (B,A) be a pair of finite simplicial complexes of dimension $\leqslant m$, such that $H^m(B,A;Z_2)=0$. Then B collapses simplicially to A union a subcomplex of the $(m-1)$-skeleton $B^{(m-1)}$ of B.* Hint: $H_m(B,A;Z_2)=0$, and this implies that, if $\dim(B-A)=m$, some m-simplex S of $B-A$ has a $(m-1)$-face T not in A nor in another m-simplex of B; then deletion of $\mathrm{int}T \cup \mathrm{int}S$ is a collapse, preserving the homological condition. ■

(b) Finiteness lemma. *If A is a compactum in a locally compact, locally connected and connected Hausdorff space M, then each open neighborhood U of A contains all but finitely many of the connected components of $M-A$.* Hint: We can assume that closure $\bar{U}$ in M is compact; then the compact frontier $\delta\bar{U}$ can meet only finitely many of the (disjoint and open!) components of $M-A$. ■

† Reprinted in this volume.

Definition : Given M as above and a closed subset A , any connected component of M–A that is bounded (i.e. has compact closure in M) is called a **hole** in A . •

(c) Drainage lemma . *In the setting of (b) if* $B \supset A$ *and* $V \subset M$ *is an open set , each connected component of which meets* $M-B$ *, then the holes in* $(A-V)$ *filled by* B *(i.e. contained in* B*) are precisely those holes of* A *filled by* B *which fail to meet* V. ■

(d) Detection lemma . *Let* (B, A) *be a closed pair in a manifold* M^m *without boundary . Then* B *fills no hole in* A *if and only if* $\check{H}^m(B, A; Z_2) = 0$ *(for Čech cohomology)*. Hint : By a simple injective limit argument using taughtness (continuity) of Čech groups [Sp , p.316 , p.334] , it suffices to prove the case $B = M$. Now use the (sadly neglected) Poincaré duality isomorphism $\check{H}^m(M, A; Z_2) \cong$ $\cong H_0^{\ell f}(M-A; Z_2)$ where $H_*^{\ell f}$ denotes homology based on singular chains **locally finite in M** but possibly infinite . ■

(e) Intersection lemma . *Let* (B, A) *be a compact pair in a chart of the topological m-manifold without boundary* M^m *such that* $\check{H}^m(B, A; Z_2) = 0$. *There exists a nested sequence* $(B_1, A_1) \supset$ $\supset (B_2, A_2) \supset \cdots$ *of finite simplicial pairs satisfying* $H^m(B_i, A_i; Z_2) =$ $= 0$, *and* $(B, A) = \cap\{(B_i, A_i) \mid i=1,2,\ldots\}$.
Hint : Given (B′, A′) , a simplicial pair containing (B, A) such that B′ does fill (i.e. contain) a hole H in A′ , we have available one of two remedies : if B does not fill H delete from B′ the interior of a PL disc in H disjoint from B ; if B does fill H then using the drainage lemma delete from A′ the interior V of a connected closed PL neighborhood of a PL path running from H to the complement of B and chosen so that $V \cap A = \phi$. ■

(f) Elementary engulfing lemma . *Let* Δ *be a simplex linearly embedded in* R^m *and let* Λ *be* $\partial\Delta$ *minus a (m–1)-face . Let* V *be an open set containing* Λ *and let* W *be a neighborhood of* $\Delta - V$. *There exists an isotopy* h_t, $0 \leqslant t \leqslant 1$, *of* $h_o = id \mid R^m$ *with compact support in* W *so that* $h(V) \supset \Delta$. ■

(g) Induction lemma . *Let* (B, A) *be a compact pair in a topological m-manifold without boundary* M^m , *so that* $\check{H}^m(B, A; Z_2) = 0$. *If the closure* $Cl(B-A)$ *is covered by* k *co-ordinate charts , there exists a closed set* B' , $A \subset B' \subset B$, *so that*

$\check{H}^m(B, B'; Z_2) = \check{H}^m(B', A; Z_2) = 0$ and each of $Cl(B-B')$ and $Cl(B'-A)$ is contained in $< k$ co-ordinate charts.

Proof: Let U be one of the k charts covering Cl(B–A), and let W be the union of the rest. Choose any compactum B_o, $A \subset B_o \subset B$, so that $Cl(B_o-A) \subset W$ and $Cl(B-B_o) \subset U$. To B_o add every hole in B_o that is filled by B and lies in W. In view of the finiteness lemma, the result, still denoted B_o, is compact, satisfies the same conditions, and has only finitely many holes $H_1, \ldots, H_s$ filled by B.

We shall get rid of these holes using the drainage lemma. As B contains no component of M–A, there exists, for each i, a path $\gamma_i : I \to (M-A)$ from $\gamma_i(0) \in H_i$ to $\gamma_i(1) \in (M-B)$. Applying a chart by chart engulfing process to these paths regarded as a homotopy of the set $\{\gamma_i(1) \mid i=1, \ldots, s\}$ into U we can find new paths in M–A, say γ_i', $i=1, \ldots, s$, joining the same points, and an isotopy of $id \mid M$ with compact support in $M-Cl(B-B_o)$ to a homeomorphism h of M such that

$$h(U) \supset \cup_i \{\gamma_i'(I) \mid i=1,2, \ldots, s\} \cup Cl(B-B_o).$$

This h(U) is a new chart, constructed using (f) for 1-simplices. Let V_i be a connected open neighborhood of the continuum $\gamma_i'(I)$ so that $Cl(V_i)$ is a compactum lying in h(U)–A. We define B′ to be $B_o - \cup\{V_i \mid i=1, \ldots, s\}$. Now $Cl(B'-A) \subset W$ a fortiori; and $Cl(B-B') \subset$ $\subset hU$ because $Cl(B-B_o) \cup Cl(V_i) \subset h(U)$. ∎

(g) Filtration lemma. *Let (B, A) closed pair in a m-manifold M^m without boundary such that $\check{H}^m(B, A; Z_2) = 0$. Then B admits a filtration*† *$A = A_1 \subset A_2 \subset A_3 \subset \ldots$ by closed subsets such that, for all i, $A_{i+1} - A_i$ is bounded in M (= relatively compact) while $\check{H}^m(B, A_i; Z_2) = 0$, and (hence) $\check{H}^m(A_{i+1}, A_i; Z_2) = 0$.* Hint: If A′ is any closed set, $A \subset A' \subset B$ with A′–A bounded, the finiteness lemma lets one simply add to A′ every hole in A′ filled by B to obtain a closed set A″ with A″–A still bounded, which, by the detection lemma, satisfies $\check{H}^m(B, A''; Z_2) = 0$. ∎

† It is understood that $\cup \mathring{A}_i = B$ where $\mathring{A}_i$ is the interior of A_i in B (not in M).

Consider a contravariant functor Φ from the category of inclusions of open subsets of a topological space M to the category of quasi-spaces†, which carries to projective limit of sets any injective limit expressed as a union of open sets‡. This Φ is nothing but a **sheaf of quasi-spaces on** M. We extend Φ to arbitrary subsets $A \subset M$ by setting $\Phi(A) = \varinjlim \{\Phi(U)\}$ where U varies over open neighborhoods of A in M.

In the examples we have in mind ([Si_{10}, §4] and this essay §1.4.1), M is a manifold and Φ associates to each open set $U \subset M$ a suitably defined space of structures on U. The sheaf condition merely expresses the fact that structures piece together uniquely. In these examples there is a second such functor Φ' present and a transformation $d : \Phi \to \Phi'$. It happens to be obvious that

† Intuitively a quasi-space is a sort of 'space' of which we want to know only the sets of maps to it of certain specified pleasant spaces. The definition is always set up so that projective and injective limits of quasi-spaces exist because they exist in the category of sets. And one wants to allow enough pleasant spaces and maps between them in order to do some homotopy theory for quasi-spaces. Classically a quasi-space Y is defined as a contravariant function from the category C of Hausdorff compacta and continuous maps to the category of sets, which takes union squares (pushout squares) to fiber product squares. Thus if we write $Y : X \mapsto [X,Y]$ for X compact, and $X = X_1 \cup X_2$ is a union of compacta, then the square

$$\begin{array}{ccc} [X,Y] & \to & [X_1,Y] \\ \downarrow & & \downarrow \\ [X_2,Y] & \to & [X_1 \cap X_2,Y] \end{array}$$

of inclusion-induced 'restriction' maps is a fiber product square of sets; in other words, for each pair $x_i \in [X_i,Y]$, $i = 1,2$ mapping to the same element in $[X_1 \cap X_2,Y]$, there is a unique element $x \in [X,Y]$ mapping to both x_1 and x_2. We shall write $x \mid X_1 = x_1$, $x \mid X_2 = x_2$. This definition suffices for [Si_{10}, §4]. However, for §1.4.1 of this essay we replace C by the category of piecewise-linear maps of compact polyhedra, and in §1.6(A) we need still another definition. If we replace C by the category of ordered simplicial complexes and order preserving simplicial maps, then the corresponding quasi-spaces are nothing more nor less than css sets (as used in this essay).

‡ This amounts to saying that if U_α, $\alpha \in J$, is a collection of open sets then for any allowable X (see previous note) and any selection of elements $x_\alpha \in [X, \Phi(U_\alpha)]$, $\alpha \in J$ such that for each pair $\alpha, \beta \in J$ one has $x_\alpha|(U_\alpha \cap U_\beta) = x_\beta|(U_\alpha \cap U_\beta)$ in $[X, \Phi(U_\alpha \cap U_\beta)]$, there exists exactly one element x in $[X, \Phi(U_\alpha \cap U_\beta)]$ such that $x|U_\alpha = x_\alpha$ for each $\alpha \in J$.

$d_A : \Phi(A) \to \Phi'(A)$ is a weak homotopy equivalence† if A is a point, and the task of the immersion theoretic method is to prove if possible that $d_M : \Phi(M) \to \Phi'(M)$ is also one.

For an inclusion $i : A \to B$ of subsets of M consider the 'restriction' map $\Phi(i) : \Phi(B) \to \Phi(A)$. If $\Phi(i)$ has the [weak]‡ covering homotopy property for maps of a compact object, we say that the pair (B,A) is **[weakly] Φ-flexible** or that Φ is [weakly] flexible for (B,A). Again, if $\Phi(i)$ has the covering micro-homotopy property in the sense that some initial interval of any homotopy of a compact object can be lifted, we say (B, A) is **Φ-microflexible** (**microgibki** in Russian).

Recall that in the examples we have in mind, the only immediately provable property was Φ-microflexibility for compact pairs.

A cardinal lemma of the immersion theoretic method asserts that if M^m is an open manifold and Φ enjoys some extra functoriality‡‡, (as it often does) then Φ-microflexibility implies Φ-flexibility for any simplicial pair of *dimension* $< m$ in a co-ordinate chart. See [Gr, §3.3.1] [Ha_1]; we will not repeat the proof.

The same extra functoriality lets one quickly show, using the elementary engulfing lemma, that in case $A \hookrightarrow B$ is a finite simplicial expansion the restriction $\Phi(B) \to \Phi(A)$ is both a *weak* fibration and a weak homotopy equivalence. (Compare §1.4, property (2)(a).)

With this much motivation we prove two theorems.

THEOREM A.1.

Let $\Phi : U \to \Phi(U)$ be a sheaf of quasi-spaces on a topological m-manifold without boundary M^m. Suppose it is known that Φ is [weakly] flexible for every compact pair (B, A) such that B is a simplex linearly imbedded in some chart of M and either (i) or (ii) holds:

† A weak homotopy equivalence gives isomorphisms of homotopy groups for all base points. Avoid confusion with the usage of 'weak' in 'weak covering homotopy' or 'weakly flexible'.

‡ This property is just the covering homotopy property for homotopies that are constant on an initial interval of time.

‡‡ The general nonsense formulation of this extra functoriality seems to run as follows. For each pair of open sets U, V in M, there is a map of quasi-spaces $\Phi_{UV} : \mathrm{Imb}(U,V) \to \mathrm{Map}(\Phi(V), \Phi(U))$ where Imb(U,V) is the quasi space of open embeddings $U \to V$, and these Φ_{UV} together constituting an extension of Φ to a quasi-continuous contravariant functor of quasi-topologized categories.

(i) $\dim B < m$ *and* $A = \partial B$.
(ii) A *is* ∂B *less one principal face (so that* $\partial B \hookrightarrow A$ *is an elementary expansion).*

Then Φ *is [weakly] flexible for every closed pair* (B, A) *in* M *such that* $\check{H}^m(B, A; Z_2) = 0$.

THEOREM A.2.

Let $d : \Phi \to \Phi'$ *be a natural transformation between two sheaves of quasi-spaces on a topological m-manifold without boundary, such that:*

(1) $d_A : \Phi(A) \to \Phi'(A)$ *is a weak homotopy equivalence if* A *is a single point.*

(2) For any finite simplicial expansion $i : A \hookrightarrow B$, *the restrictions* $\Phi(i)$ *and* $\Phi'(i)$ *are weak homotopy equivalences.*

(3) Every closed pair (B, A) *verifying* $\check{H}^m(B, A; Z_2) = 0$ *is weakly flexible for both* Φ *and* Φ'.

Then $d_A : \Phi(A) \to \Phi'(A)$ *is a weak homotopy equivalence for every closed subset* $A \subset M$ *verifying* $\check{H}^m(A; Z_2) = 0$.

Remark. Suppose in A.1 and A.2 we delete the conditions $\check{H}^m(\ldots,\ldots; Z_2) = 0$ from hypotheses and conclusions, and in A.1 we suppress the condition $< m$; then the modified statements are valid, classical, and far easier to prove – much as under § 1.4 of this essay.

Proofs of A.1 and A.2.

Φ and Φ' send any square

$$\begin{array}{ccc} A_1 \cup A_2 & \underset{j'}{\leftarrow} & A_1 \\ i' \uparrow & & \uparrow i \\ A_2 & \overset{j}{\leftarrow} & A_1 \cap A_2 \end{array}$$

of inclusions to a fiber product square (= projective limit square). Hence if $\Phi(i)$ is a has a covering homotopy property so has $\Phi(i')$ and $\Phi(j')$ maps the fibers isomorphically. Also if $\Phi(i)$, $\Phi'(i)$ have the weak covering homotopy property and if d_X is a weak equivalence for $X = A_1$, A_2, and $A_1 \cap A_2$, then, for $X = A_1 \cup A_2$, the map d_X is also a weak equivalence, by application of the 5-lemma.

Every simplicial complex mentioned below is understood to be finite and simplex-wise linearly imbedded in some co-ordinate chart .

Now we can verify in order seven cases of A.1 . In this discussion let it be agreed that (B, A) is a closed pair in M verifying the condition $\check{H}^m(B, A; Z_2) = 0$.

(α) *(B, A) is finite simplicial of dimension* $< m$. Proof : Apply (i) .

(β) *(B, A) is finite simplicial and* $A \hookrightarrow B$ *is an expansion* . Proof : Apply (ii) .

(γ) *(B, A) is finite simplicial* . Proof : Apply (α) , (β) , and the collapsing lemma .

(δ) *(B, A) is compact in a co-ordinate chart* . Proof : Apply (γ) and the intersection lemma .

(ε) *(B, A) is compact* . Proof : Apply (δ) and the induction lemma .

(η) *(B, A) is such that B–A has compact closure in M* . Proof : Apply (ε) and the opening remarks about a square of inclusions .

(ζ) *The general case* . Proof : Apply (η) the filtration lemma and an infinite induction to construct the liftings desired , (cf. last step for A.2) . This completes the proof of A.1 . ■

Finally we go through the proof of A.2 in six cases . Here A is understood to be a closed subset of M verifying $\check{H}^m(A; Z_2) = 0$.

(α) *A is a simplex* . Proof : Apply (1) and (2) .

(β) *A is a finite simplicial complex of dimension* $< m$. Proof : Use (α) , (3) and the preliminary remarks in an induction on the number of simplices .

(γ) *A is a finite simplicial complex* . Proof : Apply the collapsing lemma and (2) to reduce to case (β) .

(δ) *A is a compactum in a co-ordinate chart* . Proof : Apply the intersection lemma and (γ) .

(ε) *A is a compactum* . Proof : Apply (δ) and the induction lemma together with the preliminary remarks .

(η) *The general case* . Proof : The filtration lemma applied to the pair (A, ϕ) yields a filtration $A_1 \subset A_2 \subset \cdots$ of A such that in the commutative ladder

$$\begin{array}{ccccccc}
\Phi(A_1) & \leftarrow & \Phi(A_2) & \leftarrow\cdots\leftarrow & \varprojlim \Phi(A_i) & = & \Phi(A) \\
\downarrow d_1 & & \downarrow d_2 & & & & \downarrow d_A \\
\Phi'(A_1) & \leftarrow & \Phi'(A_2) & \leftarrow\cdots\leftarrow & \varprojlim \Phi'(A_i) & = & \Phi'(A)
\end{array}$$

the horizontals are weak fibrations by (3) while the veriticals are weak equivalences by (ε) .

(Note that the equalities on the right are implied by the sheaf property and the fact from the filtration lemma that $A = \cup \mathring{A}_i$, whereas $A = \cup A_i$ would not imply them !) This ladder will permit the reader to check that d_A maps all homotopy groups isomorphically for all base points , as required to prove A.2 . ∎

Concluding remark : The above results A.1 , A.2 and their proofs , framed for TOP manifolds , apply verbatim to PL and DIFF manifolds , and to other sorts . We did need to engulf a finite 1-complex chart by chart in the induction lemma (g) .

Microflexibility implies flexibility (some indications) .

Above A.1 we bypassed the cardinal lemma [Gr, §3.3.1] [Ha_1] asserting that microflexibility of Φ implies flexibility for simplicial pairs in M^m of dimension $< m$. To help the reader toward an understanding of it, we shall conclude by isolating one idea in it:

Tucking Lemma: *With the data above A.1 , let $B \subset M^m$ be a (linear) simplex. Given a standard collar C of ∂B in B , and an open neighborhood V of C in B , consider two maps $f_0 , f_1 : P \to \Phi(V)$ that (co-)restrict both to the same map $P \to \Phi(\partial B)$. Then, if $\dim B < m$, there exists a homotopy $F : P \times I \to \Phi(V)$ of $f_0 = F_0$, constant when restricted outside of a compactum in V , to a map $F_1 : P \to \Phi(V)$ equal to f_1 near the collar C .*

Proof: It uses just the extra functoriality mentioned above A.1 , not the microflexibility. It is perhaps sufficient to illustrate the proof in the classical case $M = R^2$, $\Phi(U) = \{$ open immersions $U \to R^2 \}$, with B = 1-simplex , and P = point – as follows:

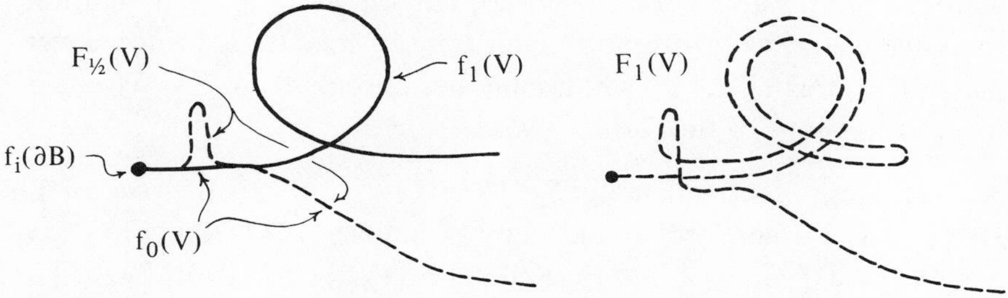

One combines this lemma with a vigorous use of microflexibility ($P \times I$ appears where one might expect P) . It becomes easy to lift homotopies after inserting finitely many pauses in their time flow. (The pauses give time to tuck.) This *weaker* flexibility is just as serviceable as flexibility or weak flexibility, e.g. for A.1 , A.2 ; but in fact, with a little more effort, one establishes genuine flexibility.

We remind the reader that, in the classical case just cited, when B is a 2-disc, flexibility fails utterly (even the weaker flexibility), because  does not bound an immersed 2-disc whereas the regularly homotopic immersion O does.

Appendix B.
CLASSIFICATION OF HOMOTOPY TORI: THE NECESSARY CALCULATIONS

This surgical classification of A. Casson, W.C. Hsiang & J. Shaneson, and C.T.C. Wall was used to settle the Hauptvermutung in $[KS_1]$. One should recall that it was first applied in $[Ki_1]$ to prove the Stable Homeomorphism Theorem in dimensions $\geqslant 5$. The latter's role in studying topological manifolds (in Essay III for example) is in itself so vital that this monograph would be incomplete without an account of the classification. The reader may wish to compare the original articles on the PL classification $[HS_1]$ $[Wa_3]$ $[HS_2]$ $[Wa_1]$, which we do not follow in all details.

In this appendix, we establish just that part of the classification needed elsewhere in these essays, namely the fact that the set $S_o(I^k \times T^n \text{ rel } \partial)$ of CAT (= DIFF or PL) homotopy tori invariant under passage (transfer) to standard finite coverings is $\pi_k(K(Z_2,3))$ provided $k+n \geqslant 5$ and $k \leqslant 4$. For precise definitions see §5.4 and below.

Notations: For typographical reasons the symbol $\mathcal{S}^*$ of §5.4 becomes S in this appendix. A subscripted 'oh' as in S_o will continue to indicate a (sub)set invariant under transfer to standard finite coverings. Henceforth ' rel ∂ ' will usually be suppressed for brevity's sake from expressions of the form $S(X \text{ rel } \partial)$.

The case $k = 0$, namely $S_o(T^n) = 0$, $n \geqslant 5$, is the one used in $[Ki_1]$. The other cases are needed to establish $\pi_k(TOP/CAT) = \pi_k(K(Z_2,3))$, $k \leqslant 4$, in as $[KS_1]$ or [IV, §10.12] or [V, §5] .

To complete the picture, we shall go on in Appendix C to give a full calculation of the homotopy-CAT structure sets $S(I^k \times T^n)$, $k+n \geqslant 5$, in all three categories DIFF , PL and TOP .

What follows is a tour-de-force of the non-simply connected surgery theory of Wall ; at the same time the strategy of the calculation (a splitting procedure) can be traced back to W. Browder's article $[Br_0]$.

A good deal of surgical machinery must be brought into view to make the discussion understandable; we do this in finer print.

Let X henceforth be a compact oriented CAT m-manifold with boundary ∂X (that is possibly empty, and possibly with corners if CAT = DIFF). An element of the set $S(X)$ of **homotopy**-CAT **structures** on X rel ∂ is represented by a homotopy equivalence $f : M \to X$ defined on a compact CAT m-manifold M^m , such that f is a CAT isomorphism near ∂X (i.e. on the preimage of a neighborhood of ∂X). Another such equivalence $f' : M' \to X$ represents the same element $[f'] = [f]$ in $S(X)$ precisely if there exists an h-cobordism $(W;M,M')$ from M to M' rel boundary, and a homotopy equivalence $F : W \to I \times X$ that is a product with I near $I \times \partial X$ while $F(x) = (f(x),0)$ for $x \in M$, and $F(x) = (f'(x),1)$ for $x \in M'$ after the identification $\partial W - F^{-1}(\mathring{I} \times \partial X) \cong M \amalg M'$ is made. When $m \geqslant 6$, and $\pi_1 X$ is free abelian, every h-cobordism is a product cobordism (cf. [I,§ 1]), so that $[f] = [f']$ then simply means there exists a CAT isomorphism $G : M \to M'$ with $f'G \simeq f$ rel ∂ .

The first tool in the surgical analysis of $S(X)$ is a map τ to a rather similarly defined set $N(X)$ of **normal invariants** (with our conventions 'tangential invariants' would be more appropriate terminology !). An element of $N(X)$ is represented by a pair (f,φ) called a **normal map** to X rel ∂ , where $f : M^m \to X$ is a degree ± 1 map of a compact CAT manifold that is a CAT isomorphism near ∂X , and $\varphi : \tau(M) \to \xi$ is a CAT stable bundle map over f from the tangent (micro-)bundle of M to a CAT bundle ξ over X . Another such pair (f',φ') represents the same element $[f,\varphi] = [f',\varphi'] \in N(X)$ precisely if there exists a normal map, called a **normal cobordism ,** (F, Φ) to $I \times X$ rel $I \times \partial X$ whose restrictions over $0 \times X$ and $1 \times X$ are identified to (f,φ) and (f',φ') respectively. It is plain to see that given $[f] \in S(X)$ there exists a unique element $[f,\varphi] \in N(X)$; this rule $[f] \mapsto [f,\varphi]$ defines

$$\tau \; : \; S(X) \; \to \; N(X) \quad .$$

The map τ lends itself to computation because there is a natural bijection first perceived by D. Sullivan (thesis Princeton 1965) $N(X) \to [X \text{ rel } \partial , G/CAT]$ to the π_0 of the space of those maps of X to G/CAT carrying ∂X to the base point. It arises from a rule to derive from $[f,\varphi] \in N(X)$ a homotopy trivialization $\alpha : \eta^k \to R^k$ of the bundle $\eta^k = \xi \oplus \nu(X)$ so that α is a CAT bundle trivialization near ∂X (here $\nu(X)$ is the normal bundle to X in euclidean space and ξ is the target of φ). Since G/CAT classifies homotopy trivialized CAT bundles, α determines a class $[\alpha] \in [X \text{ rel } \partial , G/CAT]$. Now the recipe for α . Approximate $f : M \to X$ rel ∂ by a CAT imbedding $g : M \to E(\xi \oplus \nu(X))$. There is a trivialization of the normal bundle ϵ to $g(M)$ in this total space such that the inclusion $E(\epsilon) \hookrightarrow E(\xi \oplus \nu(X))$ of total spaces induces, on their stable tangent bundles, the original stable CAT bundle map $\varphi : \tau(M) \to \xi$ up to CAT bundle homotopy . Stabilizing $\xi \oplus \nu(X)$ once, we can assume (see [III, § 4]) that $\xi \oplus \nu(X) = \eta^k$ contains a k-disc bundle η_1 , and that (after a standard move), g(M) lies with trivialized normal bundle in the boundary sphere bundle $\delta\eta_1$. By the (pre-Thom) Pontrjagin construction, one deduces a map $E(\delta\eta_1) \to S^{k-1}$ of degree ± 1 on each fiber, and then, by coning, a fiber homotopy equivalence $\alpha : \eta \to R^k$ to the trivial bundle over a point. Keeping the construction standard near ∂ we find that α is a CAT trivialization near ∂X. It is not difficult to show that this rule $[f,\varphi] \mapsto [\alpha]$ is well-defined and bijective; the inverse rule comes from making a homotopy trivialization $\alpha : \eta^k \to R^k$ suitably transverse to $0 \in R^k$. Compare [Wa_1, § 10].

We can now enunciate

PROPOSITION B.1. *If* $k \leqslant 4$ *, the normal invariant mapping* $\tau : \mathcal{S}_0(I^k \times T^n) \to [I^k \times T^n \text{ rel } \partial, G/CAT]$ *is zero.*

But the proof requires more basic theory.

G/CAT carries a Hopf-space structure representing Whitney sum, and making the functor $A \mapsto [A, G/CAT]$ a contravariant functor to abelian groups, cf. [Ad_1].

This H-space structure on G/CAT, with the connectivity of G/CAT, implies that it does not matter whether the set of homotopy classes $[A, G/CAT]$ is defined using base point preserving maps or arbitrary maps $A \to G/CAT$.

In case $(Y, \partial Y)$ is a CAT submanifold of $(X, \partial X)$ with trivial normal bundle ϵ, there is a simple rule to express the restriction map $[X \text{ rel } \partial, G/CAT] \to$ $\to [Y \text{ rel } \partial, G/CAT]$ in terms of normal maps and transversality. Given $[f,\varphi] \in$ $\in \mathcal{N}(X \text{ rel } \partial)$, deform $f : M \to X$ rel ∂ to be transverse to Y (for ϵ), deforming φ correspondingly. Set $N = f^{-1}(Y)$, $g = f|N$, $\eta = \xi|Y$. Then $\tau(M)|N = \tau(N)$ stably, and thus φ gives a CAT stable bundle map $\psi : \tau(N) \to \eta$ over g. Now $[g, \psi] \in \mathcal{N}(Y)$ is the advertised restriction of $[f, \varphi] \in \mathcal{N}(X)$.

Wall has constructed by surgery a (usually non-additive) mapping of pointed sets

$$\theta : \mathcal{N}(X^m) \longrightarrow L_m(\pi_1 X)$$

to an algebraically defined abelian group $L_m(\pi_1 X)$ that depends on m only modulo 4. The map θ depends up to sign on the choice of homology orientation class in $H_m(X, \partial X)$. (In the so-called non-orientable case where $w_1(X) \neq 0$ one must use $w_1(X)$-twisted integer coefficients for $H_m(X, \partial X)$; also $L_m(\pi_1 X)$ depends then on $w_1(X) : \pi_1(X) \to Z_2$ as well as on $\pi_1(X)$.) In this situation the main theorem of Wall's surgery states that the kernel of θ in $\mathcal{N}(X)$ is the image $\tau(\mathcal{S}(X))$, provided $m = \dim X \geqslant 5$.

We shall make extensive use of Wall's geometric periodicity theorem [Wa_1, § 9.9] asserting that producting with $P^8 = (CP_2)^2$, or with any $(CP_2)^k$, does not alter surgery obstructions, i.e. the square

$$\begin{array}{ccc} \mathcal{N}(X) & \xrightarrow{\theta} & L_m(\pi_1 X) \\ {\scriptstyle P\times}\downarrow & & \| \\ \mathcal{N}(P^8 \times X) & \xrightarrow{\theta} & L_{m+8}(\pi_1 X) \end{array}$$

is commutative. (See [Morg$_1$][Ran] for more general P .) The service rendered by periodicity in this appendix is the technical one of lifting our calculations into high dimensions out of the way of notorious difficulties met by surgery and handlebody theory in dimensions 3 , 4 , and 5 , cf. [Si$_8$] .

As starting material, we shall need information about θ for $\pi_1(X) = 0$. The group $L_m(0)$ is $0, Z_2, 0, Z$ for $m = 1,2,3,4$ respectively, see [Br$_2$][Ser$_3$][MiH]. Applied to $[f,\varphi] \in \mathcal{N}(X^m)$, $f : M^m \to X$, $m=0 \bmod 4$, the map θ yields 1/8 of the signature of the intersection form of M restricted to the kernel of $f_* : H_*(M) \to H_*(X)$; for $m = 2 \bmod 4$, the map θ yields the Arf invariant (from φ) of this form (for Z_2 coefficients), cf. [RSu] .

We shall gradually need to know that $\mathcal{N}(S^k) = \pi_k(G/CAT)$ (CAT = DIFF or PL) is 0 for k=0 , and is isomorphic to $L_k(0)$ for $k = 1,2,3,4$, while θ is bijective for $k = 2$ and is *multiplication by* 2 for $k = 4$. For CAT = DIFF these results are quickly deduced by using the fibration $G/DIFF \to B_O \to B_G$ and the basic information $\pi_k O = Z_2, Z_2, 0, Z$ for $k = 0,1,2,3$ and $\pi_k(G) = Z_2, Z_2, Z_2, Z_{24}, 0$ for $k = 0,1,2,3,4$, cf. [Hus] [Ser$_2$] [MiK] . The reader should pause to do this most basic calculation weaving in the following facts: (a) w_2 comes from B_G , (b) the 2-torus, with standard trivialization of its tangent bundle, represents an element x of $\pi_2(G/DIFF)$ with Arf invariant $\theta(x) \neq 0$, (c) $\pi_3 O \to \pi_3 G$ is onto [Toda] , equivalently any spin 3-manifold is spin nul-cobordant, cf. [Lick] , (d) the least signature of a closed DIFF spin 4-manifold is 16 (*Rohlin's theorem*) [MiK] , cf. [IV, Appendix B] .

The same result for CAT = PL is deduced from the DIFF calculation and a map $\varphi : \mathcal{N}_{DIFF}(S^k) \to \mathcal{N}_{PL}(S^k)$, $k \leq 4$, constructed by Whitehead's triangulation theorems. This φ turns out to be bijective for $k \leq 4$ by classical smoothing theory; compare [IV, Appendix B] . Fortunately Cerf's difficult result $\Gamma_4 = 0$ [Ce$_5$] is not needed, the injectivity for k=4 being assured by signatures. Also triangulation and smoothing of bundles is not strictly necessary; it is enough to triangulate and smooth total spaces and use some normal bundle theorems in [IV, Appendix A] .

For CAT = PL , the map $\theta : \pi_k(G/CAT) \to L_k(0)$ is bijective for all $k \geq 5$. One uses the PL Poincaré theorem (beware k = 5 [IV, Appendix B]), plus (for surjectivity) Milnor's plumbing [Br$_2$] and (for injectivity) the main theorem of surgery.

For CAT = TOP , it turns out (*after* our calculation of $\pi_*(TOP/CAT)$) that $\theta : \pi_k(G/TOP) \to L_k(0)$ is bijective for all k , cf. [Si$_{10}$, § 5, § 13] , and C.1 below.

The calculation of $[I^k \times T^n \text{ rel } \partial, G/CAT]$ is quite elementary. Let H be any Hopf-space, such as G/CAT , for which the rule $Y \mapsto [Y,H]$ is a functor to abelian groups. Applying this functor to the Puppe cofibration sequence $X/\partial \to (X \times T^1)/\partial \to (X \times I)/\partial \to \Sigma(X/\partial) \to \Sigma(X \times T^1 / \partial) \to \ldots$, we get an exact sequence of abelian groups

$$[X \text{ rel } \partial, H] \xleftarrow{\alpha} [X \times T^1 \text{ rel } \partial, H] \xleftarrow{\beta} [X \times I \text{ rel } \partial, H] \xleftarrow{\gamma} \ldots .$$

Note that α is a retraction since $(X \times T^1)/\partial$ retracts canonically onto X/∂. For a similar reason γ is zero and so β is injective, and we have a canonical decomposition

$$(\dagger) \qquad [X \times T^1 \text{ rel } \partial , H] = [X \text{ rel } \partial , H] \oplus [X \times I \text{ rel } \partial , H] \quad .$$

It is easily checked that the 'transfer' map in $[X \times T^1 \text{ rel } \partial , H]$ induced by multiplication by a positive integer λ in T^1 respects the summands on the right, fixing the first while giving multiplication by λ in the second.

We use $(\dagger)$ n times in succession to establish

$$\textbf{(B.2)} \quad [I^k \times T^n \text{ rel } \partial , H] = \bigoplus_i \binom{n}{i} [I^{k+i} \text{ rel } \partial , H] = \bigoplus_i \binom{n}{i} \pi_{k+i}(H) = \bigoplus_i \binom{n}{i} [S^{k+i} , H] \quad .$$

After s applications there are 2^s summands each of which will split into two at the next stage. The first application for example gives

$$[I^k \times T^n \text{ rel } \partial , H] = [I^k \times T^{n-1} \text{ rel } \partial , H] \oplus [I^{k+1} \times T^{n-1} \text{ rel } \partial , H] .$$

Inspection shows that each one of these $2^n = \sum_i \binom{n}{i}$ summands $[S^{k+i} , H]$ corresponds naturally to a standard subtorus T^i of T^n, and that the composite map $(I^k \times T^n)/\partial \xrightarrow{\alpha} (I^k \times T^i)/\partial \xrightarrow{\beta} S^{k+i}$ induces the injection of this $[S^{k+i} , H]$. Here α comes from projection $T^n \to T^i$ and β from collapsing the $(i-1)$-skeleton of T^i.

Passage to the standard λ-fold covering along any *one* factor T^1 of T^n corresponds to the map on the right in (B.2) that is multiplication by λ in any summand $[S^{k+i}, H]$ corresponding to a subtorus T^i containing this T^1 and to multiplication by 1 in the remaining summands. Hence the subgroup, called $[I^k \times T^n \text{ rel } \partial , H]_o$, invariant in (B.2) under passage to standard finite coverings is just the summand $[S^k , H]$.

After all these preliminaries we can give

Proof of Proposition B.1 .

We have observed, for $H = G/CAT$, that the restriction map $[I^k \times T^n \text{ rel } \partial , H]_o \to [I^k \text{ rel } \partial , H] = [S^k, H]$ is an isomorphism, and also that $\theta : [S^k , H] \to L_k(0)$ is injective, $k \leqslant 4$. It follows, using geometric periodicity and the discussion of normal maps, that this injection θ can now be expressed as a composition

$$N_o(I^k \times T^n) \xrightarrow{P\times} N(P\times I^k \times T^n) \xrightarrow{\rho} N(P\times I^k) \xrightarrow{\theta} L_k(0) \quad ,$$

where $P = (CP_2)^2$, and ρ is restriction (expressed via transversality).

Consider now $[f,\varphi] \in N(I^k \times T^n)$ where $f : M \to I^k \times T^n$ is a homotopy equivalence. By Farrell's fibering theorem [Fa] (cf. [IV, Appendix B.3]), we can deform $P\times f : P\times M \to$ $\to P\times I^k \times T^n$ rel ∂ (and $P\times\varphi$ correspondingly) to a bundle map of CAT bundles over T^n . The deformed map then gives, by restriction over $P\times I^k \times 0$, a normal map (g,ψ) where $g : N^{8+k} \to P\times I^k$ is a *homotopy equivalence* of fibers. This implies $\theta[g,\psi] =$ $= 0 \in L_{8+k}(0) = L_k(0)$, and we conclude, in view of our preliminary remarks, that $\theta[f,\varphi] = 0$. The proposition is proved. We repeat that P merely served to jack up dimensions high enough to apply Farrell's result. ∎

We now need the full five-term Sullivan-Wall exact sequence

(B.3) $$N(I\times X) \xrightarrow{\theta'} L(X) \xrightarrow{d} S(X) \xrightarrow{\tau} N(X) \xrightarrow{\theta} L_m(X) \quad .$$

An element of $L(X)$ is represented by a normal map (f,φ) to $I\times X$ rel $0\times X \cup I\times\partial X$ that gives a homotopy equivalence of m-manifolds over $1\times X^m$. The equivalence relation on these normal maps defining $L(X)$ is normal cobordism rel $0\times X \cup I\times\partial X$ giving an h-cobordism of m-manifolds over $1\times X^m$. Thus, restriction over $1\times X$ yields a map d as indicated. Clearly there is a forgetting map θ' as indicated. And it is easy to see the resulting sequence of pointed sets is exact.

Surgery defines a map $\theta'' : L(X) \to L_{m+1}(\pi_1 X)$ which Wall [Wa_1, § 5.8, § 6.5] has shown to be bijective for $m = \dim X \geqslant 5$. (His argument is strongly analogous to Stallings' classification of s-cobordisms with initial end X .) This imposes an abelian group structure on $L(X^m)$ for $m \geqslant 5$.

One readily verifies that the abelian group law in $[I\times X \text{ rel } \partial, G/CAT] = N(I\times X)$ arising from Whitney sum of bundles corresponds to fitting together two copies of $I\times X$. Thus, for once, the addition can be seen in terms of normal maps; it is end to end concatenation. An elementary additivity property of the surgery obstruction then reveals that θ' is additive.

It is easy to check then that d defines a bijection $\mathrm{Cokernel}(\theta') \cong \mathrm{Image}(d)$. Wishing to know the latter we may as well calculate the former.

If $\pi_1(X)$ is free abelian, a map $p : \mathcal{L}(X \times T^1) \to \mathcal{L}(X)$ is defined geometrically for $\dim X \geqslant 5$ as follows. Given $[F,\Phi] \in \mathcal{L}(X \times T^1)$, where $F : W \to I \times X \times T^1$, adjust F rel $0 \times X \cup I \times \partial X$ using first Farrell's fibering theorem over $1 \times X \times T^1$, then mere transversality, in such a way that F becomes transverse to $I \times X \times 0$ giving a homotopy equivalence over $1 \times X \times 0$. After this adjustment $p[F,\Phi] = [G,\Psi] \in \mathcal{L}(X)$, where (G,Ψ) is simply the restriction of (F,Φ) over $I \times X \times 0 = I \times X$.

A left inverse to p is obtained by crossing with T^1. It is readily seen that both these maps are additive, once addition in $\mathcal{L}$ is suitably expressed as a concatenation.

Proposition B.4 . (Shaneson [Sh], Wall [Wa_1, §13A.8])

$$\mathcal{L}(X \times T^1) \cong \mathcal{L}(X) \oplus \mathcal{L}(X \times I) \quad , \dim X \geqslant 5 ,$$

where the injection of $\mathcal{L}(X)$ *comes from crossing with* T^1 *and that of* $\mathcal{L}(X \times I)$ *comes from matching end points of* I .

Proof of B.4.

The specified map $\mathcal{L}(X \times I) \to \mathcal{L}(X \times T^1)$ is injective as it corresponds to the obvious group injection $\pi_1 X \to \pi_1(X \times T^1)$, which has a right inverse making $L_{m+1}(\pi_1 X)$ a retract of $L_{m+1}(\pi_1(X \times T^1))$ (The rule $\pi \leftrightarrow L_*(\pi)$ is covariantly functorial !)

Clearly $\mathcal{L}(X \times I) \subset \text{kernel}(p)$. In fact equality holds since $p[F,\Phi] = 0$, $[F,\Phi] \in \mathcal{L}(X \times T^1)$, means that, in the transversal situation discussed above, the restriction (G,Ψ) of (F,Ψ) over $I \times X \times 0$ is normally cobordant (in the restricted sense defining $[G,\Psi]$) to the trivial normal map involving id $|(I \times X \times 0)$. Such a normal cobordism of (G,φ), when thickened laterally along T^1, provides a normal cobordism $(F,\Phi) \sim (F',\Phi')$ (of the restricted sort) so that F' and Φ' are the identity over $I \times X \times 0$. Then clearly $[F,\Phi] \in \text{Image}(\mathcal{L}(X \times I))$.
■

With a little more work, one shows that the splitting $L_{m+1}(Z^{n+1}) = L_m(Z^n) \oplus L_{m+1}(Z^n)$ provided by B.4 (with $\pi_1 X = Z^n$) is independent of the geometry used to construct it; but we do not use this fact.

We note that there is a parallel decomposition $\mathcal{N}(I \times X \times T^1) = \mathcal{N}(I \times X) \oplus \mathcal{N}(I \times X \times I)$, dim X arbitrary, which can be established by a parallel argument (that is easier, as Farrell's theorem is not needed over $1 \times X \times T^1$). It is then clear that these decompositions correspond under the forgetting maps θ' of (B.3) . Incidentally, one readily checks that this decomposition corresponds to the one of $[I \times X \times T^1 \text{ rel } \partial, G/CAT]$ that we have used to prove Proposition B.2 .

The transfer map in $L(X \times T^1)$ corresponding to multiplication by λ in T^1 is readily seen to respect both summands, fix $L(X)$ and multiply by λ in $L(X \times I)$. Similarly in $\mathcal{N}(I \times X \times T^1)$.

We tend to regard the above result of Shaneson and Wall as an integral part of the calculation of $\mathcal{S}_0(I^k \times T^n)$. It permits us to complete the calculation apace.

Applying the above results n times in succession to $L(P \times I^k \times T^n)$, just as in proving B.2 , we decompose:

$$L(P \times I^k \times T^n) = \bigoplus_i \binom{n}{i} L(P \times I^{k+i}) = \bigoplus_i \binom{n}{i} L_{1+k+i}(0) \quad ;$$

and similarly:

$$\mathcal{N}(P \times I^{1+k} \times T^n) = \bigoplus_i \binom{n}{i} \mathcal{N}(P \times I^{1+k+i}) \quad ,$$

$$\mathcal{N}(I^{1+k} \times T^n) = \bigoplus_i \binom{n}{i} \mathcal{N}(I^{1+k+i}) = \bigoplus_i \binom{n}{i} \pi_{1+k+i}(G/CAT)$$

Thereby, the left and bottom maps in the following commutative square are expressed as direct sums

$$\begin{array}{ccc} \mathcal{N}(I^{1+k} \times T^n) & \xrightarrow{\ \theta'\ } & L(I^k \times T^n) \\ {\scriptstyle P\times}\downarrow & & {\scriptstyle P\times}\downarrow \cong \text{ periodicity} \\ \mathcal{N}(P \times I^{1+k} \times T^n) & \xrightarrow{\ \theta'\ } & L(P \times I^k \times T^n) \end{array} \quad .$$

The diagonal map, which is isomorphic to the top map θ' , is thus the direct sum of 2^n surgery maps:

$$\theta : \pi_{1+k+i}(G/CAT) \to L_{1+k+i}(0)$$

whose kernel and cokernel we henceforth denote K_{1+k+i} and K^{1+k+i} respectively. We have proved

Proposition B.5 . *The cokernel of* $\theta' : \mathcal{N}(I^{1+k} \times T^n) \to L(I^k \times T^n)$. $k+n \geq 5$ *,.is the sum* $\bigoplus_i \binom{n}{i} K^{1+k+i}$ *, of which the subgroup invariant under transfer is* K^{1+k} .

In the Sullivan-Wall sequence (B.3) for $X = I^k \times T^n$, this is equally the image of d or the kernel of τ . Also the transfer in the sets N , L and S clearly correspond under the maps and actions in (B.3) . Thus, combining B.1 and B.5 , we get

THEOREM B.6 .

For CAT = DIFF or PL , we have $S_o(I^k \times T^n \text{ rel } \partial) = K^{1+k} = \pi_k(K(Z_2,3))$ provided $k \leqslant 4$, $k+n \geqslant 5$.

The task of this appendix is accomplished.

Our budget of prequisites.

(i) It may be helpful to sum up our prerequisites for the result $S_o(T^n) = 0$, $n \geqslant 5$, that is used to prove the Stable Homeomorphism Theorem [Ki_1] . The basics of non-simply connected surgery are essential, including the geometric periodicity theorem [Wa_1, §5.8, §6.5] . Thus most of the first 112 pages of [Wa_1] must in the end be understood.† The one algebraic result about L-groups required is $L_1(0) = 0$, for which a pleasant proof is given in [Br_2] . To be sure, handlebody theory is needed (as in Essay I), plus Farrell's fibering theorem [Fa] based on [Si_1] . The PL result here requires no classical smoothing theory. Either the PL or the DIFF result suffices to prove the Stable Homeomorphism Theorem as in [Ki_1] .

(ii) Concerning the prerequisites for proving $S_o(I^k \times T^n) = \pi_k(K(Z_2,3))$, $k \leqslant 4$, we can say much the same for CAT = DIFF except that it is the calculation of $L_{k+1}(0)$ that is needed. When k = 3 Rohlin's Theorem [MiK] is vital. For CAT = PL , we must add classical smoothing theory (at present 200 pages of arduous reading: we suggest [Mu_3] , then [Hi_3] , then [HiM_2,I]) plus $\Gamma_i = 0$ for $i \leqslant 3$ (we suggest [V, §3.4, §5.8] or [Ce_4, pp.127-131] or again a recent improvement of the latter by A. Douady [Lau, Appendix]).

† Unfortunately for beginners, Wall's exposition is couched in the language of (n+1)-ads. In learning this language note that $[0,\infty)^n = R^n_\square$ is naturally a (n+1)-ad (the i-th of n subsets being defined by vanishing of the i-th co-ordinate); it can play a role of 'model' analogous to that of the standard geometric n-simplex in semi-simplicial work.

Appendix C. SOME TOPOLOGICAL SURGERY

In contrast to Appendix B which carries out calculations vital to these essays, the present appendix is optional material that exploits what has gone before to enlarge the reader's understanding of surgery as it relates to topological manifolds. In particular it relies all on the surgical apparatus set out in Appendix B, and on all the tools for the TOP version of surgery that have been forged in Essay III.

Staying initially in the elementary line of argument of Appendix B, we complete the classification of DIFF, PL, or TOP homotopy tori. A delightful isomorphism $\theta : \pi_k(G/TOP) \cong L_k(0)$, $k > 0$, lets one quickly calculate that $\mathcal{S}^{TOP}(I^k \times T^n) = 0$, $k+n \geqslant 5$; then structure theory (essay IV or V) reveals that $\mathcal{S}^{CAT}(I^k \times T^n) = [\Sigma^k T^n, TOP/CAT]$, $k+n \geqslant 5$. (Recall that these homotopy-CAT structure sets are *rel boundary.*)

This done, we adopt F. Quinn's semi-simplicial formulation [Qn1] of Wall's surgery, and proceed to explain some far-reaching consequences of the isomorphism $\pi_*(G/TOP) = L_*(0)$. To begin, we derive the periodicity of Sullivan and Casson $G/TOP \simeq \Omega^4(G/TOP)$ from the periodicity $L_k(\pi) \cong L_{k+4}(\pi)$ of algebraic surgery obstruction groups. Then, having paid due attention to fundamental group, we can observe a corresponding periodicity $\mathcal{S}^{TOP}(X) \cong \mathcal{S}^{TOP}(I^4 \times X)$ in homotopy-TOP structure sets (rel boundary), where X is any compact TOP manifold of dimension $\geqslant 5$. This makes the whole TOP Sullivan-Wall exact sequence satisfy periodicity.

We close with a few applications of these ideas.

(i) $\mathcal{S}^{TOP}(I^k \times T^n) = 0$ for $k \geqslant 5$, by induction on n via a simple splitting argument; hence $\mathcal{S}^{TOP}(I^k \times T^n) = 0$, $k+n \geqslant 5$. And this new proof applies immediately to significant generalizations by Wall and Quinn that let one calculate many groups $L_*(\pi)$.

(ii) A PL analog of all the foregoing yields a rapid, if ultra sophisticated, PL proof ab initio that $\mathcal{S}^{PL}(I^k \times T^n) = [\Sigma^k T^n, K(Z_2,3)]$, $k+n \geqslant 5$. (It must have much in common with A. Casson's unpublished semi-simplicial one dating from 1967-8.)

(iii) We mention a long standing general program involving fibrations, to reduce DIFF surgery problems to the study of the map $G/DIFF \to G/TOP$ and to the parallel TOP surgery problems, which enjoy periodicity.

We begin with

THEOREM C.1 . *The surgery map* $\theta_k : \pi_k(G/TOP) \to L_k(0)$ *is an isomorphism for all* $k \geqslant 1$.

Proof of C.1 .

For $k \geqslant 5$, argue as for PL in Appendix B .

For $k < 4$ the fibration $TOP/CAT \to G/CAT \to G/TOP$ together with the equation $\pi_k(TOP/CAT) = \pi_k(K(Z_2,3))$ ($k \leqslant 4$, and CAT = DIFF or PL , see [IV, §10.11] [V, §5]) show that θ_k is an isomorphism for TOP since (Appendix B) it is one for CAT .

Coming to $k = 4$, there is even some trouble defining θ_4 , since there is (still !) no TOP microbundle transversality theorem when the expected preimage would have dimension 4 . (See [III, §1] ; one *can* in fact use M. Scharlemann's transversality theorem [Sch] .) Avoiding this problem, we define θ_4 to make the following square commute, where $P = (CP_2)^2$:

$$\begin{array}{ccc} \pi_4(G/TOP) = [S^4, G/TOP] & \xrightarrow{\theta_4} & L_4(0) \\ {\scriptstyle P\times} \downarrow & & \| \\ [P \times S^4, G/TOP] & \xrightarrow{\theta} & L_{12}(0) \end{array} .$$

The exact sequence $0 \to \pi_4(G/CAT) \to \pi_4(G/TOP) \to$ $\to \pi_3(TOP/CAT) \to 0$ leaves just two possibilities: either θ_4 is an isomorphism, or $\pi_4(G/TOP)$ is $Z \oplus Z_2$ and has the same image $2Z$ in $L_4(0) = Z$ as $\pi_4(G/CAT)$. We thus complete the proof by showing θ_4 is onto.

The discussion in Appendix B (TOP version) identifies θ_4 to the geometric forgetting map

$$\theta' : \mathcal{N}_o^{TOP}(I \times X) \longrightarrow \mathcal{L}_o^{TOP}(X) \cong L_4(0) \quad ,$$

where $X = I^3 \times T^n$, with $3+n = 6$ (say), and subscripted 'oh' continues to indicate a (sub)set invariant under transfer to standard finite coverings. The surjectivity required now follows from the commutative diagram of forgetting maps with exact rows (see B.3).

$$\begin{array}{ccccc}
\mathcal{N}_o^{TOP}(I\times X) & \xrightarrow{\theta_4=\theta'} & \mathcal{L}_o^{TOP}(X) & \xrightarrow{d} & \mathcal{S}_o^{TOP}(X) \\
\uparrow & & \cong\uparrow & & \text{zero}\uparrow \text{(by [5,Proof of 5.3], cf. [IV, §10.11])} \\
\mathcal{N}_o^{CAT}(I\times X) & \xrightarrow{\theta'} & \mathcal{L}_o^{CAT}(X) & \xrightarrow[\text{onto}]{d} & \mathcal{S}_o^{CAT}(X) = Z_2 \text{ (by B.6)}
\end{array}$$

To verify the exactness observe that the square at left is a summand of the similar square without subscripted oh's. ∎

For an alternative proof see [Si_{10}, §13.4] .

The classification of Appendix B is enlarged and clarified by

THEOREM C.2 . *(with notations from Appendix B)*

(i) $\mathcal{S}^{TOP}(I^k \times T^n) = 0$, *for* $k+n \geqslant 5$.

(ii) $\mathcal{S}^{CAT}(I^k \times T^n) = [I^k \times T^n \text{ rel } \partial \text{ , TOP/CAT}]$

$$= \oplus_i \binom{n}{i} \pi_{k+i}(\mathrm{TOP/CAT}) \quad ,$$

for CAT = DIFF *or* PL , *and* $k+n \geqslant 5$.

Proof of C.2 part (i) : TOP homotopy tori are trivial.

Imitating Appendix B , we apply the Sullivan-Wall exact sequence (TOP version) :

(B.3) $\mathcal{N}(I\times X) \xrightarrow{\theta'} \mathcal{L}(X) \xrightarrow{d} \mathcal{S}(X) \xrightarrow{\tau} \mathcal{N}(X) \xrightarrow{\theta} L_m(\pi_1 X)$,

with $X = I^k \times T^n$, $m = k+n \geqslant 5$. The analysis above B.5 shows that the homomorphism θ' decomposes as a sum of 2^n surgery maps $\theta_j\colon \pi_j(G/TOP) \to L_j(0)$. By C.1 each θ_j is an isomorphism, so θ' is an isomorphism.

At this point $\mathcal{S}(X) = 0$ amounts to showing that θ has zero kernel.

For $k \geqslant 1$, and $k+n \geqslant 6$, we have a proof in hand since θ factorises (see below B.3) as

$$\mathcal{N}(I^k \times T^n) \xrightarrow{\theta'} \mathcal{L}(I^{k-1}\times T^n) \xrightarrow[\cong]{\theta''} L_m(Z^n) \; ,$$

while (as above) θ' is an isomorphism. If $k \geqslant 1$, and $k+n = 5$, the same argument works via the detour of producting with T^1 . ∎

There remains the case $k = 0$, which is troublesome because

$\theta : N(T^n) \to L_n(Z^n)$ is not additive. (For example the composed map $[T^8,G/TOP] \cong N(T^8) \to L_8(Z^8) \to L_8(0)$ given by signature is not additive as the Hirzebruch genus for signature shows; the non zero cup products in T^8 cause the trouble).

J. Levine found the following elementary proof that $\tau : S(T^n) \to N(T^n)$ has image zero, see [HS_2] . It entirely obviates discussion of θ .

For each of the $\binom{n}{i}$ subtori T^i of T^n we have a composed map $N(T^n) \xrightarrow{PX} N(P \times T^n) \xrightarrow{\rho} N(P \times T^i) \xrightarrow{\theta} L_i(0)$ where ρ is restriction. Summing these we get a map $\sigma : N(T^n) \to \oplus_i \binom{n}{i} L_i(0)$, which could in fact be identified to θ . Avoiding even this identification we propose to show that $\sigma\tau = 0$ and then that $\sigma^{-1}(0) = 0$.

To show that $\sigma\tau = 0$, consider any standard subtorus T^i and the corresponding component $S(T^n) \to L_i(0)$ of $\sigma\tau$, which is the composition $S(T^n) \xrightarrow{\tau} N(T^n) \xrightarrow{PX} N(P \times T^n) \xrightarrow{\rho} N(P \times T^i) \xrightarrow{\theta}$ $\xrightarrow{\theta} L_i(Z^i) \to L_i(0)$. Farrell's fibering theorem shows that the image of any $[f] \in S(T^n)$ is represented in $N(P \times T^i)$ by a normal map that is a homotopy equivalence, and therefore goes to zero in $L_i(Z^i)$.

To show that $\sigma^{-1}(0) = 0$, we induct upwards on n (J. Levine's trick). If $\sigma(x) = 0$, inductive hypothesis assures that the restriction of x for each standard $T^{n-1} \subset T^n$ is zero. Then the analysis of $[T^n,G/TOP]$ (see B.2) shows that x lies in the injected subgroup $\pi_n(G/TOP)$ on which σ is clearly the injective surgery map $\pi_n(G/TOP) \to L_n(0)$. This completes the proof that τ is zero, and therewith the proof of Theorem C.2 . ■

Proof of C.2 part (ii) from part (i) .

Since the set $[I^k \times T^n \text{ rel } \partial , TOP/CAT]$ has been shown in §5 (or in Essay [IV]) to be in bijective correspondence with the set $\mathcal{S}(I^k \times T^n)$ of concordance classes rel boundary of CAT manifold structures on $I^k \times T^n$, we have only to prove:

LEMMA C.3. *Let* X *be a compact* CAT *manifold such that* $S^{TOP}(X) = 0 = S^{TOP}(I \times X)$. *Then the forgetting map* $\mathcal{S}(X) \to$ $\to S^{CAT}(X)$ *is bijective.*

Proof of lemma . Surjectivity clearly follows from $S^{TOP}(X) = 0$; injectivity from $S^{TOP}(I \times X) = 0$. ■

PERIODICITY IN TOPOLOGICAL SURGERY

This is one of the distinguishing features of the topological category as one can see by reading [Si_{10}, §14, §15] [MorgL] [CapS] .

THEOREM C.4 .

There exists a 'periodicity' map $\Pi : G/TOP \to \Omega^4(G/TOP)$ *giving a homotopy equivalence to the identity component of* $\Omega^4(G/TOP)$ *, and a new (homotopy everything !)* † *H-space structure on* G/TOP *, such that for any compact TOP manifold* X^m *(with twisted orientation class) the resulting square (see below B.1 for* θ *)*

$$(\square)\qquad \begin{array}{ccc} [X \text{ rel } \partial, G/TOP] & \xrightarrow{\theta} & L_m(\pi_1 X) \\ \Pi \Big\downarrow \cong & & = \Big\downarrow \\ [I^4 \times X \text{ rel } \partial, G/TOP] & \xrightarrow{\theta} & L_{m+4}(\pi_1 X) \end{array}$$

is a commutative square of group homomorphisms.

Remarks.

a) The proof as we explain it below is a typical application of F. Quinn's semi-simplicial formulation of Wall's surgery [Qn_1] .

b) The periodicity and H-space structure we shall describe in fact coincide with those defined rather differently by D. Sullivan using his Characteristic Variety Theorem [Sull_1] . Further, J. Morgan is able to verify the commutativity and additivity of the square ($\square$) using his unpublished generalization of the Characteristic Variety Theorem for non simply connected manifolds. (I am grateful to him for explaining to me this line of proof.)

c) It should be clear that this result can be of great assistance in calcu-

† All H-space structures we shall meet are homotopy associative and homotopy commutative. What is more, I believe that one could interpret H-space structure henceforth to mean homotopy everything H-space structure, which is a much more complicated and refined sort of structure, [BoV], cf. [Seg] [May_2] ; but perhaps little would be gained thereby.

lating with the Sullivan-Wall surgery exact sequence. For example we could have used it to abbreviate the classification of TOP homotopy tori above (obviating J. Levine's argument).

d) A *distinct* periodicity $\Gamma: G/TOP \to \Omega^4(G/TOP)$, (called the Casson-Sullivan periodicity in [Si_{10}, §14]) , has become well known through C. Rourkes's exposition [Ro_0] (which presents a corresponding PL *near*-periodicity). Beware that Γ *fails* in general to make the square (□) of C.4 commute. For an example close to our heart, take $X = T^8$ and observe using Hirzebruch L-genus that even the square

$$\begin{array}{ccc} [T^8, G/TOP] & \xrightarrow{\theta} & L_8(0) \\ \Gamma \downarrow & & \downarrow = \\ [I^4 \times T^8, G/TOP] & \xrightarrow{\theta} & L_{12}(0) \end{array}$$

is non-commutative. The trouble arises because the construction of Γ involves Whitney sum; incidentally we have already observed that the upper θ is not additive when Whitney sum is used in $[T^8, G/TOP]$. In short, Π has superior properties.

In what follows, m is a fixed integer $\geqslant 0$, and π is a fixed finitely presented group equipped with a fixed 'orientation' homomorphism $w : \pi \to Z_2$. The symbols X, M, W, etc. stand for (varying) compact TOP manifolds each equipped with an orientation class in w_1-twisted integral homology.

We shall use the s-version of surgery involving *simple* rather than ordinary homotopy equivalence; so the L-groups appearing are the L^s-groups [Wa_1, p.249] Thus we rely on the result of [III, §5] that every $(X, \partial X)$ is a simple Poincaré pair. There is no pressing reason for switching from the slightly simpler h-version we have used so far, but the result of C.4 is clearly a little stronger in the s-version.

We shall need to use numerous semi-simplicial sets that lack degeneracy operations; these are called Δ-sets, see [RS_1, I] .

The proof of C.4 will follow Quinn's article [Qn_1] rather closely. We break it into seven steps.

1) $\mathbf{L}_m(\pi)$ is defined to be the Δ-set of which a typical zero-simplex is a referenced normal map f, φ, g , where the pair f, φ constitutes a degree 1 normal map over $f: (M^m, \partial M) \to (X^m, \partial X)$ such that $f: \partial M \to \partial X$ is a simple homotopy equivalence, and where $g: X \to K(\pi,1)$ is a 'reference' map such that $wg_* : \pi_1 X \to Z_2$ is the orientation homomorphism of X . A 1-simplex is a referenced normal cobordism F, Φ, G between two such referenced normal maps f, φ, g and f', φ', g' . Thus in particular $G : X^{m+1} \to \Delta^1 \times K(\pi,1)$ is a bordism from g to g' , and $F : M^{m+1} \to$ $\to X^{m+1}$ is a bordism from f to f' (in X^{m+1}) and F gives a simple equivalence on that compact part of boundary not accounted for by f and f' .

Continuing in the obvious way, one defines a k-simplex of $\mathbf{L}_m(\pi)$ as a triple F, Φ, G where for example $F : M^{m+k} \to X^{m+k}$ is a degree one map of compact topological manifold poly-ads, whose model poly-ad is $B^m \times \Delta^k$ with k+2 given subsets $B^m \times \partial_i \Delta^k$ and $(\partial B^m) \times \Delta^k$, so that, on that part of boundary coming from the boundary model poly-ad $(\partial B^m) \times \Delta^k$, the map F gives a simple homotopy equivalence of poly-ads. (The *mock-bundle* notion of [BuRS] would be helpful here.) Finally, the boundary maps in $\mathbf{L}_m(\pi)$ are defined by restriction.

The resulting Δ-set clearly verifies the Kan extension condition, there being no trouble with corners. The same holds for all Δ-sets below (ordered simplicial complexes excepted). Thus a well-behaved homotopy theory of these Δ-sets is available, cf. [RS_1, I] ; for example one can regard them as quasi-spaces based on the category of inclusion maps of ordered simplicial complexes (compare footnote in Appendix A).

2) $\mathbf{L}_m(\pi)$ is canonically an H-space. The multiplication $\mathbf{L}_m(\pi) \times \mathbf{L}_m(\pi) \to \mathbf{L}_m(\pi)$ is just disjoint sum. There is a canonical homotopy-inverse operation, namely the switch of orientations. There is a natural identity given by the empty normal map and the empty (Δ^k-fold) normal cobordisms.

3) There is a canonical group homomorphism $\theta : \pi_i \mathbf{L}_m(\pi)) \to L_{m+i}(\pi)$ to the algebraic Wall group, and it is an isomorphism if $m+i \geqslant 5$. One represents $x \in \pi_i$ by a (quasi-space) map $S^i \to \mathbf{L}_m(\pi)$, where S^i is the i-sphere triangulated as an ordered simplicial complex; then one *assembles* all the corresponding normal maps and cobordisms to get a well-defined normal map in dimension m+i and a reference map from its target to $S^i \times K(\pi,1)$. Its surgery obstruction, mapped thereby into $L_{m+1}(\pi)$, is $\theta(x)$. Wall provides the nontrivial proof of isomorphism in [Wa_1, §9] .

4) $\mathbf{L}_m(\pi) \xrightarrow{\times P} \mathbf{L}_{m+4p}(\pi)$ the map defined by producting everything with $P^{4p} = (CP_2)^p$ is a homotopy equivalence if $m \geqslant 5$. Indeed it is an isomorphism on homotopy groups by step 3 and Wall's geometric periodicity theorem [Wa_1, §9.9] . This map strictly respects H-space structure.

5) There is a composed homotopy equivalence of Δ-sets
$\theta_{k,p} : \Omega^k(G/TOP) \xrightarrow{\alpha} \{\Omega^k(G/TOP)\}^P \xrightarrow{\beta} \mathbf{N}(I^k \times P \text{ rel } \partial) \xrightarrow{\gamma} \mathbf{L}_{k+4p}(0)$,
provided $k \geqslant 0$, $k+4p \geqslant 5$, and provided $\mathbf{L}_{k+4p}(0)$ is replaced by its identity component when $k = 0$. Here $\Omega^k(G/TOP)$ is the singular complex of the usual k-th loop-space of maps $(I^k, \partial) \to (G/TOP, *)$; similarly for the second Δ-complex ; α comes from producting with P ; the Δ-complex $\mathbf{N}(I^k \times P \text{ rel } \partial)$ has as typical 0-simplex a normal map to the fixed target $I^k \times P$ rel $\partial I^k \times P$ (cf. Appendix B) ; β comes from applying TOP transversality simplex by simplex (this requires $k+4p \geqslant 5$) ; finally, γ is essentially a forgetting map as we may canonically install reference maps using $K(0,1) = \text{point}$. On the i-th homotopy groups the composed map $\theta_{k,p}$ is readily seen to give the surgery obstruction map
$\theta : \pi_{i+k}(G/TOP) \to L_{i+k+4p}(0) = L_{i+k}(0)$, which is an isomorphism by C.1 if $i+k > 0$, and zero if $i+k = 0$.

6) The periodicity Π is now defined to make the square

$$\begin{array}{ccc} G/TOP & \xrightarrow[\simeq]{\theta_{0,p}} & \mathbf{L}_{4p}(0) \\ \Pi \downarrow & & \downarrow \times CP_2 \\ \Omega^4(G/TOP) & \xrightarrow[\simeq]{\theta_{4,p}} & \mathbf{L}_{4+4p}(0) \end{array}$$

homotopy commute. Here $p \geqslant 2$, for convenience, but Π is clearly independent of p. The desired H-space structure on G/TOP is the one transported from that of $\mathbf{L}_{4p}(0)$ by $\theta_{0,p}$; we could in fact replace G/TOP by $\mathbf{L}_{4p}(0)$ using $\theta_{0,p}$!

Assertion: With this new H-space structure on G/TOP, the periodicity Π *is an H-map, i.e., commutes up to homotopy with the multiplication laws.*

Proof: This clearly amounts to the existence of an equivalence $\varphi_4 : \mathbf{L}_{4+4p} \xrightarrow{\simeq} \Omega^4 \mathbf{L}_{4p}$ preserving H-space structures so that $\varphi_4\theta_{4,p} \simeq \Omega^4(\theta_{0,p})$. To prove this we can use four times over the fact that a standard equivalence $\varphi : \mathbf{L}_n(\pi) \to \Omega\mathbf{L}_{n-1}(\pi)$ exists preserving H-space structure and verifies $\varphi\theta_{k+1,p} \simeq \Omega\theta_{k,p}$. To see this last fact form a strictly commutative diagram

$$\begin{array}{ccccc} & \Omega^k(G/TOP) & \longrightarrow & \mathbf{L}_{k+4p}(0) & \\ & \uparrow & & \uparrow & \\ (\text{point}) \simeq & \Lambda\Omega^k(G/TOP) & \xrightarrow{\Lambda\theta} & \Lambda\mathbf{L}_{k+4p}(0) & \simeq (\text{point}) \\ & \uparrow & & \uparrow & \\ & \Omega^{k+1}(G/TOP) & \longrightarrow & \mathbf{L}_{k+1+4p}(0) & \xrightarrow[\varphi]{\simeq} \Omega\mathbf{L}_{k+4p}(0) \end{array}$$

in which the columns are ad hoc Kan fibration sequences (cf. [BuRS, II, §5]) and the maps on the right respect the disjoint sum operations, while the top and bottom maps are (homotopic to) $\theta_{k,p}$ and $\theta_{k+1,p}$ respectively. ∎

7) For any compact TOP manifold X there is a homotopy commutative diagram

$$(7) \quad \begin{array}{ccccc} \Delta(X_+, G/TOP) & \xrightarrow{\theta_{0,2}} & \Delta(X_+, \mathbf{L}_8(0)) & \xrightarrow{\sigma} & \mathbf{L}_{m+8}(\pi_1 X) \\ \Pi \downarrow \simeq & & \simeq \downarrow \times CP_2 & & \simeq \downarrow \times CP_2 \\ \Delta(X_+, \Omega^4(G/TOP)) & \xrightarrow{\theta_{4,2}} & \Delta(X_+, \mathbf{L}_{12}(0)) & \xrightarrow{\sigma} & \mathbf{L}_{m+12}(\pi_1 X) \end{array}$$

where the left hand square comes from the definition of Π and the maps σ are defined using the assembly operation first met in step 3. We must specify what we mean by $\Delta(X_+, \ )$.

In case X is triangulable, let X_+ be an ordered triangulation and ∂_+X the induced triangulation of the boundary. Then $\Delta(X_+,?)$ is defined to be the Δ-set of which a k-simplex is a (quasi-space) map $(\Delta^k \times X_+)' \to ?$ sending $(\Delta^k \times \partial_+ X)'$ to the base point; here prime indicates a standard ordered triangulation which the reader can supply. The right hand square commutes strictly and strictly respects H-space multiplication. Thus, applying π_0 to the outer rectangle, we get a commutative square. A straight forward pursuit of definitions reveals that this is our square (□).

In case X_+ is not triangulable, let the space underlying X_+ be a stable normal disc bundle to X in a euclidean space R^n, and let $\partial_+ X$ be its restriction over the boundary ∂X. The pair $(X_+, \partial_+ X)$ is readily given a triangulation using the Product Structure Theorem of Essay I, (cf. [III, §4, §5]). This restores meaning to the left hand square. But unfortunately the assembly procedure no longer carries us into $\mathbf{L}_{m+8}(\pi_1 X)$ and $\mathbf{L}_{m+12}(\pi_1 X)$.

To remedy this simply, one can follow up the assembly procedure by an inductive TOP microbundle transversality procedure, as now described. (Compare the more sure-footed but tiresome arguments to resolve such technical difficulties in §2.) Discussing the upper σ first, observe that assembly canonically creates for each k-simplex of $\Delta(X_+, \mathbf{L}_8(0))$ an assembled normal map f, φ and an assembled reference map g_o, where $M^{k+n+8} \xrightarrow{f} N^{k+n+8} \xrightarrow{g_o} \Delta^k \times X_+^n$. Assuming the transversality process carried out for simplices of dimension $< k$, we apply the first TOP transversality theorem of [III, §1] in two steps. First make g_o transverse to $\Delta^k \times X^m$ in $\Delta^k \times X_+^n$ (for the given normal bundle); second make f transverse to the new-found $g_o^{-1}(\Delta^k \times X^m)$ adjusting φ accordingly; both steps are accomplished by homotopies fixing boundary which we then extend to higher dimensional simplices with this face. After this inductive procedure is completed, we restrict over $\Delta^k \times X^m$, for each such k-simplex, and reference via a fixed map $X \to K(\pi_1(X), 1)$ so as to land up in $\mathbf{L}_{8+m}(\pi_1 X)$ as desired. Doing the same (in relative fashion) for the lower map σ, we make the right hand square commutative; σ commutes with H-space multiplication at least up to homotopy. Then applying π_0 to the outside homotopy commutative rectangle we again get the square $(\square)$; this is amplified by step 8 below. ∎

Our first target, the proof of the periodicity theorem C.4 has been reached. We now restate it in a sharper semi-simplicial form and go on to establish the geometric periodicity $S(X) \cong S(I^4 \times X)$, $\dim X \geqslant 5$.

8) The outer rectangle of (7) is naturally homotopy equivalent to the homotopy commutative square

$$(8) \qquad \begin{array}{ccc} (G/TOP)^{(X/\partial)} & \xrightarrow{\theta} & \mathbf{L}_{m+8}(\pi_1 X) \\ \Pi \downarrow & & \downarrow \times CP_2 \\ \Omega^4 (G/TOP)^{(X/\partial)} & \xrightarrow{\theta} & \mathbf{L}_{m+12}(\pi_1 X) \end{array} .$$

The bottom θ for example is prescribed on a typical k-simplex as follows: regard it as a map $\Delta^k \times I^4 \times X \to G/TOP$; use this map to pull back the universal bundle; cross with P^8; apply transversality to get a normal map to $\Delta^k \times I^4 \times X^m \times P^8$ rel $\Delta^k \times \partial(I^4 \times X \times P)$; and lastly regard this as a simplex of $\mathbf{L}_{m+12}(\pi_1 X)$ using the obvious reference map to $\Delta^k \times K(\pi_1 X, 1)$. ∎

9) When $\dim X = m \geqslant 5$, the square (8) is naturally equivalent to a square

$$(9) \qquad \begin{array}{ccc} \mathbf{N}(X) & \xrightarrow{\theta} & \mathbf{L}_m(\pi_1 X) \\ \Pi \downarrow \simeq & & \simeq \downarrow \times CP_2 \\ \mathbf{N}(I^4 \times X) & \xrightarrow{\theta} & \mathbf{L}_{m+4}(\pi_1 X) \end{array} .$$

$\mathbf{N}(X)$ is the *Δ-set of normal maps to* X (rel ∂) : a typical k-simplex is a normal map to $\Delta^k \times X$ rel $\Delta^k \times \partial X$, (of course given in blocks, one over each $\Delta^k \times X$, so that face maps are defined by restriction). The maps θ just add the standard reference map to $\Delta^k \times K(\pi_1 X,1)$. The equivalence with (8) is given by a homotopy commutative diagram:

$$\begin{array}{ccccccc} \circ & \xrightarrow[\simeq]{\alpha} & \circ & \longrightarrow & \circ & \xrightarrow[\simeq]{\times P} & \circ \\ \Pi \downarrow & & \Pi \downarrow & (9) & \downarrow & & \downarrow \\ \circ & \xrightarrow[\alpha]{\simeq} & \circ & \longrightarrow & \circ & \xrightarrow[\times P]{\simeq} & \circ \end{array}$$

where the outer rectangle is (8) and α is defined using TOP transversality. ■

10) $\mathbf{S}(X) \xrightarrow{\tau} \mathbf{N}(X) \xrightarrow{\theta} \mathbf{L}_m(\pi_1 X)$ is a homotopy fibration for $\dim X = m \geqslant 5$, called the *structure fibration*. $\mathbf{S}(X)$ is here the *Δ-set of homotopy-TOP structures on* X *rel* ∂ defined in analogy with $\mathbf{N}(X)$ so that $\pi_0 \mathbf{S}(X) = S(X)$ and more generally $\pi_k \mathbf{S}(X)) = S(I^k \times X)$. The map τ is defined by adding structure to convert simple equivalences into normal maps (cf. Appendix B) .

The fibration property is proved by considering the Serre fiber $\mathbf{F}(X)$ of $\theta : \mathbf{N}(X) \to \mathbf{L}'_m(\pi_1 X)$ where $\mathbf{L}'$ is the sub Δ-set of $\mathbf{L}$ consisting of all simplices for which the reference map is a 1-equivalence (on all faces). Using [Wa_1, §9] , one shows that $\mathbf{L}'_m \cong \mathbf{L}_m$ by inclusion for $m \geqslant 5$. There is a natural map $\mathbf{S}(X) \to \mathbf{F}(X)$ which one readily shows to be a homotopy equivalence, see [Wa_1, §9], by using the obstruction-free π-π surgery theorem of [Wa_1 , §4] .

It is not difficult to show that the long exact homotopy sequence of the structure fibration is the Sullivan-Wall structure sequence.

What we have said here in (10) applies to DIFF and PL manifolds as well, and even to Poincaré spaces, cf. [J] [LLM] [Qn_2] .

11) There is a periodicity $\Pi : \mathbf{S}(X) \to \mathbf{S}(I^4 \times X) \simeq \Omega^4 \mathbf{S}(X)$ of homotopy-TOP structure Δ-sets for $\dim X \geqslant 5$. It is defined to be the induced homotopy equivalence of the homotopy fibers of the maps θ in the square (9) . Thus we have a homotopy commutative diagram

$$(11) \qquad \begin{array}{ccccc} \mathbf{S}(X) & \xrightarrow{\tau} & \mathbf{N}(X) & \xrightarrow{\theta} & \mathbf{L}_m(\pi_1 X) \\ \Pi \downarrow \simeq & & \Pi \downarrow \simeq & (9) & \simeq \downarrow \times CP_2 \\ \mathbf{S}(I^4 \times X) & \xrightarrow{\tau} & \mathbf{N}(I^4 \times X) & \xrightarrow{\theta} & \mathbf{L}_{m+4}(\pi_1 X) \end{array}$$

To specify $\Pi : \mathbf{S}(X) \to \mathbf{S}(I^4 \times X)$ in a preferred homotopy class, we need to *choose* a homotopy making square (9) commute; this is best built in $\mathbf{L}_{m+12}(\pi_1 X)$ using TOP transversality, cf. (7) .

To then verify that the homotopy class of $\Pi : \mathbf{S}(X) \to \mathbf{S}(I^4 \times X)$ is thereby well-defined, independent of the choices involved, we can rely on the *extendibility* of its construction as we have specified it. Thus, having specified constructions of this Π for $X \times 0$ and again for $X \times 1$ (call the results Π_0 and Π_1) , we can extend to a construction of an equivalence $\Pi : \mathbf{S}(\check{X}) \to \mathbf{S}(I^4 \times \check{X})$ where $\check{X}$ is the *triad* $X \times (I;0,1)$, cf. [Qn$_1$] . We conclude that $\Pi_0 \simeq \Pi_1$ using the four equivalences by restriction $\mathbf{S}(\check{X}) \simeq \mathbf{S}(X \times i)$, and $\mathbf{S}(I^4 \times \check{X}) \simeq \mathbf{S}(I^4 \times X \times i)$, $i = 0 , 1$.

Here are some salient conclusions that require no semi-simplicial language.

PERIODICITY THEOREM FOR STRUCTURES C.5 .

For any compact TOP manifold X^m, $m \geqslant 5$, *the Sullivan-Wall long exact structure sequence is a long exact sequence of abelian groups, and it is canonically isomorphic to the one for* $I^4 \times X^m$. *In particular* $\mathcal{S}(X) \cong \mathcal{S}(I^4 \times X)$.

As Wall remarks [Wa$_1$, § 10] , there is no sign that for DIFF or PL manifolds (in place of TOP) the structure sets should all naturally be groups.

The periodicity $\mathcal{S}(X) \cong \mathcal{S}(I^4 \times X)$ should also exist for noncompact TOP manifolds X^m , $m \geqslant 5$, cf. [Mau$_2$] .

The periodicity C.5 is surely an attractive result; but my treatment of it evokes a rueful song:

Maybe I'm doin' it wrong.
It just don't move me the way that it should ...
Sometimes I throw off a good one.
Least I think it is.
No, I know it is ...
Maybe I'm doin' it wrong ...

Randy Newman

APPLICATIONS OF PERIODICITY

I) Generalized Klein Bottles.

This is a reedition of $[Wa_1, \S\ 15B]$ with an improved proof.

A group G is **poly-Z of rank n** if it has a filtration $0 = G_0 \subset G_1 \subset ... \subset G_n = G$ so that G_k is normal in G_{k+1} with quotient $G_{k+1}/G_k \cong Z$ (infinite cyclic), $0 \leqslant k < n$. If $t \in G$ generates G/G_{n-1}, the inner automorphism $\theta(x) = t^{-1}xt$ of G_{n-1} can be realized by a homotopy equivalence $\theta' : K(G_{n-1},1) \to K(G_{n-1},1)$ and one observes that the mapping torus of θ' is a $K(G,1)$.

A **generalized Klein bottle** is a closed manifold $X \simeq K(G,1)$ with G poly-Z. For homological reasons $\dim X = \operatorname{rank} G$. The Klein bottle is the simplest one that is not a torus.

For any poly-Z group G, Farrell and Hsiang $[FH_2]$ have shown that $Wh(G) = 0$. Also, we can see by induction on rank that $K(G,1)$ has finite homotopy type. This will let us use Farrell's fibration theorem [Fa] (cf. [IV, Appendix B.3]).

HANDLEBODY LEMMA C.6 .

Let X^m *, be a compact* $CAT(= PL$ *or* $TOP)$ *manifold with boundary, homotopy equivalent to* $K(G,1)$ *where* G *is poly-Z of rank* n *. Suppose* $m - n \geqslant 6$ *and* $\pi_i(\partial X) \cong \pi_i(X)$ *by inclusion,* $i = 0, 1$ *.*

Then the set $\mathcal{S}(X^m)$ *of homotopy* CAT *structures on* X^m *rel* ∂ *is zero.*

Proof of Lemma C.6 by induction on n .

The map $g : X \to T^1$ corresponding to $G \to G/G_{n-1} = \pi_1 T^1$ has homotopy fiber $K(G_{n-1},1)$, and so Farrell's fibration theorem (see [III, §3] for its TOP version) tells us that g can be deformed first on the boundary then in the interior to become a CAT bundle projection with fiber $X_1 \simeq K(G_{n-1},1)$ over $0 \in T^1$ say, so that $\pi_1 \partial X_1 \cong \pi_1 X_1$ by inclusion.

Remark: The necessary fact that the homotopy fiber of $g | \partial X$ above has finite type follows from use of Poincaré duality, with $Z[G_{n-1}]$ coefficients, in the infinite cyclic covering. At the cost of assuming that

$m \gg \operatorname{rank} G$, one can see that both X and ∂X split because X is a regular neighborhood of a split $K(G,1)$, e.g. of one that is a mapping torus as was described. This last argument adapts best to the generalizations mentioned below.

Consider now any $[f] \in S(X^m)$, $f : (M^m, \partial) \to (X, \partial)$. Applying Farrell's theorem rel ∂ to the map gf , we can deform gf rel ∂ to a CAT bundle projection with fiber M_1^{m-1} and deform f rel ∂ to respect fibers. Then f is split at an equivalence $f_1 : M_1 \to X_1$, representing $[f_1] \in S(X_1)$ zero by induction. We can thus further deform f fixing ∂ so that in addition f_1 is an isomorphism. Cutting X and M open along X_1 and M_1 , we get from f a well defined homotopy equivalence $f_0 : (M_0^m, \partial) \to (X_0^m, \partial)$ representing $[f_0] \in S(X_0) = 0$, which again is zero by induction. Then we can finally deform f fixing $(\partial M) \cup M_1$ to an isomorphism. Thus $[f] = 0$.

The induction starts with $S(B^m) = 0$, $m \geqslant 5$. This is the one part of the argument that fails for DIFF manifolds. ∎

UNIQUENESS THEOREM C.7 [Wa$_1$, § 15B] .

Let X^m , $m \geqslant 5$, *be a compact TOP manifold that is a* $K(G,1)$ *with G poly-Z (rank* $\leqslant m$ *). Then* $S^{TOP}(X) = 0$.

Proof of C.7 : $S(X) \cong S(I^8 \times X)$ by TOP periodicity C.5 . And $S(I^8 \times X) = 0$ by C.6 . ∎

Remark. Wall's proof is rather different (more computational), although at the crucial point he envisages the use of periodicity in G/TOP (indeed C.4 exactly fits his needs). Is there a proof not strongly dependent on periodicity? Wall suggests there is one [Wa$_1$, p. 229] .

EXISTENCE THEOREM C.8 .

Given any poly-Z group G *of rank m , there exists a closed TOP manifold* $X^m \cong K(G,1)$.

Proof of C.8 (Wall's argument): Consider the filtration $G_0 \subset G_1 \subset \ldots \subset G_m = G$ for G and suppose we have a TOP manifold $X^n \cong K(G_n,1)$, $n > m$. Then we have a homotopy equivalence $f : X^n \to X^n$ whose mapping torus $T(f)$ is a $K(G_{n+1},1)$. For $n \neq 4$

we can deform f to a homeomorphism and set $T(f) = X^{n+1}$. This deformation is possible for $n = 3$ by [St₁][Neu] (as X^3 will be irreducible by construction). For $n \geqslant 5$ it is possible by C.7 .

Thus the only remaining problem is to construct X^5 . As $\mathcal{S}(X^4 \times T^1) = 0$, we can build $M^6 \simeq K(G_5,1) \times T^1$ as above, then fiber it over T^1 with fiber $X^5 \simeq K(G_5,1)$. ■

STRUCTURE THEOREM C.9 .

If π *is poly-Z of rank* m *then* $\theta : [K(\pi,1), \Omega^k(G/TOP)] \xrightarrow{\cong} L_{m+k}(\pi,w_1)$ *is an isomorphism for* $k > 0$ *, where* w_1 *is the orientation homomorphism for* $K(\pi,1)$.

Proof: Immediate from C.7 , C.8 and the Sullivan-Wall structure sequence. ■

GENERALIZATIONS C.10 (by F. Quinn [$Qn_{1,2}$] using [$Cap_{1,2,3}$])

(i) Let g be Waldhausen's simplest class of 'accessible' finitely presented torsion-free groups, namely the one generated from the trivial group by successive construction of free product with amalgamation and and its one-sided analog (= HNN extension) . These include poly-Z groups, free groups, classical knot groups.

Theorem: *In the s- or the h-surgery theory, for any TOP manifold* $N^m \simeq K(\pi,1)$, $m \geqslant 5$, *with* $\pi \in g$, *the group* $\mathcal{S}(X)$ *is 2-primary.* [*Hints:* One can imitate the proof above for the poly-Z case using, in place of Farrell's result, the strongest splitting theorem of S. Cappell [$Cap_{1,2}$] , which lets one split *up to h-cobordism* provided two 2-primary obstructions vanish (the more mysterious one comes from Cappell's UNil functor and may indeed be non-zero [Cap_3]) . For technical convenience in extending the handlebody lemma C.6 , one can initially suppose that $m \gg \dim K(\pi,1)$ and that X is a product with [0,1] and suitably split † . This factor [0,1] of X permits geometric doubling in $\mathcal{S}(X)$, which eventually kills Cappell's 2-primary obstructions, and also any distinction between the h- and s-theories. In a number of important special cases, e.g. *surface groups*

† See the remark in the handlebody lemma C.6 . Since $Wh(\pi)$ is not known to be zero for $\pi \in g$, one must add the elementary observation (for the h- or the s-theory), that $\mathcal{S}(X) \cong \mathcal{S}(X')$ if X' is X with an h-cobordism added along ∂X .

and *classical fibered knot groups*, a closer examinations reveals that $S(X) = 0$ since Cappell's obstructions and Whitehead groups vanish, see [Cap$_1$] [Wald] .

(ii) We conclude that θ is an isomorphism mod 2-primary groups from $[X^m \text{ rel } \partial , G/TOP]$ to $L_m(\pi_1 X)$, whenever $X \simeq K(\pi,1)$, $\pi \in g$. And it is a strict isomorphism in the more favorable cases mentioned.

(iii) According to Sullivan [Sull$_1$] the group $[X^m \text{ rel } \partial , G/TOP]$ is isomorphic mod 2-primary groups to $KO^0(X^m \text{ rel } \partial)$, and isomorphic rationally to $\oplus_k \{H^{4k}(X,\partial;Q)$; ; $k \in Z\}$, all this for X connected and $\partial X \neq \phi$.

(iv) In case $w_1 X^m = 0$, and X^m is chosen in R^m , one has an (extraordinary) Poincaré duality isomorphism, from each of these three (graduation 0) cohomology groups, to the corresponding m-th (extraordinary) homology group of X . Thus we have (writing $\pi_1 X = \pi$) :

$$\theta : H_m(X;G/TOP) \rightarrow L_m(\pi) \qquad \dagger$$
$$\theta : KO_m(X) \otimes Z[1/2] \rightarrow L_m(\pi) \otimes Z[1/2]$$
$$\theta : \oplus_k H_{m+4k}(X;Q) \rightarrow L_m(\pi) \otimes Q \quad .$$

By (ii) the last two are isomorphisms if $X = K(\pi,1)$ with $\pi \in g$. For the non-orientable case (with $w_1 : \pi \rightarrow Z_2$ non zero), see [Qn$_{1,2}$] .

II) PL homotopy tori by Casson's approach.

One can carry out the discussion of periodicity for PL manifolds in place of TOP , because $\theta : \pi_k(G/PL) \rightarrow L_k(0)$ is an isomorphism at least for $k > 4$. Indeed periodicity was first exploited by Sullivan and Casson in the PL context. We can pretend it is 1967 . The one, notorious failure of θ to be an isomorphism (it is $Z \xrightarrow{\times 2} Z$ for $k = 4$) prevents the PL periodicity maps from being true homotopy equivalences; in fact $\Pi : G/PL \rightarrow \Omega^4(G/PL)$ is easily seen to have fiber $K(Z_2,3)$.

We propose to use the PL version of diagram (11) (preceding C.5), with X a compact PL manifold of dimension $\geqslant 5$. Applied

† The spectrum for G/TOP is of course the periodic Ω-spectrum ..., Ω^4, Ω^3 , Ω^2 , Ω^1 , ... where $\Omega^k = \Omega^k(G/TOP)$.

twice, it gives a homotopy commutative diagram with rows homotopy fibrations:

$$\begin{array}{ccccc} \mathbf{S}(X) & \xrightarrow{\quad\tau\quad} & (G/PL)^{(X/\partial)} & \xrightarrow{\theta} & \mathbf{L}_m(\pi_1 X) \\ \Pi_8 \big\downarrow & & \Pi_8 \big\downarrow & & \simeq \big\downarrow \times(CP_2)^2 \\ \mathbf{S}(I^8 \times X) & \xrightarrow{\quad\tau'\quad} & \Omega^8 (G/PL)^{(X/\partial)} & \xrightarrow{\theta'} & \mathbf{L}_{m+8}(\pi_1 X) \end{array}$$

Here $\Pi_8 = (\Omega^4 \Pi) \circ \Pi$. Note that the homotopy fiber of $\Pi_8 : G/PL \to \Omega^8(G/PL)$ is just $K(Z_2,3)$. Also note that the left hand square is a (homotopy) fiber product square since an equivalence is induced from the fiber $\Omega \mathbf{L}_m$ of τ to the fiber $\Omega \mathbf{L}_{m+8}$ of τ' , namely the looping of $\times(CP_2)^2$ in the mapping of Puppe fibration sequences. *It follows that the fibers of the two maps* Π_8 *are homotopy equivalent.*

The fiber of the Π_8 at left is $\mathbf{S}(X)$ for $X = K(G,1)$ with G *poly-Z* : indeed $S(I^{8+k} \times X) = 0$, $k \geqslant 0$, by the (PL) handlebody lemma C.6 , thus $\mathbf{S}(I^8 \times X)$ is contractible, and so the fiber is $\mathbf{S}(X)$ itself !

The fiber of the Π_8 at center is $K(Z_2,3)^{(X,\partial)}$.

Therefore we have a natural homotopy equivalence $\mathbf{S}(X) \simeq K(Z_2,3)^{(X/\partial)}$ (for $\dim X \geqslant 5$, $X \simeq K(G,1)$, G poly-Z), whence $S(I^k \times X) \cong H^{3-k}(X,\partial;Z_2)$, $k \geqslant 0$.

This elegant calculation must be a close approximation to A. Casson's unpublished semi-simplicial classification of PL homotopy tori dating from 1967-68 . (It is not clear whether he took advantage of the handlebody lemma C.6). I have not given it greater emphasis because of the sophistication required and equally because it would seem to deny access to TOP manifolds to those equipped with no more than DIFF techniques.

III) Related problems .

Contemplate the following homotopy commutative diagram for X a compact CAT m-manifold, $m \geqslant 5$.

$$\begin{array}{ccccc}
\mathbf{S}^{T/C}(X) & \xrightarrow[\simeq]{\gamma} & (TOP/CAT)^{(X/\partial)} & & \\
\downarrow & & \downarrow & & \\
\mathbf{S}^{C}(X) & \longrightarrow & (G/CAT)^{(X/\partial)} & \longrightarrow & \mathbf{L}^{C}_{m}(\pi_1 X) \\
\downarrow & & \downarrow & & \downarrow \simeq \\
\mathbf{S}^{T}(X) & \xrightarrow{\tau} & (G/TOP)^{(X/\partial)} & \xrightarrow{\theta} & \mathbf{L}^{T}_{m}(\pi_1 X)
\end{array}$$

The bottom row is the structure fibration (10) for TOP ; the middle row is its analogue for CAT ; all vertical maps forget. $\mathbf{S}^{T/C}(X)$ is the pointed Δ-set of CAT manifold structures on X so defined that its k-th homotopy group is naturally isomorphic to the set of concordance classes (rel ∂) of CAT manifold structures on $I^k \times X$ standard near ∂ as discussed in Essay IV . There is a natural map from $\mathbf{S}^{T/C}(X)$ to a Serre-type fiber of $\mathbf{S}^{C}(X) \to \mathbf{S}^{T}(X)$ and one requires only the TOP s-cobordism theorem to show that it is a homotopy equivalence.

Since the lower left square is a fiber product square (argue as in **II**) we can obtain a 'classifying' homotopy equivalence γ as indicated. One can show that, on arc components π_0 , it is precisely the classification of [IV, §10] (for X compact). To deal also with the *existence* problem for CAT structures one can modify the middle column (use spaces of liftings !).

This inserts the classification of [IV] into a larger context. Although the proof of it provided thereby is not really easier than that in [IV] , there are analogous situations where this approach involving surgery seems the most workable one. This is the case for the problem of classifying up to concordance *arbitrary* triangulations (perhaps not PL homogeneous) of a given TOP manifold. For this we refer to forthcoming articles by T. Matumoto and D. Galewski & R. Stern.

For a given CAT manifold a primary surgical problem is to understand the vertical fibration on the left with base space $\mathbf{S}^{T}(X)$. This has been accomplished (as Quinn notes [Qn_1]) when CAT = PL and X is a torus (or PL generalized Klein bottle) ; indeed $\mathbf{S}^{T/C}(X) \simeq \mathbf{S}^{C}(X) \simeq K(Z_2,3)^X$. In general, note that the vertical homotopy fibration on the left is the pull-back by τ of the middle fibration. But this τ is the fiber of θ . Hence, if one can fully analyze the middle vertical homotopy fibration and θ (or τ), the job will be done. Much of Sullivan's work on surgery [$Sull_{1,2}$] has been directed to understanding the middle fibration beginning with an exhaustive study of G/TOP , (see also [Qn_3] [MadM]) .

Annex A. Reprinted with permission, from
Annals of Mathematics, Vol. 89, No. 3, May 1969, pp. 575-582.

Stable homeomorphisms and the annulus conjecture*

By ROBION C. KIRBY

A homeomorphism h of R^n to R^n is stable if it can be written as a finite composition of homeomorphisms, each of which is somewhere the identity, that is, $h = h_1h_2 \cdots h_r$ and $h_i \mid U_i =$ identity for each i where U_i is open in R^n.

Stable Homeomorphism Conjecture, SHC$_n$: All orientation preserving homeomorphisms of R^n are stable.

Stable homeomorphisms are particularly interesting because (see [3]) SHC$_n \Rightarrow$ AC$_n$, and AC$_k$ for all $k \leqq n \Rightarrow$ SHC$_n$ where AC$_n$ is the

Annulus Conjecture, AC$_n$: Let $f, g : S^{n-1} \rightarrow R^n$ be disjoint, locally flat imbeddings with $f(S^{n-1})$ inside the bounded component of $R^n - g(S^{n-1})$. Then the closed region A bounded by $f(S^{n-1})$ and $g(S^{n-1})$ is homeomorphic to $S^{n-1} \times [0, 1]$.

Numerous attempts on these conjectures have been made ; for example, it is known that an orientation preserving homeomorphism is stable if it is differentiable at one point [10] [12], if it can be approximated by a PL homeomorphism [6], or if it is $(n-2)$-stable [4]. "Stable" versions of AC$_n$ are known ; $A \times [0, 1)$ is homeomorphic to $S^{n-1} \times I \times [0, 1)$, $A \times R$ is $S^{n-1} \times I \times R$, and $A \times S^k$ is $S^{n-1} \times I \times S^k$ if k is odd (see [7] and [13]). A counter-example to AC$_n$ would provide a non-triangulable n-manifold [3].

Here we reduce these conjectures to the following problem in PL theory. Let T^n be the cartesian product of n circles.

Hauptvermutung for Tori, HT$_n$: Let T^n and τ^n be homeomorphic PL n-manifolds. Then T^n and τ^n are PL homeomorphic.

THEOREM 1. *If $n \geqq 6$, then* HT$_n \Rightarrow$ SHC$_n$.

(*Added December* 1, 1968. It can now be shown that SHC$_n$ is true for $n \neq 4$. If $n \leqslant 3$, this is a classical result. Theorem 1 also holds for $n = 5$, since Wall [19, p. 67] has shown that an end which is homeomorphic to $S^4 \times R$ is also PL homeomorphic to $S^4 \times R$.

In the proof of Theorem 1, a homeomorphism $f: T^n \rightarrow \tau^n$ is constructed. If $\hat{f}: \hat{T}^n \rightarrow \hat{\tau}^n$ is any covering of $f: T^n \rightarrow \tau^n$, then clearly f is stable if and only if

* Partially supported by NSF Grant GP 6530.

$\hat{f}$ is stable. Using only the fact that f is a simple homotopy equivalence, Wall's non-simply connected surgery techniques [15] provide an "obstruction" in $H^3(T^n; Z_2)$ to finding a PL homeomorphism between T^n and τ^n. It is Siebenmann's idea to investigate the behavior of this obstruction under lifting $f: T^n \to \tau^n$ to a 2^n-fold cover; he suggested that the obstruction would become zero. Wall [16] and Hsiang and Shaneson [17] have proved that this is the case; that is, if $\hat{\tau}^n$ is the 2^n-fold cover of a homotopy torus τ^n, $n \geqq 5$, then $\hat{\tau}^n$ is PL homeomorphic to $T^n (= \hat{T}^n)$. Therefore, following the proof of Theorem 1, $\hat{f}: \hat{T}^n \to \hat{\tau}^n$ is stable, so f is stable, and thus SHC_n holds for $n \neq 4$. Hence the annulus conjecture AC_n holds for $n \neq 4$.)

(*Added April* 15, 1969. Siebenmann has found a beautiful and surprising counter-example which leads to non-existence and non-uniqueness of triangulations of manifolds. In particular HT_n is false for $n \geqq 5$, so it is necessary to take the 2^n-fold covers, as above. One may then use the fact that $\hat{f}: T^n \to \hat{\tau}^n$ is homotopic to a PL homeomorphism to show that $f: T \to \tau^n$ was actually isotopic to a PL homeomorphism. Thus, although there are homeomorphisms between T^n and another PL manifold which are not even homotopic to PL homeomorphisms, they cannot be constructed as in Theorem 1. Details will appear in a forthcoming paper by Siebenmann and the author. See also R. C. Kirby and L. C. Siebenmann, *On the triangulation of manifolds and the Hauptvermutung*, to appear in Bull. Amer. Math. Soc.)

Let $\mathcal{H}(M^n)$ denote the space (with the compact-open topology) of orientation preserving homeomorphisms of an oriented stable n-manifold M, and let $\mathcal{SH}(M^n)$ denote the subspace of stable homeomorphisms.

THEOREM 2. *$\mathcal{SH}(R^n)$ is both open and closed in $\mathcal{H}(R^n)$.*

Since a stable homeomorphism of R^n is isotopic to the identity, we have the

COROLLARY. *$\mathcal{SH}(R^n)$ is exactly the component of the identity in $\mathcal{H}(R^n)$.*

COROLLARY. *A homeomorphism of R^n is stable if and only if it is isotopic to the identity.*

THEOREM 3. *If M^n is a stable manifold, then $\mathcal{SH}(M^n)$ contains the identity component of $\mathcal{H}(M^n)$.*

In general this does not imply that the identity component is arcwise connected (as it does for $M^n = R^n$ or S^n), but arcwise connectivity does follow from the remarkable result of Cernavskii [5] that $\mathcal{H}(M^n)$ is locally contractible if M^n is compact and closed or $M^n = R^n$. From the techniques in this paper, we have an easy proof of the last case.

THEOREM 4. *$\mathcal{H}(R^n)$ is locally contractible.*

We now give some definitions, then a few elementary propositions, the crucial lemma, and finally the proofs of Theorems 1 – 4 in succession.

The following definitions may be found in Brown and Gluck [3], a good source for material on stable homeomorphisms. A homeomorphism h between open subsets U and V of R^n is called stable if each point $x \in U$ has a neighborhood $W_x \subset U$ such that $h \mid W_x$ extends to a stable homeomorphism of R^n. Then we may define stable manifolds and stable homeomorphisms between stable manifolds in the same way as is usually done in the PL and differential categories. Whenever it makes sense, we assume that a stable structure on a manifold is inherited from the PL or differential structure. Homeomorphisms will always be assumed to preserve orientation.

PROPOSITION 1. *A homeomorphism of R^n is stable if it agrees with a stable homeomorphism on some open set.*

PROPOSITION 2. *Let $h \in \mathcal{H}(R^n)$ and suppose there exists a constant $M > 0$ so that $| h(x) - x | \leqq M$ for all $x \in R^n$. Then h is stable.*

PROOF. This is Lemma 5 of [6].

Letting rB^n be the n-ball of radius r, we may consider $5D^n = i(5B^n)$ as a subset of T^n, *via* some fixed differentiable imbedding $i: 5B^n \to T^n$.

PROPOSITION 3. *There exists an immersion $\alpha: T^n - D^n \to R^n$.*

PROOF. Since $T^n - D^n$ is open and has a trivial tangent bundle, this follows from [8, Th. 4. 7].

PROPOSITION 4. *If A is an $n \times n$ matrix of integers with determinant one, then there exists a diffeomorphism $f: T^n \to T^n$ such that $f_* = A$ where $f_*: \pi_1(T^n, t_0) \to \pi_1(T^n, t_0)$.*

PROOF. A can be written as a product of elementary matrices with integer entries, and these can be represented by diffeomorphisms.

PROPOSITION 5. *A homeomorphism of a connected stable manifold is stable if its restriction to some open set is stable.*

For the proof, see [3, p. 35]

PROPOSITION 6. *Let $f: S^{n-1} \times [-1, 1] \to R^n$ be an imbedding which contains S^{n-1} in its interior. Then $f \mid S^{n-1} \times 0$ extends canonically to an imbedding of B^n in R^n.*

PROOF. This is shown in [9]. However, there is a simple proof; one just re-proves the necessary part of [2] in a canonical way. This sort of canonical construction is done carefully in the proof of Theorem 1 of [11].

The key to the paper is the following observation.

LEMMA. *Every homeomorphism of T^n is stable.*

PROOF. Let $e: R^n \longrightarrow T^n$ be the usual covering map defined by

$$e(x_1, \cdots, x_n) = (e^{2\pi i x_1}, \cdots, e^{2\pi i x_n})$$

and let $t_0 = (1, 1, \cdots, 1) = e(0, \cdots, 0)$. e fixes a differential and hence stable structure on T^n.

Let h be a homeomorphism of T^n, and assume at first that $h(t_0) = t_0$ and $h_*: \pi_1(T^n, t_0) \longrightarrow \pi_1(T^n, t_0)$ is the identity matrix. h lifts to a homeomorphism $\hat{h}: R^n \longrightarrow R^n$ so that the following diagram commutes.

$$\begin{array}{ccc} R^n & \xrightarrow{\hat{h}} & R^n \\ \downarrow{\scriptstyle e} & & \downarrow{\scriptstyle e} \\ T^n & \xrightarrow{h} & T^n \end{array}$$

Since $I^n = [0, 1] \times \cdots \times [0, 1]$ is compact,

$$M = \sup\{|\hat{h}(x) - x| \mid x \in I^n\}$$

exists. The condition $h_* =$ identity implies that $\hat{h}$ fixes all lattice points with integer coordinates. Thus $\hat{h}$ moves any other unit n-cube with vertices in this lattice in the "same" way it moves I^n; in particular $|\hat{h}(x) - x| < M$ for all $x \in R^n$. By Proposition 2, $\hat{h}$ is stable. e provides the coordinate patches on T^n, so h is stable because $e^{-1}he \mid e^{-1}$ (patch) extends to the stable homeomorphism $\hat{h}$ for all patches.

Given any homeomorphism h of T^n, we may compose with a diffe-omorphism g so that $gh(t_0) = t_0$. If $A = (gh)_*^{-1}$, then Proposition 4 provides a diffeomorphism f with $f_* = A = (gh)_*^{-1}$, so $(fgh)_* =$ identity. We proved above that fgh was stable so $h = g^{-1}f^{-1}(fgh)$ is the product of stable home-omorphisms and therefore stable.

PROOF OF THEOREM 1. Let g be a homeomorphism of R^n. $g\alpha$ induces a new differentiable structure on $T^n - D^n$, and we call this differential manifold $\widetilde{T^n - D^n}$. We have the following commutative diagram,

$$\begin{array}{ccc} T^n - D^n & \xrightarrow{\text{id}} & \widetilde{T^n - D^n} \\ \downarrow{\scriptstyle \alpha} & & \downarrow{\scriptstyle g\alpha} \\ R^n & \xrightarrow{g} & R^n \end{array} .$$

α and $g\alpha$ are differentiable and therefore stable, so g is stable if and only if the identity is stable (use Proposition 1).

Since $\widetilde{T^n - D^n}$ has one end, which is homeomorphic to $S^{n-1} \times R$, and $n \geqq 6$, there is no difficulty in adding a differentiable boundary [1]. Since the boundary is clearly a homotopy $(n-1)$-sphere, we can take a C^1-triangulation and use the PL h-cobordism theorem to see that the boundary is a PL $(n-1)$-sphere. To be precise, there is a proper PL imbedding $\beta: S^{n-1} \times [0, 1) \to \widetilde{T^n - D^n}$, and we add the boundary by taking the union $\widetilde{T^n - D^n} \cup_\beta S^{n=1} \times [0, 1]$ over the map β.

Finally we add B^n to this union, *via* the identity map on the boundaries, to obtain a closed PL manifold τ^n.

We can assume that $\partial 2D^n$ lies in $\beta(S^{n-1} \times [0, 1))$. Thus $\partial 2D^n$ lies in an n-ball of τ^n and, since it is locally flat, bounds an n-ball by the topological Schoenflies theorem [2]. Now, we may extend the $id \mid T^n - 2D^n$, by coning on $\partial 2D^n$, to a homeomorphism $f: T^n \to \tau^n$.

Using HT_n, we have a PL (hence stable) homeomorphism $h: T^n \to \tau^n$. By the Lemma, $h^{-1}f: T^n \to T^n$ is stable, so $f = h(h^{-1}f)$ is stable, $f \mid T^n - 2D^n =$ identity is stable, and finally g is stable.

Note that it is only necessary that HT_n gives a stable homeomorphism h.

PROOF OF THEOREM 2. We shall show that a neighborhood of the identity consists of stable homeomorphisms. But then by translation in the topological group $\mathcal{H}(R^n)$, any stable homeomorphism has a neighborhood of stable homeomorphisms, so $\mathcal{SH}(R^n)$ is open. Now it is well known that an open subgroup is also closed (for a coset of $\mathcal{SH}(R^n)$ in $\mathcal{H}(R^n)$ is open, so the union of all cosets of $\mathcal{SH}(R^n)$ is open and is also the complement of $\mathcal{SH}(R^n)$, which is therefore closed).

If C is a compact subset of R^n and $\varepsilon > 0$, then it is easily verified that $N(C, \varepsilon) = \{h \in \mathcal{H}(R^n) \mid |h(x) - x| < \varepsilon \text{ for all } x \in C\}$ is an open set in the CO-topology. Let C be a compact set containing $\alpha(T^n - D^n)$. If $\varepsilon > 0$ is chosen small enough, then

$$\overline{h\alpha(T^n - 5D^n)} \subset \alpha(T - 4D^n) \subset \overline{\alpha(T^n - 4D^n)} \subset h\alpha(T^n - 3D^n)$$
$$\subset \overline{h\alpha(T^n - 2D^n)} \subset \alpha(T^n - D^n)$$

for any $h \in N(C, \varepsilon)$. There exists an imbedding $\hat{h}$, which "lifts" h so that the following diagram commutes.

$$\begin{array}{ccc} T^n - 2D^n & \xrightarrow{\hat{h}} & T^n - D^n \\ \downarrow \alpha & & \downarrow \alpha \\ R^n & \xrightarrow{h} & R^n \end{array}$$

To define $\hat{h}$, first we cover C with finitely many open sets $\{U_i\}$, $i = 1, \cdots, k$, so that α is an imbedding on each component of $\alpha^{-1}(U_i)$, $i = 1, \cdots, k$. Let $\{V_i\}$, $i = 1, \cdots, k$, be a refinement of $\{U_i\}$. If ε was chosen small enough, then $h(V_i) \subset U_i$. Let $W_i = U_i \cap \alpha(T^n - D^n)$ and $X_i = V_i \cap \alpha(T^n - 2D^n)$. Since $h\alpha(T^n - 2D^n) \subset \alpha(T^n - D^n)$, we have $h(X_i) \subset W_i$, $i = 1, \cdots, k$. Let $W_{i,j}$, $j = 1, \cdots, u_i$ be the components of $\alpha^{-1}(W_i)$, let $X_{i,j} = W_{i,j} \cap T^n - 2D^n$, and let $\alpha_{i,j} = \alpha \mid W_{i,j}$ for all i and j. Now we can define $\hat{h}$ by

$$\hat{h} \mid X_{i,j} = (\alpha_{i,j})^{-1} h\alpha \mid X_{i,j} \qquad \text{for all } i \text{ and } j.$$

Clearly $\hat{h}$ is an imbedding.

$\alpha(T^n - 4D^n) \subset h\alpha(T^n - 3D^n)$ which implies that $\alpha(4D^n - D^n) \supset h\alpha(\partial 3D^n)$, so $\hat{h}(\partial 3D^n) \subset 4D^n$ and hence $\hat{h}(\partial 3D^n)$ bounds an n-ball in $4D^n$. By coning, we extend $\hat{h} \mid T^n - 3D^n$ to a homeomorphism $H: T^n \to T^n$. H is stable by the lemma, so $\hat{h}$ is stable and h is stable. Hence $N(C, \varepsilon)$ is a neighborhood of the identity consisting of stable homeomorphisms, finishing the proof of Theorem 2.

PROOF OF THEOREM 3. As in the proof of Theorem 2, it suffices to show that a neighborhood of the identity consists of stable homeomorphisms; then $\mathcal{SH}(M^n)$ is both open and closed and therefore contains the identity component.

Let $j: R^n \to M$ be a coordinate patch. Let $\varepsilon > 0$ and $r > 0$ be chosen so that $N(rB^n, \varepsilon) \subset \mathcal{H}(R^n)$ consists of stable homeomorphisms. Then there exists a $\delta < 0$ such that if $h \in N(j(rB^n), \delta) \subset \mathcal{H}(M^n)$, then $hj(2rB^n) \subset j(R^n)$ $j^{-1}hj \mid 2rB^n \in N(rB^n, \varepsilon)$. We may isotope $j^{-1}hj \mid 2rB^n$ to a homeomorphism H of R^n with $H = j^{-1}hj$ on rB^n and therefore $H \in N(rB^n, \varepsilon) \subset \mathcal{H}(R^n)$. Thus H is stable and so $j^{-1}hj \mid 2rB^n$ is stable. By Proposition 5, h is stable, and hence $N(j(rB^n), \delta)$ is our required neighborhood of the identity.

PROOF OF THEOREM 4. We will observe that Theorem 2 can be proved in a "canonical" fashion; that is, if h varies continuously in $\mathcal{H}(R^n)$, then H varies continuously in $\mathcal{H}(T^n)$. First note that $\mathcal{H}(R^n)$ may be contracted onto $\mathcal{H}_0(R^n)$, the homeomorphisms fixing the origin. The immersion $\alpha: T^n - D^n \to R^n$ can be chosen so that $\alpha e = \text{id}$ on $1/4B^n$. Pick a compact set C and $\varepsilon > 0$ as in the proof of Theorem 2 and let $h \in N(C, \varepsilon)$. h lifts canonically to $\hat{h}: T^n - 2D^n \to T^n - D^n$. Since $\hat{h}(\text{int } 5D^n - 2D^n)$ contains $\partial 4D^n$, it follows from Proposition 6 that $\hat{h}(\partial 3D^n)$ bounds a canonical n-ball in $4D^n$. Then $\hat{h} \mid T^n - 3D^n$ extends by coning to $H: T^n \to T^n$.

Clearly $H(t_0) = t_0$ and $H_* = \text{identity}$ so H lifts uniquely to a homeomorphism $g: R^n \to R^n$, with $|g(x) - x| < \text{constant}$ for all $x \in R$, (see the lemma). We have the commutative diagram

$$\begin{array}{ccc}
R^n & \xrightarrow{g} & R^n \\
e \downarrow & & \downarrow e \\
T^n & \xrightarrow{H} & T^n \\
\cup & & \cup \\
T^n - 3D^n & \xrightarrow{\hat{h}} & T^n - 2D^n \\
\alpha \downarrow & & \downarrow \alpha \\
R^n & \xrightarrow{h} & R^n
\end{array}$$

Since $e(1/4B^n) \cap 4D^n = \varnothing$ and $\alpha e = \text{id}$ on $1/4B^n$, it follows that $g = h$ on $1/4B^n$. The construction of g being canonical means that the map $\psi: \mathcal{H}_0(R^n) \to \mathcal{H}_0(R^n)$, defined by $\psi(h) = g$, is continuous.

Let $P_t: R^n \to R^n$, $t \in [0, 1]$, be the isotopy with $P_0 = h$ and $P_1 = g$ defined by

$$P_t(x) = g\left\{\frac{1}{1-t} \cdot \left[g^{-1}h((1-t)x)\right]\right\} \qquad \text{if } t < 1, \text{ and } P_1 = g\,.$$

Let $Q_t: R^n \to R^n$, $t \in [0, 1]$ be the isotopy with $Q_0 = g$ and $Q_1 = \text{identity}$ defined by

$$Q_t(x) = (1-t) \cdot g\left(\frac{1}{1-t} \cdot x\right) \qquad \text{if } t < 1, \text{ and } Q_1 = \text{identity}.$$

Now let $h_t: R^n \to R^n$, $t \in [0, 1]$ be defined by

$$h_t(x) = \begin{cases} P_{2t}(x) & \text{if } 0 \leqq t \leqq 1/2 \\ Q_{2t-1}(x) & \text{if } 1/2 \leqq t \leqq 1\,. \end{cases}$$

It can be verified that h_t is an isotopy of h to the identity which varies continuously with respect to h. Then $H_t: N(C, \varepsilon) \to \mathcal{H}_0(R^n)$, $t \in [0, 1]$ defined by $H_t(h) = h_t$ is a contraction of $N(C, \varepsilon)$ to the identity where $H_t(\text{identity}) = \text{identity}$ for all $t \in [0, 1]$.

This proof can be easily modified to show that if a neighborhood V of the identity in $\mathcal{H}_0(R^n)$ is given, then C and ε may be chosen so that $N(C, \varepsilon)$ contracts to the identity and the contraction takes place in V. To see this, pick $r > 0$ and δ so that $N(rB^n, \delta) \subset V$. Then we may re-define α and e so that $\alpha e = \text{identity}$ on rB^n. If $h \in N(rB^n, \delta)$, then $P_t \in N(rB^n, \delta)$, and if ε is chosen small enough (with respect to δ), then $h \in N(rB^n, \varepsilon)$ implies that $Q_t \in N(rB^n, \delta)$. Therefore $N(rB^n, \varepsilon)$ contracts in V.

UNIVERSITY OF CALIFORNIA, LOS ANGELES, AND
INSTITUTE FOR ADVANCED STUDY

REFERENCES

[1] W. BROWDER, J. Levine, and G. R. Livesay, *Finding a boundary for an open manifold*, Amer. J. Math. **87** (1965), 1017-1028.

[2] MORTON BROWN, *A proof of the generalized Schoenflies theorem*, Bull. Amer. Math. Soc. **66** (1960), 74-76.

[3] ——— and Herman Gluck, *Stable structures on manifolds*, I, II, III, Ann. of Math. **79** (1964), 1-58.

[4] A. V. CERNAVSKII, *The k-stability of homeomorphisms and the union of cells*, Soviet Math. **9** (1968), 729-732.

[5] ———, Local contractibility of groups of homeomorphisms of a manifold, to appear.

[6] E. H. CONNELL, *Approximating stable homeomorphisms by piecewise linear ones*; Ann. of Math. **78** (1963), 326-338.

[7] A. C. CONNOR, *A stable solution to the annulus conjecture*, Notices Amer. Math. Soc. **13** (1966), 620, No. 66 T-338.

[8] MORRIS W. HIRSCH, *On embedding differentiable manifolds in euclidean space*, Ann. of Math. **73** (1961), 566-571.

[9] WILLIAM HUEBSCH and MARSTON MORSE, *The dependence of the Schoenflies extension on an accessory parameter (the topological case)*, Proc. Nat. Acad. Sci. **50** (1963), 1036-1037.

[10] R. C. KIRBY, *On the annulus conjecture*, Proc. Amer. Math. Soc. **17** (1966), 178-185.

[11] J. M. KISTER, *Microbundles are fiber bundles*, Ann. of Math. **80** (1964), 190-199.

[12] W. A. LABACH, *Note on the annulus conjecture*, Proc. Amer. Math. Soc. **18** (1967), 1079.

[13] L. SIEBENMANN, *Pseudo-annuli and invertible cobordisms*, to appear.

[14] ———, *A total Whitehead torsion obstruction to fibering over the circle*, to appear.

[15] C. T. C. WALL, *Surgery on compact manifolds*, to appear.

[16] ———, *On homotopy tori and the annulus theorem*, to appear.

[17] W. C. HSIANG and J. L. SHANESON, *Fake tori, the annulus conjecture, and the conjectures of Kirby*, to appear.

[18] J. L. SHANESON, *Embeddings with codimension two of spheres in spheres and h-cobordisms of* $S^1 \times S^3$, Bull. Amer. Math. Soc. **74** (1968), 972-974.

[19] C. T. C. WALL, *On bundles over a sphere with a fibre euclidean space*, Fund. Math. **LXI** (1967), 57-72.

(Received October 29, 1968)

Annex B. Reprinted with permission, from
Bulletin of the American Mathematical Society
July, 1969, Vol. 75, No. 4, pp. 742-749.

ON THE TRIANGULATION OF MANIFOLDS AND THE HAUPTVERMUTUNG

BY R. C. KIRBY[1] AND L. C. SIEBENMANN[2]

Communicated by William Browder, December 26, 1968

1. The first author's solution of the stable homeomorphism conjecture [**5**] leads naturally to a new method for deciding whether or not every topological manifold of high dimension supports a piecewise linear manifold structure (triangulation problem) that is essentially unique (Hauptvermutung) cf. Sullivan [**14**]. At this time a single obstacle remains[3]—namely to decide whether the homotopy group $\pi_3(\mathrm{TOP}/\mathrm{PL})$ is 0 or Z_2. The positive results we obtain in spite of this obstacle are, in brief, these four: any (metrizable) topological manifold M of dimension $\geqq 6$ is triangulable, i.e. homeomorphic to a piecewise linear (=PL) manifold, provided $H^4(M; Z_2)=0$; a homeomorphism $h: M_1 \to M_2$ of PL manifolds of dimension $\geqq 6$ is isotopic to a PL homeomorphism provided $H^3(M; Z_2)=0$; any compact topological manifold has the homotopy type of a finite complex (with no proviso); any (topological) homeomorphism of compact PL manifolds is a simple homotopy equivalence (again with no proviso).

R. Lashof and M. Rothenberg have proved some of the results of this paper, [**9**] and [**10**]. Our work is independent of [**10**]; on the other hand, Lashof's paper [**9**] was helpful to us in that it showed the relevance of Lees' immersion theorem [**11**] to our work and reinforced our suspicions that the *Classification theorem* below was correct.

We have divided our main result into a *Classification theorem* and a *Structure theorem.*

(I) CLASSIFICATION THEOREM. *Let M^m be any topological manifold of dimension $m \geqq 6$ (or $\geqq 5$ if the boundary ∂M is empty). There is a natural one-to-one correspondence between isotopy classes of PL structures on M and equivalence classes of stable reductions of the tangent microbundle $\tau(M)$ of M to PL microbundle.*

(There are good relative versions of this classification. See [7] and proofs in §2.)

Explanations. Two PL structures Σ and Σ' on M, each defined by a PL compatible atlas of charts, are said to be *isotopic* if there exists a

[1] Partially supported by NSF Grant GP 6530.

[2] Partially supported by NSF Grant GP 7952X.

[3] See note added in proof at end of article.

topological isotopy h_t, $0 \leq t \leq 1$, of 1_M so that h_1 is a PL homeomorphism of (M, Σ) with (M, Σ'). If M has a metric d and ϵ is a continuous function $M \to (0, \infty)$, then h_t is called an ϵ-isotopy provided $d(x, h_t(x)) < \epsilon(x)$ for all $x \in M$ and all $t \in [0, 1]$. To (I) we can add: *Isotopic PL structures are ϵ-isotopic for any ϵ.*

By n-microbundle one can, by the Kister-Mazur theorem, understand simply a locally-product bundle with fiber Euclidean n-space R^n, and zero-section. If ξ is then a TOP (= topological) microbundle, over a locally finite simplicial complex X, with fiber R^n, a reduction of ξ to PL microbundle is given by a triangulation of ξ as a PL microbundle over X. Two such triangulations of ξ give equivalent reductions if the identity of ξ is bundle isotopic to a PL isomorphism from the one PL microbundle structure to the other. The notion of stable reduction differs in allowing addition of a trivial bundle at any moment. Since M is not a priori triangulable, one should, to define reductions, first pull back $\tau(M)$ to a homotopy-equivalent simplicial complex. The total space of a normal microbundle of $M \cup \{$collar on $\partial M\}$ in R^{m+k} (k large) is convenient. This technicality obscures, but does not destroy, the pleasant properties of the notion of reduction.

If TOP_m/PL_m is the fiber of the map, $BPL_m \to BTOP_m$ of classifying spaces for microbundles, define TOP/PL as the telescope of the sequence $TOP_1/PL_1 \to TOP_2/PL_2 \to TOP_3/PL_3 \to \cdots$ arising from stabilization of bundles.

(II) Structure theorem (Partially announced in [8]). *$\pi_k(TOP/PL)$ is 0 if $i \neq 3$ and Z_2 or 0 if $i = 3$. Also $\pi_k(TOP_m/PL_m) \cong \pi_k(TOP/PL)$ by stabilization, for $k < m$, $m \geq 5$.*

When it became known that, with Wall, we had shown that π_k(TOP/PL) is 0 for $k \neq 3$ and $\leq Z_2$ for $k = 3$, we [8] and Lashof and Rothenberg [10] independently noticed that Lees' immersion theorem [11] gives the corresponding nonstable results above. This was of critical importance to Lashof's triangulation theorem [9].

The equivalence classes of stable reductions of $\tau(M)$ can be put in one-to-one correspondence with vertical homotopy classes of sections of a bundle over M with fiber TOP/PL, namely the pull-back by a classifying map $M \to B_{TOP}$ for $\tau(M)$ of the fibration $TOP/PL \to B_{PL} \to B_{TOP}$. Combining (I) and (II) we find

(1) *There is just one well-defined obstruction in $H^4(M; \pi_3(TOP/PL))$ to imposing a PL structure on M.*

(2) *Given one PL structure on M the isotopy classes of PL structures on M are in (1-1)-correspondence with the elements of $H^3(M; \pi_3(TOP/PL))$.*

As applications of (I) alone consider:

(a) The total space E of any normal k-disc bundle [**3**] of M^n in R^{n+k}, $n+k \geqq 6$, is triangulable as a PL manifold, since $\tau(E)$ is trivial.

(b) If $h: E \to E'$ is a homeomorphism of parallelizable PL m-manifolds, there exists a topological disc-bundle automorphism $\alpha: E \times D^s \to E \times D^s$ over E (s large) so that $(h \times id) \circ \alpha: E \times D^s \to E' \times D^s$ is topologically isotopic to a PL homeomorphism.

PROOF OF (b). The PL reduction of $\tau(E)$ given by h is classified by an element $y \in [E, \mathrm{TOP}_m/\mathrm{PL}_m]$. Since $\tau(E)$ and $\tau(E')$ are trivial bundles, y comes from $x \in [E, \mathrm{TOP}_m]$. Represent $-x$ by an automorphism $\beta: E \times R^m \to E \times R^m$ of the trivial R^m bundle. Then, up to bundle isotopy $\beta \times 1_R$ extends [**3**] to a disc bundle automorphism $\alpha: E \times D^{m+1} \to E \times D^{m+1}$, where $R^{m+1} = \mathrm{int} D^{m+1}$. The PL reductions of $\tau(E \times D^{m+1})$ given by $(h \times id) \circ \alpha$ and by $id | E \times D^{m+1}$ are stably the same; so (b) follows from (I).

These seemingly innocent observations readily affirm two important conjectures (cf. [**13**]).

(III) FINITENESS OF COMPACT TYPES. *Every compact topological manifold has the homotopy type of a finite complex—even if it be nontriangulable.*

(IV) TOPOLOGICAL INVARIANCE OF TORSIONS. *Every topological manifold M has a well-defined simple homotopy type*[4], *namely the type of its normal disc bundles triangulated as PL manifolds. In particular, if $h: M \to M'$ is a homeomorophism of compact connected PL manifolds, the Whitehead torsion $\tau(h) \in \mathrm{Wh}(\pi_1 M)$ of h is zero.*

2. We now sketch the proof of (I) and (II). Important elements of it were announced in [**7**], [**8**]. An important role is played by Lees' recent classification theorem for topological immersions incodimension zero [**11**], and by Wall's surgery of nonsimply connected manifolds [**15**]; we suspect that one or both could be eliminated from the proof of (I), but they are essential in the proof of (II). In this regard see the weaker triangulation theorems of Lashof [**9**] and Lees [**11**] proved before (I).

HANDLE STRAIGHTENING PROBLEM $P(h)$. *Consider a homeomorphism $h: B^k \times R^n \to V^m$, $k+n=m$, (where B^k = standard PL k-ball in R^k) onto a PL manifold V so that $h | \partial B^k \times R^n$ is a PL homeomorphism. Can one find a topological isotopy $h_t: B^k \times R^n \to V^n$, $0 \leqq t \leqq 1$, of $h = h_0$ such that*

(1) $h_1 | B^k \times B^n$ is PL,

(2) $h_t = h$, $0 \leqq t \leqq 1$, on $\partial B^k \times R^n \cup B^k \times (R^n - rB^n)$ for some r ?

---

[4] This makes good sense even when M is noncompact, cf. [**13**, p. 74].

One should think of $B^k \times R^n$ as an open PL handle with core $B^k \times 0$. The analysis of $P(h)$ is based on the *main diagram* below, which for $k=0$, originated in [5].

MAIN DIAGRAM

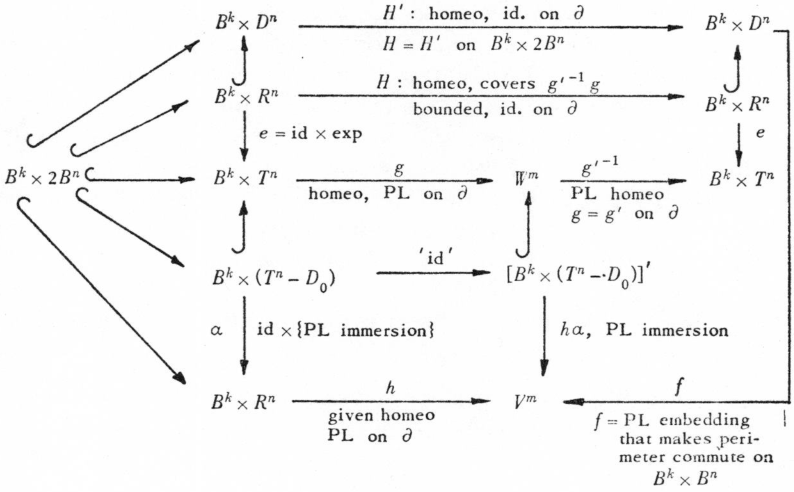

$m = n+k \geqq 5$, ∂ indicates boundary, B^k = standard PL k-ball in R^k.

D^n = a PL n-ball with int$D^n = R^n$ = euclidean n-space, so that bounded maps $R^n \to R^n$ extend by the identity to D^n.

T^n = n-torus, the n-fold product of circles; D_0 = a PL n-ball, collared in T^n.

Arrange that h is PL near $\partial B^k \times R^n$ and successively construct e, α, W, g, g' (when possible), H, H', f.

Explanations. α, e and the inclusions of $B^k \times 2B^n$ are chosen so that the triangles at left commute; $B^k \times (T^n - D_0)$ with the unique PL structure making $h\alpha$ PL is by definition $[B^k \times (T^n - D_0)]'$; W is PL and "caps it off". g extends *id*. Finding g' is the

TORUS PROBLEM $Q(g)$. *To extend* $g | \partial B^k \times T^n$ *to a PL homeomorphism homotopic to* g.

Supposing g' solves $Q(g)$, form H and 'squeeze' to allow extension to H'; then define f by engulfing to extend the natural PL identification $H'(B^k \times B^n) \to h(B^k \times B^n)$. Since f embeds the ball $B^k \times D^n$, for any isotopy H'_t of H' fixing ∂ (= boundary), there exists an isotopy h_t of h fixing ∂ and $B^k \times R^n - h^{-1}f(B^k \times D^n)$ so that, for each t, $h_t = fH'_t$ on $B^k \times B^n$. Thus, if H'_t is the Alexander isotopy of H' to the identity, h_t solves $P(h)$!!

PROPOSITION 1. *If g is derived from h as above, $Q(g)$ is solvable$\Leftrightarrow P(h)$ is solvable.*

PROOF. It remains to show $\Leftarrow$. Using a solution h_t of $P(h)$ form an isotopy $\bar{h}: I\times B^k\times R^n \to I\times V$, fixing ∂, from h to a PL homeomorphism. One can extend the construction of g from h to get from the isotopy $\bar{h}$ a homeomorphism $\bar{g}: I\times B^k\times T^n \to W'$ with $\bar{g}|\{0\}\times B^k\times T^n = g$, that is PL on ∂ minus $\{0\}\times \text{int}B^k\times T^n$. Applying the s-cobordism theorem to W' one gets a solution of $Q(g)$. ■

Given the problem $P(h)$ consider the tangent bundle map $h_*: \tau(B^k\times R^n) \to \tau(V^m)$. Both bundles are PL trivial. Restricting over $B^k\times\{0\}$ in the source and projecting to the fiber in the target we get a map $(B^k, \partial B^k) \to (\text{TOP}_m, \text{PL}_m)$ (to be understood semisimplicially [12, §2]). Call its class $d(h) \in \pi_k(\text{TOP}_m, \text{PL}_m) \cong \pi_k(\text{TOP}_m/\text{PL}_m)$. Denote this group by $\Pi_k(m)$.

PROPOSITION 2. *$P(h)$ is solvable$\Leftrightarrow d(h)=0$.*

PROOF. $\Rightarrow$ is trivial. If $d(h)=0$, Lees' immersion theorem [**11**], together with its familiar PL counterpart due to Haefliger and Poenaru, provides a regular homotopy (modulo ∂) from h to a PL immersion. The argument under Proposition 1 applies to solve $Q(g)$ and hence $P(h)$. ■

PROPOSITION 2′. *For each $x\in\Pi_k(m)$, $k<m$, there exists a problem $P(h)$ with $d(h)=x$.*

PROOF. Immersion theory here provides an immersion $h_0: B^k\times R^n \to R^m$, PL on ∂, such that if V^m is $B^k\times R^n$ with the PL structure making h_0 PL, then $h=id: B^k\times R^n \to V^m$ has $d(h)=x$. ■

For any solution h_t of $P(h)$, the induced tangent bundle map gives a map $f: I\times B^k \to \text{TOP}_m$ sending $I\times\partial B^k \cup \{1\}\times B^k$ into PL_m. Thus, given two solutions h_t, h_t' of $P(h)$ we can piece together f, f' to get a 'difference' class $\delta(h_t, h_t') \in \pi_{k+1}(\text{TOP}_m, \text{PL}_m) = \pi_{R+1}(m)$.

PROPOSITION 3. *Let h_t solve $P(h)$ for $h: B^k\times R^n \to V^m$, $k<m$. Given any y in $\Pi_{k+1}(m+1)$ there exists a solution h_t' of $P(h)$ such that $s\delta(h_t, h_t') = y$ (s denotes stabilization which by* (II) *is an isomorphism).*

INDICATION OF PROOF. One can reduce to the special case where h_t is the identity solution of $P(id|B^k\times R^n)$. Proposition 2′ provides a problem $P(h_1)$, $h_1: B^{k+1}\times R^n \to V^{m+1}$ with $d(h_1)=y$, which yields a torus problem $Q(g_1)$, $g_1: B^{k+1}\times T^n \to W^{m+1}$. Impose on $B^{k+1}\times T^n = I\times B^k\times T^n$ the PL structure Σ making g_1 PL. Σ is standard on ∂. Apply the s-cobordism theorem to $(I; 0, 1)\times B^k\times T^n$ with structure

Σ to derive a PL automorphism θ of $\{1\} \times B^k \times T^n$ that fixes ∂. Solve $P(id \mid B^k \times R^n)$ using θ (for g' of the main diagram) in place of id, and call the solution h_t'. Then $s\delta(id, h_t') = y$.

PROOF OF (II). (a) *For $m > k > 3$ there is an elementary proof that* $\Pi_k(m) = 0$. It is an induction on $n = m - k \geqq 1$ exploiting the main diagram. Also, surgery can be used as in (*b*).

(b) $\Pi_k(m) = 0$ *for* $m \geqq 5$ *and* $k = 0, 1, 2$. Represent $x \in \Pi_k(m)$ by a problem $P(h)$ and pass to the torus problem $Q(g)$. There is an obstruction $[g] \in H^{3-k}(T^n; Z_2)$ to solving $Q(g)$. For $[g]$ to exist, g need only be a homotopy equivalence which is a PL homeomorphism on ∂, and, strictly speaking, $[g]$ is the obstruction to homotoping g (modulo ∂) to a PL homeomorphism. Note that to solve $P(h)$ it suffices to solve $Q(\bar{g})$ where $\bar{g}$ is the covering of g for $(2Z)^n \subset Z^n = \pi_1(T^n)$.

$$\begin{array}{ccc} B^k \times T^n & \xrightarrow{\bar{g}} & \overline{W} \\ p \Big\downarrow 2^n & & \Big\downarrow 2^n\text{-fold covering map.} \\ B^k \times T^n & \xrightarrow{g} & W \end{array}$$

To see this just add a new tier over g in the main diagram! Now $[\bar{g}] = p^*[g]$. But p^* kills $H^{3-k}(T^n; Z_2)$. Thus $P(h)$ is solvable and $x = 0$. (Also $Q(g)$ is solvable by Proposition 1, so $[g]$ was already 0!)

(c) $\Pi_3(m) \subset Z_2$ *for* $m \geqq 5$. In the above argument $p^* = id$. But by inspection of definitions $x \mapsto [g] \in H^0(T^n; Z_2)$ is additive.

(d) *Stability*: $s: \Pi_3(m) \to \Pi_3(m+1)$ *is an isomorphism*, $m \geqq 5$. That s is onto follows by the descent argument for (a). To prove s is injective we check that, if $P(h)$ gives $Q(g)$ with surgery obstruction $y \in Z_2$, then $P(h \times 1_R)$ also gives $y \in Z_2$.

Prior to this work, C. T. C. Wall, W. C. Hsiang, and J. Shaneson (jointly), and A. Casson understood the classification of homotopy tori (the case when $k = 0$ above).[5] When our specific questions (see [**6**] and [**8**]) were posed, Wall [**16**] and Hsiang and Shaneson [**4**], independently verified that $[\bar{g}] = p^*[g]$ and extended their work to the cases $k \neq 0$. It seems to us that the fibration theorem of Farrell [**1**] reworked in [**2**] and [**15**]) plays an essential role in the construction and use of $[g]$. We thank W. Browder for a very transparent definition of $[g]$.

Propositions 1, 2, 2′, 3 and the above stability permit us to prove (I).

Uniqueness (Hauptvermutung). Consider an unbounded TOP manifold M with two PL structures Σ, Σ' giving reductions ρ, ρ' of

[5] Casson apparently had examined the general case $k \geqq 0$ [**16**].

$\tau(M)$ to PL microbundle that are (stably) related by a deformation σ of reductions. Take a handle decomposition of M, and suppose that, for a subhandlebody M_0, one has found an isotopy h_t, $0 \leqq t \leqq s$, of 1_M with $h_s | M_0$ being PL, which realizes $\sigma | M_0$. If H is a k-handle attached to M_0, the obstruction in $\pi_k(\mathrm{TOP/PL})$ (Proposition 2) to prolonging h_t fixing M_0 to an isotopy h_t, $0 \leqq t \leqq s+1$, with $h_{s+1} | M_0 \cup H$ being PL, can be identified with the obstruction to extending over $M_0 \cup H$ the deformation determined by $h_t | M_0$ between the reductions given by Σ and Σ'. It is zero because $\sigma | M_0 \cup H$ exists. Proposition 3 allows us to choose h_t, $0 \leqq t \leqq s+1$, so as to realize $\sigma | M_0 \cup H$. This indicates how to construct inductively the desired isotopy. By using small handles, one can make this isotopy small. In a relative form, which this argument also yields, the case for open manifolds takes care of manifolds with boundary.

Existence (Triangulation). Once it is stated in a relative form (see [7]) we can assume that $\partial M = \emptyset$ and (by passing to charts of a locally finite cover) that M is triangulable. Again we use a handle induction, and spread the PL structure handle by handle. Proposition 2′ takes the above role of Proposition 3, so this argument for existence is simpler than that for uniqueness above.

ADDED IN PROOF. We have proved that π_3 (TOP/PL) is Z_2 not 0, see [17]. Here is an argument in outline. Surgery provides a homotopy equivalence $g: B^1 \times T^n \to W$, $n \geqq 5$, that is a PL homeomorphism on ∂ and has nonzero invariant $[g] \in H^2(T^n; Z_2)$. (Here and below compare proof of II.) Using the s-cobordism theorem (cf. Proposition 3) one derives from g a PL automorphism h of T^n, well-defined by $[g]$ up to PL pseudo-isotopy (concordance). Let g_λ, $\lambda = 1, 3, 5, \cdots$ be the standard λ^n-fold covering of g and let h_λ be a λ^n-fold covering of h derived from g_λ. For λ large, h_λ can be arbitrarily close to $id | T^n$; hence is topologically isotopic to $id | T^n$ by a result of Černavskii (Doklady 1968) provable [6] by a method of [5]. Thus if $\pi_*(\mathrm{TOP/PL})$ were zero, h_λ would be at least PL pseudo-isotopic to $id | T^n$. But it is not, since $[g\lambda] = [g] \neq 0$ as λ is odd. Therefore $\pi_3(\mathrm{TOP/PL}) = Z_2$.

This discovery leads to many striking conclusions (e.g. see Notices Amer. Math. Soc., June 1969). These will be discussed fully in a paper devoted to a careful development of our results.

REFERENCES

1. F. T. Farrell, Thesis, Yale University, New Haven, Conn., 1967.
2. F. T. Farrell and W. C. Hsiang, *Manifolds with* $\pi_1 = Z \times_\alpha G$ (to appear).
3. M. W. Hirsch, *On non-linear cell-bundles*, Ann. of Math (2) **84** (1966), 373–385.

4. W. C. Hsiang and J. L. Shaneson, *Fake tori, the annulus conjecture and the conjectures of Kirby*, Proc. Nat. Acad. Sci. U.S.A. (to appear).

5. R. C. Kirby, *Stable homeomorphisms and the annulus conjecture*, Ann. of Math. (2) **90** (1969).

6. ———, Announcement distributed with preprint of [5], 1968.

7. R. C. Kirby and L. C. Siebenmann, *A triangulation theorem*, Notices Amer. Math. Soc. **16** (1969), 433.

8. R. C. Kirby, L. C. Siebenmann and C. T. C. Wall, *The annulus conjecture and triangulation*, Notices Amer. Math. Soc. **16** (1969), 432.

9. R. K. Lashof, *Lees' immersion theorem and the triangulation of manifolds*, Bull. Amer. Math. Soc. **75** (1969), 535–538.

10. R. K. Lashof and M. Rothenberg, *Triangulation of manifolds*. I, II, Bull. Amer. Math. Soc. **75** (1969), 750–754, 755–757.

11. J. A. Lees, *Immersions and surgeries of topological manifolds*, Bull. Amer. Math. Soc. **75** (1969), 529–534.

12. C. P. Rourke and B. J. Sanderson, *On the homotopy theory of Δ-sets* (to appear).

13. L. C. Siebenmann, *On the homotopy type of compact topological manifolds*, Bull. Amer. Math. Soc. **74** (1968), 738–742.

14. D. P. Sullivan, *On the hauptvermutung for manifolds*, Bull. Amer. Math. Soc. **63** (1967), 598–600.

15. C. T. C. Wall, *Surgery on compact manifolds*, preprint, University of Liverpool.

16. ———, *On homotopy tori and the annulus theorem*, Proc. London Math. Soc. (to appear).

17. R. C. Kirby and L. C. Siebenmann, *For manifolds the Hauptvermutung and the triangulation conjecture are false*, Notices Amer. Math. Soc. **16** (1969), 695.

University of California, Los Angeles, California 90024 and
Institute for Advanced Study, Princeton, New Jersey 08540

Annex C. Reprinted with permission, from
Proceedings of the International Congress of Mathematicians
Nice, September, 1970, Gauthier-Villars, éditeur, Paris 6e,
1971, Volume 2, pp. 133-163.

TOPOLOGICAL MANIFOLDS *

by L. C. SIEBENMANN

0. Introduction.

Homeomorphisms – topological isomorphisms – have repeatedly turned up in theorems of a strikingly conceptual character. For example :

(1) (19th century). There are continuously many non-isomorphic compact Riemann surfaces, but, up to homeomorphism, only one of each genus.

(2) (B. Mazur 1959). Every smoothly embedded $(n-1)$-sphere in euclidean n-space R^n bounds a topological n-ball.

(3) (R. Thom and J. Mather, recent work). Among smooth maps of one compact smooth manifold to another the topologically stable ones form a dense open set.

In these examples and many others, homeomorphisms serve to reveal basic relationships by conveniently erasing some finer distinctions.

In this important role, PL (= piecewise-linear)(**) homeomorphisms of simplicial complexes have until recently been favored because homeomorphisms in general seemed intractable. However, PL homeomorphisms have limitations, some of them obvious ; to illustrate, the smooth, non-singular self-homeomorphism $f : R \to R$ of the line given by $f(x) = x + \frac{1}{4} \exp(-1/x^2) \sin(1/x)$ can in no way be regarded as a PL self-homeomorphism since it has infinitely many isolated fixed points near the origin.

Developments that have intervened since 1966 fortunately have vastly increased our understanding of homeomorphisms and of their natural home, the category of (finite dimensional) topological manifolds(***). I will describe just a few of them below. One can expect that mathematicians will consequently come to use freely the notions of homeomorphism and topological manifold untroubled by the frustrating difficulties that worried their early history.

(*) This report is based on theorems concerning homeomorphisms and topological manifolds [44] [45] [46] [46 A] developed with R.C. Kirby as a sequel to [42]. I have reviewed some contiguous material and included a collection of examples related to my observation that $\pi_3(\mathrm{TOP/PL}) \neq 0$. My oral report was largely devoted to results now adequately described in [81], [82].

(**) A continuous map $f : X \to Y$ of (locally finite) simplicial complexes is called PL if there exists a simplicial complex X' and a homeomorphism $s : X' \to X$ such that s and fs each map each simplex of X' (affine) linearly into some simplex.

(***) In some situations one can comfortably go beyond manifolds [82]. Also, there has been dramatic progress with infinite dimensional topological manifolds (see [48]).

1. **History.**

A topological (= TOP) *m*-manifold M^m (with boundary) is a metrizable topological space in which each point has an open neighborhood U that admits an open embedding (called a *chart*) $f : U \to R^m_+ = \{(x_1, \ldots, x_m) \in R^m | x_1 \geqslant 0\}$, giving a homeomorphism $U \approx f(U)$.

From Poincaré's day until the last decade, the lack of techniques for working with homeomorphisms in euclidean space R^m (m large) forced topologists to restrict attention to manifolds M^m equipped with atlases of charts $f_\alpha : U_\alpha \to R^m_+$, $\cup U_\alpha = M$, (α varing in some index set), in which the maps $f_\beta f_\alpha^{-1}$ (where defined) are especially tractable, for example all DIFF (infinitely differentiable), or all PL (piecewise linear). Maximal such atlases are called respectively DIFF or PL manifold structures. Poincaré, for one, was emphatic about the importance of the naked homeomorphism – when writing philosophically [68, §§ 1, 2] – yet his memoirs treat DIFF or PL manifolds only.

Until 1956 the study of TOP manifolds as such was restricted to sporadic attempts to prove existence of a PL atlas (= *triangulation conjecture*) and its essential uniqueness (= *Hauptvermutung*). For $m = 2$, Rado proved existence, 1924 [70] (Kerékjártó's classification 1923 [38] implied uniqueness up to isomorphism). For $m = 3$, Moise proved existence and uniqueness, 1952 [62], cf. a misproof of Furch 1924 [21].

A PL manifold is easily shown to be PL homeomorphic to a simplicial complex that is a so-called combinatorial manifold [37]. So the *triangulation conjecture* is that any TOP manifold M^m admits a homeomorphism $h : M \to N$ to a combinatorial manifold. The *Hauptvermutung* conjectures that if h and $h' : M \to N'$ are two such, then the homeomorphism $h'h^{-1} : N \to N'$ can be replaced by a PL homeomorphism $g : N \to N'$. One might reasonably demand that g be *topologically isotopic* to $h'h^{-1}$, or again *homotopic* to it. These variants of the Hauptvermutung will reappear in §5 and §15.

The Hauptvermutung was first formultated in print by Steinitz 1907 (see [85]). Around 1930, after homology groups had been proved to be topological invariants without it, H. Kneser and J.W. Alexander began to advertise the Hauptvermutung for its own sake, and the triangulation conjecture as well [47] [2]. Only a misproof of Nöebling [66] (for any m) ensued in the 1930's. Soberingly delicate proofs of triangulability of DIFF manifolds by Cairns and Whitehead appeared instead.

Milnor's proof (1956) that some 'well-known' S^3 bundles over S^4 are homeomorphic to S^7 but not DIFF isomorphic to S^7 strongly revived interest. It was very relevant ; indeed homotopy theory sees the failure of the Hauptvermutung (1969) as quite analogous. The latter gives the first nonzero homotopy group $\pi_3(\mathrm{TOP/O}) = Z_2$ of TOP/O ; Milnor's exotic 7-spheres form the second $\pi_7(\mathrm{TOP/O}) = Z_{28}$.

In the early 1960's, intense efforts by many mathematicians to unlock the geometric secrets of topological manifolds brought a few unqualified successes : for example the generalized Shoenflies theorem was proved by M. Brown [7] ; the tangent microbundle was developed by Milnor [60] ; the topological Poincaré conjecture in dimensions $\geqslant 5$ was proved by M.H.A. Newman [65].

Of fundamental importance to TOP manifolds were Černavskii's proof in 1968 that the homeomorphism group of a compact manifold is locally contractible [10] [11], and Kirby's proof in 1968 of the stable homeomorphism conjecture with the help of surgery [42]. Key geometric techniques were involved – a meshing idea in the former, a particularly artful torus furling and unfurling idea(*) in the latter. The disproof of the Hauptvermutung and the triangulation conjecture I sketch below uses neither, but was conceived using both. (See [44] [44 B] [46 A] for alternatives).

2. **Failure of the Hauptvermutung and the triangulation conjecture.**

This section presents the most elementary disproof I know. I constructed it for the Arbeitstagung, Bonn, 1969.

In this discussion $B^n = [-1, 1]^n \subset R^n$ is the standard PL ball ; and the sphere $S^{n-1} = \partial B^n$ is the boundary of B^n. $T^n = R^n/Z^n$ is the standard PL torus, the n-fold product of circles. The closed interval $[0, 1]$ is denoted I.

As starting material we take a certain PL automorphism α of $B^2 \times T^n, n \geqslant 3$, fixing boundary that is constructed to have two special properties (1) and (2) below. The existance of α was established by Wall, Hsiang and Shaneson, and Casson in 1968 using sophisticated surgical techniques of Wall (see [35] [95]). A rather naive construction is given in [80, § 5], which manages to avoid surgery obstruction groups entirely. To establish (1) and (2) it requires only the s-cobordism theorem and some unobstructed surgery with boundary, that works from the affine locus $Q^4 : z_1^5 + z_2^3 + z_3^2 = 1$ in C^3. This Q^4 coincides with Milnor's E_8 plumbing of dimension 4 ; it has signature 8 and a collar neighborhood of infinity $M^3 \times R$, where $M^3 = SO(3)/A_5$ is Poincaré's homology 3-sphere, cf. [61, § 9.8].

(1) *The automorphism β induced by α on the quotient T^{2+n} of $B^2 \times T^n$ (obtained by identifying opposite sides of the square B^2) has mapping torus*

$$T(\beta) = I \times T^{2+n}/\{(0, x) = (1, \beta(x))\}$$

not PL *isomorphic to T^{3+n} ; indeed there exists(**) a* PL *cobordism $(W ; T^{n+3}, T(\beta))$ and a homotopy equivalence of W to $\{I \times T^3 \# Q \cup \infty\} \times T^n$ extending the standard equivalences $T^{3+n} \simeq 0 \times T^3 \times T^n$ and $T(\beta) \simeq 1 \times T^3 \times T^n$. The symbol # indicates (interior) connected sum* [41].

(2) *For any standard covering map $p : B^2 \times T^n \to B^2 \times T^n$ the covering automorphism α_1 of α fixing boundary is* PL *pseudo-isotopic to α fixing boundary. (Covering means that $p\alpha_1 = \alpha p$). In other words, there exists a* PL *automorphism H of $(I ; 0, 1) \times B^2 \times T^n$ fixing $I \times \partial B^2 \times T^n$ such that $H | 0 \times B^2 \times T^n = 0 \times \alpha$ and $H | 1 \times B^2 \times T^n = 1 \times \alpha_1$.*

(*) Novikov first exploited a torus *furling* idea in 1965 to prove the topological invariance of rational Pontrjagin classes [67]. And this led to Sullivan's partial proof of the Hauptvermutung [88]. Kirby's *unfurling* of the torus was a fresh idea that proved revolutionary.

(**) This is the key property. It explains the exoticity of $T(\beta)$ – (see end of argument), and the property (2) – (almost, see [80, § 5]).

In (2) choose p to be the 2^n-fold covering derived from scalar multiplication by 2 in R^n. (Any integer > 1 would do as well as 2.) Let $\alpha_0 (= \alpha), \alpha_1, \alpha_2, \ldots$ be the sequence of automorphisms of $B^2 \times T^n$ fixing boundary such that α_{k+1} covers α_k, i.e. $p\alpha_{k+1} = \alpha_k p$. Similarly define $H_0 (= H), H_1, H_2, \ldots$ and note that H_k is a PL concordance fixing boundary from α_k to α_{k+1}. Next define a PL automorphism H' of $[0, 1) \times B^2 \times T^n$ by making $H' | [a_k, a_{k+1}] \times B^2 \times T^n$, where $a_k = 1 - \frac{1}{2^k}$, correspond to H_k under the (oriented) linear map of $[a_k, a_{k+1}]$ onto $[0, 1] = I$. We extend H' by the identity to $[0, 1) \times R^2 \times T^n$. Define another self-homeomorphism H'' of $[0, 1) \times B^2 \times T^n$ by $H'' = \varphi H' \varphi^{-1}$ where

$$\varphi(t, x, y) = (t, (1-t)x, y) \quad .$$

Finally extend H'' by the identity to a bijection

$$H'' : I \times B^2 \times T^n \to I \times B^2 \times T^n \quad .$$

It is also continuous, hence a homeomorphism. To prove this, consider a sequence $q_1, q_2, \ldots$ of points converging to $q = (t_0, x_0, y_0)$ in $I \times B^2 \times T^n$. Convergence $H''(q_j) \to H''(q)$ is evident except when $t_0 = 1, x_0 = 0$. In the latter case it is easy to check that $p_1 H''(q_j) \to p_1 H''(q) = 1$ and $p_2 H''(q_j) \to p_2 H''(q) = 0$ as $j \to \infty$, where p_i, $i = 1, 2, 3$ is projection to the i-th factor of $I \times B^2 \times T^n$. It is not as obvious that $p_3 H''(q_j) \to p_3 H''(q) = y_0$. To see this, let

$$\widetilde{H}_k : I \times B^2 \times R^n \to I \times B^2 \times R^n$$

be the universal covering of H_k fixing $I \times \partial B^2 \times R^n$. Now

$$\sup \{ |p_3 z - p_3 \widetilde{H}_k z| \quad ; \quad z \in [0, 1] \times B^2 \times R^n \} \equiv D_k$$

is finite, being realized on the compactum $I \times B^2 \times I^n$. And, as $\widetilde{H}_k$ is clearly $\theta_k^{-1} \widetilde{H}_0 \theta_k$, where $\theta_n(t, x, y) = (t, x, 2^n y)$, we have $D_k = \frac{1}{2^k} D_0$. Now D_k is $\geqslant$ the maximum distance of $p_3 H_k$ from p_3, for the quotient metric on $T^n = R^n / Z^n$; so $D_k \to 0$ implies $p_3 H''(q_j) \to p_3 H''(q) = y_0$, as $j \to \infty$.

As the homeomorphism H'' is the identity on $I \times \partial B^2 \times T^n$ it yields a self-homeomorphism g of the quotient $I \times T^2 \times T^n = I \times T^{2+n}$. And as

$$g | 0 \times T^{2+n} = 0 \times \beta \quad ,$$

and $g | 1 \times T^{2+n} =$ identity, g gives a homeomorphism h of $T(\beta)$ onto

$$T(\mathrm{id}) = T^1 \times T^{2+n} = T^{3+n}$$

by the rule sending points (t, z) to $g^{-1}(t, z)$ – hence $(0, z)$ to $(0, \beta^{-1}(z))$ and $(1, z)$ to $(1, z)$

The homeomorphism $h : T^{3+n} \approx T(\beta)$ *belies the Hauptvermutung.* Further, (1) offers a certain PL cobordism $(W; T^{3+n}, T(\beta))$. Identifying T^{3+n} in W to $T(\beta)$ under h we get a closed topological manifold

$$X^{4+n} \simeq \{T^1 \times T^3 \# Q \cup \infty\} \times T^n$$

($\simeq$ indicating homotopy equivalence).

If it had a PL manifold structure the fibering theorem of Farrell [19] (or the author's thesis) would produce a PL 4-manifold X^4 with $w_1(X^4) = w_2(X^4) = 0$ and signature $\sigma(X^4) \equiv \sigma(S^1 \times T^3 \# Q \cup \infty) \equiv \sigma(Q \cup \infty) \equiv 8$ mod. 16, cf. [80, § 5]. Rohlin's theorem [71] [40] cf. § 13 shows this X^4 doesn't exist. *Hence* X^{4+n} *has no* PL *manifold structure.*

Let us reflect a little on the generation of the homeomorphism $h : T(\beta) \approx T^{3+n}$. The behaviour of H'' is described in figure 2-a (which is accurate for B^1 in place of B^2 and for $n = 1$) by partitioning the fundamental domain $I \times B^2 \times I^n$ according to the behavior of H''. The letter α indicates codimension 1 cubes on which H'' is a conjugate of α.

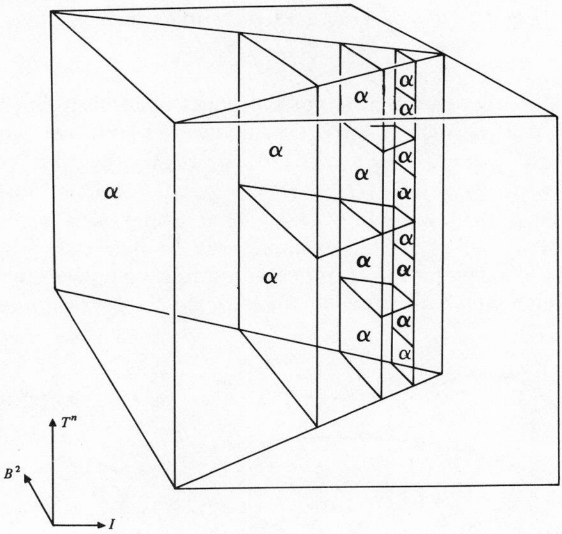

Figure 2a

Observe the infinite ramification (2^n-fold) into smaller and smaller domains converging to all of $1 \times 0 \times T^n$. In the terminology of Thom [92, figure 7] this reveals the failure of the Hauptvermutung to be a *generalized catastrophe* !

Remark 2.1. — Inspection shows that $h : T(\beta) \approx T^{3+n}$ is a Lipschitz homeomorphism and hence X^{4+n} is a Lipschitz manifold as defined by Whitehead [98] for the pseudogroup of Lipschitz homeomorphisms — see § 4. A proof that $T(\beta) \approx T^{3+n}$ (as given in [44]) using local contractibility of a homeomorphism group would not reveal this as no such theorem is known for Lipschitz homeomorphisms. Recall that a theorem of Rademacher [69] says that every Lipschitz homeomorphism of one open subset of R^m to another is almost everywhere differentiable.

3. The unrestricted triangulation conjecture.

When a topological manifold admits no PL manifold structure we know it is not homeomorphic to a simplicial complex which is a combinatorial manifold [37]. But it may be homeomorphic to *some* (less regular) simplicial complex — i.e. triangulable in an unrestricted sense, cf. [79]. For example $Q \cup \infty$ (from § 2) is triangulable and Milnor (Seattle 1963) asked if $(Q \cup \infty) \times S^1$ is a topological manifold even though $Q \cup \infty$ obviously is not one. If so, the manifold X^{4+n} of § 2 is easily triangulated.

If all TOP manifolds be triangulable, why not conjecture that that every locally triangulable metric space is triangulable ?

Here is a construction for a compactum X that is *locally triangulable* but is *non-triangulable.* Let L_1, L_2 be closed PL manifolds and

$$(W ; L_1 \times R , L_2 \times R)$$

an invertible(*) PL cobordism that is not a product cobordism. Such a W exists for instance if $\pi_1 L_i = Z_{257}$ and $L_1 \cong L_2$, compare [78]. It can cover an invertible cobordism $(W', L_1 \times S^1, L_2 \times S^1)$ [77, § 4]. To the Alexandroff compactification $W \cup \infty$ of W adjoin $\{(L_1 \times R) \cup \infty\} \times [0, 1]$ identifying each point $(x , 1)$ in the latter to the point x in $W \cup \infty$. The resulting space is X. See Figure 3-a. The properties of X and of related examples will be demonstrated in [83]. They complement Milnor's examples [57] of homeomorphic complexes that are PL (combinatorially) distinct, which disproved an *unrestricted Hauptvermutung.*

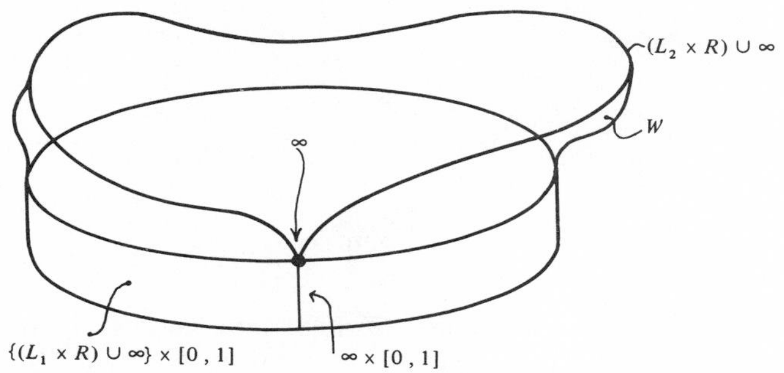

Figure 3a

(*) This means that W can be expressed as a union $W = C_1 \cup C_2$, where C_i is a closed collar neighborhood of $L_i \times R$ in W.

4. Structures on topological manifolds.

Given a TOP manifold M^m (without boundary) and a pseudo-group G of homeomorphisms(*) of one open subset of R^m to another, the problem is to find and classify G-structures on M^m. These are maximal "G-compatible" atlases $\{U_\alpha, f_\alpha\}$ of charts (= open embeddings) $f_\alpha : U_\alpha \to R^m$ so that each $f_\beta f_\alpha^{-1}$ is in G. (Cf. [29] or [48].)

One reduction of this problem to homotopy theoretic form has been given recently by Haefliger [28] [29]. Let $G(M^m)$ be the (polyhedral quasi-) space (**) of G-structures on M. A map of a compact polyhedron P to $G(M)$ is by definition a G-foliation $\mathfrak{F}$ on $P \times M$ transverse to the projection $p_1 : P \times M \to P$ (i.e. its defining submersions are transverse to p_1)(***). Thus, for each $t \in P$, $\mathfrak{F}$ restricts to a G-structure on $t \times M$ and, on each leaf of $\mathfrak{F}$, p_1 is an open embedding. Also note that $\mathfrak{F}$ gives a G_P-structure on $P \times M$ where G_P is the pseudo-group of homeomorphisms of open subsets of $P \times R^m$ locally of the form $(t, x) \to (t, g(x))$ with $g \in G$. If G consists of PL or DIFF homeomorphisms and $P = [0, 1]$, then $\mathfrak{F}$ gives (a fortiori) what is called a *sliced concordance* of PL or DIFF structures on M (see [45] [46]).

We would like to analyse $G(M^m)$ using Milnor's tangent R^m-microbundle $\tau(M)$ of M, which consists of total space $E(\tau M) = M \times M$, projection $p_1 : M \times M \to M$, and (diagonal) section $\delta : M \to M \times M$, $\delta(x) = (x, x)$. Now if ξ^m is *any* R^m microbundle over a space X we can consider $G^\perp(\xi)$ the space of G-foliations of $E(\xi)$ transverse to the fibers. A map $P \to G^\perp(\xi)$ is a G-foliation $\mathfrak{F}$ defined on an open neighborhood of the section $P \times X$ in the total space $E(P \times \xi) = P \times E(\xi)$ that is transverse to the projection to $P \times X$. Notice that there is a natural map

$$d : G(M^m) \to G^\perp(\tau M^m) \quad ,$$

which we call the *differential*. To a G-foliation $\mathfrak{F}$ of $P \times M$ transverse to p_1, it assigns the G-foliation $d\mathfrak{F}$ on $P \times M \times M = E(P \times \tau(M))$ obtained from $\mathfrak{F} \times M$

(*) e.g. the PL isomorphisms, or Lipschitz or DIFF or analytic isomorphisms. Do not confuse G with the stable monoid $G = \cup G_n$ of § 5.5.

(**) Formally such a space X is a contravariant functor $X : P \to [P, X]$ from the category of PL maps of compact polyhedra (denoted P, Q etc.) to the category of sets, which carries union to fiber product. Intuitively X is a space of which we need (or want or can) only know the maps of polyhedra to it.

(***) A *G-foliation* on a space X is a maximal G-compatible atlas $\{V_a, g_a\}$ of topological submersions $g_a : V_a \to R^m$. (See articles of Bott and Wall in these proceedings.) A map $g : V \to W$ is a *topological submersion* if it is locally a projection in the sense that for each x in V there exists an open neighborhood W_x of $g(x)$ in W a space F_x and an open embedding onto a neighborhood of x, called a *product chart* about x, $\varphi : F_x \times W_x \to V$ such that $g\varphi$ is projection $p_2 : F_x \times W_x \to W_x \subset W$. One says that g is *transverse* to another submersion $g' : V \to W'$ if for each x, φ can be chosen so that $F_x = W'_x \times F'_x$ and $g'\varphi$ is projection to W'_x an open subset of W'. This says roughly that the leaves (= fibers) of f and g intersect in general position. Above they intersect in points.

by *interchanging* the factors M. If P is a point, the leaves of $d\mathfrak{F}$ are simply

$$\{P \times M \times x \mid x \in M\} \quad .$$

Clearly $d\mathfrak{F}$ is transverse to the projection $P \times p_1$ to $P \times M$.

THEOREM 4.1. CLASSIFICATION BY FOLIATED MICROBUNDLES. — *The differential*

$$d : G(M^m) \to G^{\perp}(\tau M^m)$$

is a weak homotopy equivalence for each open (metrizable) m-manifold M^m with no compact components.

Haefliger deduces this result (or at least the bijection of components) from the topological version of the Phillips-Gromov transversality theorem classifying maps of M transverse to a TOP foliation. (See [29] and J.C. Hausmann's appendix).

As formulated here, 4.1 invites a direct proof using Gromov's distillation of immersion theory [25] [26]. This does not seem to have been pointed out before, and it seems a worthwhile observation, for I believe the transversality result adequate for 4.1 requires noticeably more geometric technicalities. In order to apply Gromov's distillation, there are two key points to check. For any $C \subset M^m$, let $G_M(C) = \text{inj lim}\,\{G(U) \mid C \subset U \text{ open in } M\}$.

(1) *For any pair $A \subset B$ of compacta in M, the restriction map $\pi : G_M(B) \to G_M(A)$ is micro-gibki* — i.e., given a homotopy $f : P \times I \to G_M(A)$ and $F_0 : P \times 0 \to G_M(B)$ with $\pi F_0 = f \mid P \times 0$ there exists $\epsilon > 0$ and $F : P \times [0\,, \epsilon] \to G_M(B)$ so that $\pi F = f \mid P \times [0\,, \epsilon]$. Chasing definitions one finds that this follows quickly from the TOP isotopy extension theorem (many-parameter version) or the relative local contractibility theorem of [10] [17].

(2) *d is a weak homotopy equivalence for $M^m = R^m$.* Indeed, one has a commutative square of weak homotopy equivalences

$$\begin{array}{ccc} G(R^m) & \xrightarrow{d} & G^{\perp}(\tau R^m) \\ \simeq\downarrow & & \downarrow\simeq \\ G_{R^m}(0) & \xleftarrow{\simeq} & G^{\perp}(\tau R^m \mid 0) \end{array}$$

in which the verticals are restrictions and the bottom comes from identifying the fiber of $\tau R^m \mid 0$ to R^m, cf. [27].

Gromov's analysis applies (1) and (2) and more obvious properties of G, $G^{\perp}$ to establish 4.1. Unfortunately, M doesn't always have a handle decomposition over which to induct ; one has to proceed more painfully chart by chart.

We can now pass quickly from a bundle theoretic to a homotopy classification of G-structures. Notice that if $f : X' \to X$ is any map and ξ^m is a R^m microbundle over X equipped with a G-folitation $\mathfrak{F}$, transverse to fibers, defined on an open neighborhood of the zero section X, then $f^*\xi$ over X' is similarly equipped with a pulled-back foliation $f^*\mathfrak{F}$. This means that equipped bundles behave much like bundles. One can use Haefliger's notion of "gamma structure" as in [29] to deduce for numerable equipped bundles the existence of a universal one $(\gamma_G^m, \mathfrak{F}_G)$ over a base space $B_{\Gamma(G)}$(*). There is a map $B_{\Gamma(G)} \to B_{\mathrm{TOP}(m)}$ classifing γ_G^m as an R^m-microbundle ; we make it a fibration. Call the fiber TOP(m)/Γ(G). One finds that there is a weak homotopy equivalence $G^{\perp}(\xi) \simeq$ Lift (f to $B_{\Gamma(G)}$), to the space of liftings to $B_{\Gamma(G)}$ of a fixed classifying map $f : X \to B_{\mathrm{TOP}(m)}$ for ξ^m. Hence one gets

THEOREM 4.2. — *For any open topological m-manifold M^m, there is a weak homotopy equivalence $G(M) \simeq$ Lift (τ to $B_{\Gamma(G)}$) from the space of G-structures $G(M)$ on M to the space of liftings to $B_{\Gamma(G)}$ of a fixed classifying map map $\tau : M \to B_{\mathrm{TOP}(m)}$ for $\tau(M)$.*

Heafliger and Milnor observe that for $G = \mathrm{CAT}^m$ the pseudo-group of CAT isomorphisms of open subsets of R^m — CAT meaning DIFF (= smooth C^∞), or PL (= piecewise linear) or TOP (= topological) — one has

$$\pi_i\,(\mathrm{CAT}(m)/\Gamma\,(\mathrm{CAT}^m)) = 0 \quad , \quad i < m \quad . \tag{4.3}$$

Indeed for CAT = TOP, 4.2 shows this amounts to the obvious fact that $\pi_0(G(S^i \times R^{m-i})) = 0$. Analogues of 4.2 with DIFF or PL in place of TOP can be proved analogously(**) and give the other cases of (4.3). Hence one has

THEOREM 4.4. — *For any open topological manifold M^m, there is a natural bijection $\pi_0 \mathrm{CAT}^m(M^m) \simeq \pi_0$ Lift (τ to $B_{\mathrm{CAT}(m)}$).*

This result comes from [44] for $m \geqslant 5$. Lashof [50] gave the first proof that was valid for $m = 4$. A stronger and technically more difficult result is sketched in [63] [45]. It asserts a weak homotopy equivalence of a "sliced concordance" variant of $\mathrm{CAT}^m(M^m)$ with Lift (τ to $B_{\mathrm{CAT}(m)}$). This is valid without the openness restriction if $m \neq 4$. For open M^m (any m), it too can be given a proof involving a micro-gibki property and Gromov's procedure.

5. The product structure theorem.

THEOREM 5.1 (*Product structure theorem*). — *Let M^m be a* TOP *manifold, C a closed subset of M and σ_0 a* CAT (= DIFF *or* PL) *structure on a neighborhood of C in M. Let Σ be a* CAT *structure on $M \times R^s$ equal $\sigma_0 \times R^s$ near $C \times R^s$. Provide that $m \geqslant 5$ and $\partial M \subset C$.*

Then M has a CAT *structure σ equal σ_0 near C. And there exists a* TOP *isotopy (as small as we please) $h_t : M_\sigma \times R^s \to (M \times R^s)_\Sigma$, $0 \leqslant t \leqslant 1$, of h_0 = identity, fixing a neighborhood of $C \times R^s$, to a* CAT *isomorphism h_1.*

It will appear presently that this result is the key to TOP handlebody theory and transversality. The idea behind such applications is to reduce TOP lemmas to their DIFF analogues.

(*) Alternatively, for our purpose, $B_{\Gamma(G)}$ can be the ordered simplicial complex having one d-simplex for each equipped bundle over the standard d-simplex that has total space in some $R^n \subset R^\infty$.

(**) The forgetful map $\varphi : B_{\Gamma(\mathrm{PL}^m)} \to B_{\mathrm{PL}(m)}$ is more delicate to define. One can make $B_{\Gamma(\mathrm{PL}^m)}$ a simplicial complex, then define φ simplex by simplex.

It seems highly desirable, therefore, to prove 5.1 as much as possible by pure geometry, without passing through a haze of formalism like that in § 4. This is done in [46]. Here is a quick sketch of proof intended to advertise [46].

First, one uses the CAT *s*-cobordism theorem (no surgery !) and the handle-straightening method of [44] to prove – without meeting obstructions –

THEOREM 5.2 (*Concordance implies isotopy*). — *Given* M *and* C *as in* 5.1, *consider a* CAT *structure* Γ *on* $M \times I$ *equal* $\sigma_0 \times I$ *near* $C \times I$, *and let* $\Gamma | M \times 0$ *be called* $\sigma \times 0$. (Γ is called a concordance of σ rel C).

There exists a TOP *isotopy (as small as we please)* $h_t : M_\sigma \times I \to (M \times I)_\Sigma$, $0 \leqslant t \leqslant 1$, *of* $h_0 =$ *identity, fixing* $M \times 0$ *and a neighborhood of* $C \times I$, *to a* CAT *isomorphism* h_1.

Granting this result, the Product Structure Theorem is deduced as follows.

In view of the relative form of 5.2 we can assume $M = R^m$. Also we can assume $s = 1$ (induct on s !). Thirdly, it suffices to build a concordance Γ (= structure on $M \times R^s \times I$) from $\sigma \times R^s$ to Σ rel $C \times R^s$. For, applying 5.2 to the concordance Γ we get the wanted isotopy. What remains to be proved can be accomplished quite elegantly. Consider Figure 5-a.

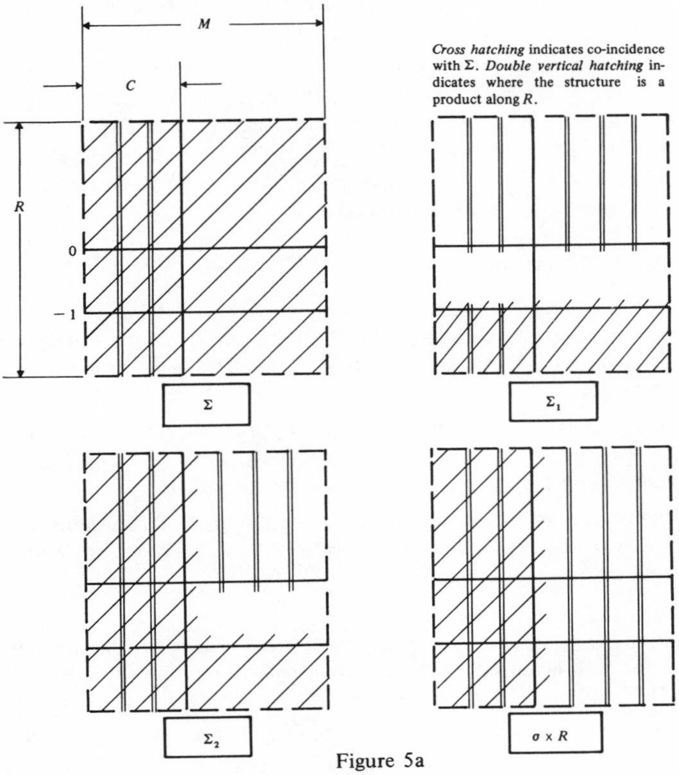

Figure 5a

We want a concordance rel $C \times R$ from Σ to $\sigma \times R$. First note it suffices to build Σ_2 with the properties indicated. Indeed Σ_2 admits standard (sliced) concordances rel $C \times R$ to $\sigma \times R$ and to Σ. The one to $\sigma \times R$ comes from sliding R over itself onto $(0, \infty)$. The region of coincidence with $\sigma \times R$ becomes total by a sort of window-blind effect. The concordance to Σ comes from sliding R over itself onto $(-\infty, -1)$. (Hint : The structure picked up from Σ_2 at the end of the slide is the same as that picked up from Σ).

It remains to construct Σ_2. Since $M \times R = R^{m+1}$, we can find a concordance (not rel $C \times R$) from Σ to the standard structure, using the STABLE homeomorphism theorem(*) [42]. Now 5.2 applied to the concordance gives Σ_1, which is still standard near $M \times [0, \infty)$. Finally an application of 5.2 to $\Sigma_1 | N \times [-1, 0]$, where N is a small neighborhood of C, yields Σ_2. The change in $\Sigma_1 | M \times 0$ (which is standard) on $N \times 0$ offered by 5.2 is extended productwise over $M \times [0, \infty)$. This completes the sketch.

It is convenient to recall here for later use one of the central results of [44]. Recall that $\mathrm{TOP}_m/\mathrm{PL}_m$ is the fiber of the forgetful map $B_{\mathrm{PL}(m)} \to B_{\mathrm{TOP}(m)}$. And TOP/PL is the fiber the similar map of stable classifying spaces $B_{\mathrm{PL}} \to B_{\mathrm{TOP}}$. Similarly one defines $\mathrm{TOP}_m/\mathrm{DIFF}_m \equiv \mathrm{TOP}_m/\mathrm{O}_m$ and TOP/DIFF $\equiv$ TOP/O.

THEOREM 5.3(**) (Structure theorem).— TOP/PL $\simeq K(Z_2, 3)$ *and*

$$\pi_k(\mathrm{TOP}_m/\mathrm{CAT}_m) = \pi_k(\mathrm{TOP/CAT})$$

for $k < m$ *and* $m \geqslant 5$. *Here* CAT = PL *or* DIFF.

Since $\pi_k(\mathrm{O}_m) = \pi_k(\mathrm{O})$ for $k < m$, we deduce that $\pi_k(\mathrm{TOP}, \mathrm{TOP}_m) = 0$ for $k < m > 5$, a weak stability for TOP_m.

Consider the second statement of 5.3 first. Theorem 4.4 says that

$$\pi_k(\mathrm{TOP}_m/\mathrm{CAT}_m) = \pi_0(\mathrm{CAT}^m(S^k \times R^{m-k})) \equiv S^m_k$$

for $k < m \geqslant 5$. Secondly, 5.1 implies $S^m_k = S^{m+1}_k = S^{m+2}_k = \cdots, m \geqslant k$. Hence $\pi_k(\mathrm{TOP}_m/\mathrm{CAT}_m) = \pi_k(\mathrm{TOP/CAT})$.

We now know that π_k(TOP/PL) is the set of isotopy classes of PL structures on S^k if $k \geqslant 5$. The latter is zero by the PL Poincaré theorem of Smale [84], combined with the stable homeomorphism theorem [42] and the Alexander isotopy. Similarly one gets $\pi_k(\mathrm{TOP/DIFF}) = \Theta_k$ for $k \geqslant 5$. Recall $\Theta_5 = \Theta_6 = 0$ [41].

The equality $\pi_k(\mathrm{TOP/PL}) = \pi_k(\mathrm{TOP/DIFF}) = \pi_k(K(Z_2, 3))$ for $k \leqslant 5$ can be deduced with ease from local contractibility of homeomorphism groups and the surgical classification [35] [95], by $H^3(T^5; Z_2)$, of homotopy 5-tori. See [43] [46 A] for details.

Combining the above with 4.4 one has a result of [44].

(*) Without this we get only a theorem about compatible CAT structures on STABLE manifolds (of Brown and Gluck [8]).

(**) For a sharper result see [63] [45], and references therein.

CLASSIFICATION THEOREM 5.4. — *For $m \geqslant 5$ a* TOP *manifold M^m (without boundary) admits a* PL *manifold structure iff an obstruction $\Delta(M)$ in $H^4(M;Z_2)$ vanishes. When a* PL *structure Σ on M is given, others are classified (up to concordance or isotopy) by elements of $H^3(M\,;Z_2)$.*

Complement. — Since π_k(TOP/DIFF) = π_k(TOP/PL) for $k < 7$ (see above calculation), the same holds for DIFF in low dimensions.

Finally we have a look at low dimensional homotopy groups involving

$$G = \lim \{G_n \,|\, n \geqslant 0\}$$

where G_n is the space of degree ± 1 maps $S^{n-1} \to S^{n-1}$. Recall that $\pi_n G = \pi_{n+k} S^k$, k large. G/CAT is the fiber of a forgetful map $B_{\mathrm{CAT}} \to B_G$, where B_G is a stable classifying space for spherical fibrations (see [15], [29]).

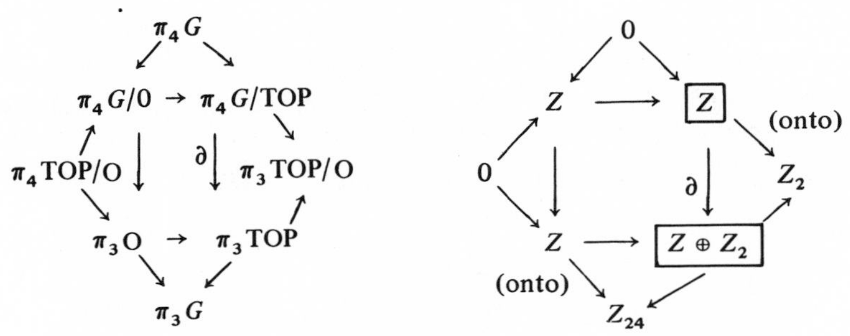

The left hand commutative diagram of natural maps is determined on the right. Only π_3TOP is unknown(*). So the exactness properties evident on the left leave no choice. Also ∂ must map a generator of $\pi_4 G/\mathrm{TOP} = Z$ to $(12, 1)$ in $Z \oplus Z/2Z = \pi_3$TOP.

The calculation with PL in place of O is the same (and follows since π_i(PL/O) = $\Gamma_i = 0$ for $i \leqslant 6$).

6. **Simple homotopy theory [44] [46 A].**

The main point is that every compact TOP manifold M (with boundary ∂M) has a preferred simple homotopy type and that two plausible ways to define it are equivalent. Specifically, a handle decomposition of M or a combinatorial triangulation of a normal disc-bundle to M give the same simple type.

The second definition is always available. Simply embed M in R^n, n large, with normal closed disc-bundle E [31]. Theorem 5.1 then provides a small homeomorphism of R^n so that $h(\partial E)$, and hence $h(E)$, is a PL submanifold.

(*) That $\pi_4 G/\mathrm{TOP}$ is Z (not $Z \oplus Z_2$) is best proved by keeping track of some normal invariants in disproving the Hauptvermutung, see [46A]. Alternatively, see 13.4 below.

Working with either of these definitions, one can see that the preferred simple type of M and that of the boundary ∂M make of $(M, \partial M)$ a finite Poincaré duality space in the sense of Wall [95], a fact vital for TOP surgery.

The Product structure theorem 5.1 makes quite unnecessary the bundle theoretic nonsense used in [44] (cf. [63]) to establish preferred simple types.

7. **Handlebody theory** (statements in [44 C] [45], proofs in [46 A]).

7.1. The main result is that handle decompositions exist in dimension $\geqslant 6$. Here is the idea of proof for a closed manifold M^m, $m \geqslant 6$. Cover M^m by finitely many compacta $A_1, \ldots, A_k$, each A_i contained in a co-ordinate chart $U_i \approx R^m$. Suppose for an inductive construction that we have built a handlebody $H \subset M$ containing $A_1 \cup \ldots \cup A_{i+1}$, $i \geqslant 0$. The Product Structure Theorem shows that $H \cap U_i$ can be a PL (or DIFF) m-submanifold of U_i after we adjust the PL (or DIFF) structure on U_i. Then we can successively add finitely many handles onto H in U_i to get a handlebody H' containing $A_1 \cup \ldots \cup A_i$. After k steps we have a handle decomposition of M.

A TOP Morse function on M^m implies a TOP handle decomposition (the converse is trivial) ; to see this one uses the TOP isotopy extension theorem to prove that a TOP Morse function without critical points is a bundle projection. (See [12] [82, 6.14] for proof in detail).

Topological handlebody theory as conceived of by Smale now works on the model of the PL or DIFF theory (either). For the sake of those familiar with either, I describe simple ways of obtaining transversality and separation (by Whitney's method) of attaching spheres and dual spheres in a level surface.

LEMMA 7.2. (Transversality). — *Let* $g : R^m \to R^m, m \geqslant 5$, *be a* STABLE *homeomorphism. In* R^m, *consider* $R^p \times 0$ *and* $0 \times R^q, p + q = m$, *with 'ideal' transverse intersection at the origin. There exists an* ϵ*-isotopy of* g *to* $h : R^m \to R^m$ *such that* $h(R^p \times 0)$ *is transverse to* $0 \times R^q$ *is the following strong sense. Near each point* $x \in h^{-1}(0 \times R^q) \cap R^p \times 0$, h *differs from a translation by at most a homeomorphism of* R^m *respecting both* $R^p \times 0$ *and* $0 \times R^q$.

Furthermore, if C *is a given closed subset of* R^m *and* g *satisfies the strong transversality condition on* h *above for points* x *of* R^m *near* C, *then* h *can equal* g *near* C.

Proof of 7.2 — For the first statement $\epsilon/2$ isotop g to diffeomorphism g' using Ed Connell's theorem [14] (or the Concordance-implies-epsilon-isotopy theorem 5.2), then $\epsilon/2$ isotop g' using standard DIFF techniques to a homeomorphism h' which will serve as h if $C = \emptyset$.

The further statement is deduced from the first using the flexibility of homeomorphisms. Find a closed neighborhood C' of C near which g is still transverse such that the frontier $\dot{C}'$ misses $g^{-1}(0 \times R^q) \cap (R^p \times 0)$ — which near C is a discrete collection of points. Next, find a closed neighborhood D of $\dot{C}'$ also missing $g^{-1}(0 \times R^q) \cap (R^p \times 0)$, and $\delta : R^m \to (0, \infty)$ so that $d(gx, 0 \times R^q) < \delta(x)$ for x in $D \cap (R^p \times 0)$. If $\epsilon : R^m \to (0, \infty)$ is sufficiently small, and h' in the first paragraph is built for ϵ, Cernavskii's local contractibility theorem [11] (also [17] and [82 , 6.3]) says that there exists a homeomorphism h equal g on C' and equal h' outside $C' \cup D$ so that $d(h', g) < \delta$. This is the wanted h.

7.3. THE WHITNEY LEMMA.

The TOP case of the Whitney process for eliminating pairs of isolated transverse intersection points (say of M^p and N^q) can be reduced to the PL case [99] [37]. The Whitney 2-disc is easily embedded and a neighborhood of it is a copy of R^m, $m = p + q$. We can arrange that either manifold, say M^p, is PL in R^m, and(*) N^q is PL near M^q in R^m. Since $5 \leqslant m = p + q$, we can assume $q \leqslant m/2$; so N^q can now be pushed to be PL in R^m by a method of T. Homma, or by one of R.T. Miller [54 A], or again by the method of [44], applied pairwise [44 A] (details in [73]). Now apply the PL Whitney lemma [37]. On can similarly reduce to the original DIFF Whitnev lemma [99].

7.4. CONCLUSION.

The s-cobordism theorem [37] [39], the boundary theorem of [76], and the splitting principle of Farrell and Hsiang [20] can now be proved in TOP with the usual dimension restrictions.

8. **Transversality** (statements in [44 C] [45], proofs in [46 A]).

If $f : M^m \to R^n$ is a continuous map of a TOP manifold without boundary to R^n and $m - n > 5$, we can homotop f to be transverse to the origin $0 \in R^n$. Here is the idea. One works from chart to chart in M to spread the transversality, much as in building handlebodies. In each chart one uses the product structure theorem 5.1 to prepare for an application of the *relative* DIFF transversality theorem of Thom.

Looking more closely one gets a *relative* transversality theorem for maps $f : M^m \to E(\xi^n)$ with target *any* TOP R^n-microbundle ξ^n over *any* space. It is parallel to Williamson's PL theorem [100], but is proved only for $m \neq 4 \neq m - n$. It is indispensible for surgery and cobordism theory.

9. **Surgery.**

Surgery of compact manifolds of dimension $\geqslant 5$ as formulated by Wall [95] can be carried out for TOP manifolds using the tools of TOP handlebody theory. The chief technical problem is to make the self-intersections of a framed TOP immersion $f : S^k \times R^k \to M^{2k}$ of S^k, $k \geqslant 3$, transverse (use Lemma 7.2 repeatedly), and then apply the Whitney lemma to find a regular homotopy of f to an embedding when Wall's self-intersection coefficient is zero.

In the simply connected case one can adapt ideas of Browder and Hirsch [4].

Of course TOP surgery constantly makes use of TOP transversality, TOP simple homotopy type and the TOP s-cobordism theorem.

(*) Use of the strong transversality of 7.2 makes this trivial in practice.

10. Cobordism theory : generalities.

Let Ω_n^{TOP} [respectively Ω_n^{STOP}] be the group of [oriented] cobordism classes of [oriented] closed n-dimensional TOP manifolds. Thom's analysis yields a homomorphism

$$\theta_n : \Omega_n^{\mathrm{TOP}} \to \pi_n(\mathrm{MTOP}) = \varinjlim_k \pi_{n+k}(\mathrm{MTOP}(k)) \quad .$$

Here $\mathrm{MTOP}(k)$ is the Thom space of the universal TOP R^k-bundle γ_{TOP}^k over $B_{\mathrm{TOP}(k)}$ – obtained, for example, by compactifying each fiber with a point (cf. [49]) and crushing these points to one. The Pontrjagin Thom definition of θ_n uses a stable relative existence theorem for normal bundles in euclidean space – say as provided by Hirsch [30] and the Kister-Mazur Theorem [49].

Similarly one gets Thom maps

$$\theta_n : \Omega_n^{\mathrm{STOP}} \to \pi_n(\mathrm{MSTOP}), \qquad \text{and} \qquad \theta_n : \Omega_n^{\mathrm{SPINTOP}} \to \pi_n(\mathrm{MSPINTOP}) \,,$$

and more produced by the usual recipe for cobordism of manifolds with a given, special, stable structure on the normal bundle [86, Chap. II].

THEOREM 10.1. — *In each case above the Thom map $\theta_n : \Omega_n \to \pi_n(M)$ is surjective for $n \neq 4$, and injective for $n \neq 3$.*

This follows immediately from the transversality theorem.

PROPOSITION 10.2. — $B_{\mathrm{SO}} \otimes Q \simeq B_{\mathrm{STOP}} \otimes Q$, *where Q denotes the rational numbers.*

Proof. — $\pi_i(\mathrm{STOP/SO}) = \pi_i(\mathrm{TOP/O})$ is finite for all i by [40] [44] cf. § 5, STOP/SO being fiber of $B_{\mathrm{SO}} \to B_{\mathrm{STOP}}$. (See § 15 or [90] for definition of $\otimes Q$).

PROPOSITION 10.3. — $\pi_* \mathrm{MSO} \otimes Q \cong \pi_* \mathrm{MSTOP} \otimes Q$.

Proof. — From 10.2 and the Thom isomorphism we have

$$H_*(\mathrm{MSO}\,; Q) \cong H_*(\mathrm{MSTOP}\,; Q) \quad .$$

Now use the Hurewicz isomorphism (Serre's from [75]).

PROPOSITION 10.4. — $\Omega_*^{\mathrm{SO}} \otimes Q \cong \Omega_*^{\mathrm{STOP}} \otimes Q$ *each being therefore the polynomial algebra freely generated by* CP_{2n}, $n \geqslant 1$.

Proof of 10.4. — The uncertainty about dimensions 3 and 4 in 10.1 cannot prevent this following from 10.2. Indeed, $\Omega_3^{\mathrm{STOP}} \to \pi_3 \mathrm{MSTOP}$ is injective because every TOP 3-manifold is smoothable (by Moise et al., cf. [80, § 5]). And

$$\Omega_4^{\mathrm{STOP}} \to \pi_4 \mathrm{MSTOP}$$

is rationally onto because $\Omega_4^{\mathrm{SO}} \to \pi_4 \mathrm{MSTOP}$ is rationally onto.

Since $\pi_i(\mathrm{STOP/SPL}) = \pi_i(\mathrm{TOP/PL})$ is Z_2 for $i = 3$ and zero for $i \neq 3$ the above three propositions can be repeated with SPL in place of SO and dyadic rationals $Z[\frac{1}{2}]$ in place of Q. The third becomes :

PROPOSITION 10.5. — $\Omega_*^{\mathrm{SPL}} \otimes Z[\frac{1}{2}] \cong \Omega_*^{\mathrm{STOP}} \otimes Z[\frac{1}{2}]$.

Next we recall

PROPOSITION 10.6. — (S.P. Novikov). $\Omega_*^{\mathrm{SO}} \to \Omega_*^{\mathrm{STOP}}$ *is injective.*

This is so because every element of Ω_*^{SO} is detected by its Stiefel-Whitney numbers (homotopy invariants) and its Pontrjagin numbers (which are topological invariants by 10.2).

In view of 10.2 we have canonical Pontrjagin characteristic classes p_k in

$$H^{4k}(B_{\mathrm{STOP}} ; Q) = H^{4k}(B_{\mathrm{SO}} ; Q)$$

and the related Hirzebruch classes $L_k = L_k(p_1, \ldots, p_k) \in H^{4k}$. Hirzebruch showed that $L_k : \Omega_{4k}^{\mathrm{SO}} \otimes Q \to Q$ sending a $4k$-manifold M^{4k} to its characteristic number $L_k(M^{4k}) = L_k(\tau(M^{4k}))[M^{4k}] \in Q$ is the signature (index) homomorphism. From 10.2 and 10.4, it follows that the same holds for STOP in place of SO. Hence we have

PROPOSITION 10.7. — *For any closed oriented* TOP *$4k$-manifold M^{4k} the signature $\sigma(M^{4k})$ of the rational cohomology cup product pairing $H^{2k} \otimes H^{2k} \to H^{4k} = Q$ is given by $\sigma(M^{4k}) = L_k(\tau M^{4k})[M^{4k}] \in Z$.*

11. **Oriented cobordism.**

The first few cobordism groups are fun to compute geometrically — by elementary surgical methods, and the next few pages are devoted to this.

THEOREM 11.1. — $\Omega_n^{\mathrm{STOP}} \simeq \Omega_n^{\mathrm{SO}} \oplus R_n$ *for $n \leqslant 7$, and we have $R_n = 0$ for $n \leqslant 3$, $R_4 \leqslant Z_2, R_5 = 0$, $R_6 = Z_2$, $R_7 \leqslant Z_2$.*

Proof of 11.1. — For $n = 1, 2, 3$, $\Omega_n^{\mathrm{STOP}} = \Omega_n^{\mathrm{SO}} = 0$ is seen by smoothing.

For $n = 4$, first observe that $Z = \Omega_4^{\mathrm{SO}} \to \Omega_4^{\mathrm{STOP}}$ maps Z to a summand because the signature of a generator CP_2 is 1 which is indivisible. Next consider the Z_2 characteristic number of the first stable obstruction $\Delta \in H^4(B_{\mathrm{STOP}} ; Z_2)$ to smoothing. It gives a homomorphism $\Omega_4^{\mathrm{STOP}} \to Z_4$ killing Ω_4^{SO}. If

$$\Delta(M^4) \equiv \Delta(\tau(M^4))[M^4] = 0 \quad ,$$

then, by 5.3, $M^4 \times R$ has a DIFF structure Σ. Push the projection $(M^4 \times R)_\Sigma \to R$ to be transversal over $0 \in R$ at a DIFF submanifold M' and behold a TOP oriented cobordism M to M'. Thus $R_4 \leqslant 0$.

For $n \geqslant 5$ note that any oriented TOP manifold M^n is oriented cobordant to a simply connected one M' by a finite sequence of 0 and 1-dimensional surgeries. But, for $n = 5$, $H^4(M' ; Z_2) \cong H_1(M' ; Z_2) = 0$ so M' is smoothable. Hence $R_5 = 0$.

For $n = 6$ we prove

PROPOSITION 11.2. — *The characteristic number $\Delta w_2 : \Omega_6^{\mathrm{STOP}} \to Z_2$ is an isomorphism.*

Proof. — It is clearly non-zero on any non-smoothable manifold $M^6 \simeq CP_3$, since $w_2(M^6) = w_2(CP_3) \neq 0$, and we will show that such a M^6 exists in 15.7 below.

Since $\Omega_6^{SO} = 0$ it remains to prove that Δw_2 is injective. Suppose $\Delta w_2(M^6) = 0$ for oriented M^6. As we have observed, we can assume M is simply connected. Consider the Poincaré dual $D\Delta$ of $\Delta = \Delta(\tau M)$ in

$$H_2(M^6 ; Z_2) = H_2(M^6 ; Z) \otimes Z_2 = \pi_2(M^6) \otimes Z_2$$

and observe that it can be represented by a locally flatly embedded 2-sphere $S \subset M^6$. (Hints : Use [24], or find an immersion of $S^2 \times R^4$ [52] and use the idea of Lemma 7.1).

Note that $\Delta | (M - S) = \Delta(M - S)$ is zero because $\Delta[x] = x \cdot D\Delta$ (the Z_2 intersection number) for all $x \in H_2(M ; Z_2)$. Thus $M - S$ is smoothable.

A neighborhood of S is smoothed, there being no obstruction to this ; and S is made a DIFF submanifold of it. Let N be an open DIFF tubular neighborhood of S. Now $0 = \Delta w_2[M] = w_2[D\Delta] = w_2[S]$ means that $w_2(\tau M)|S$ is zero. Hence $N = S^2 \times R^4$. Killing S by surgery we produce M', oriented cobordant to M, so that, writing $M_0 = M - S^2 \times \overset{\circ}{B}{}^4$, we have $M' = M_0 + B^3 \times S^3$ (union with boundaries identified). Now M' is smoothable since M_0 is and there is no further obstruction. As $\Omega_6^{SO} = 0$, Proposition 11.2, is established.

Proposition 11.3. — *The characteristic number* $(\beta\Delta) w_2 : \Omega_7^{STOP} \to Z_2$ *is injective, where* $\beta = Sq^1$.

Proof of 11.3. — We show the $(\beta\Delta) w_2[M] = 0$ implies M^7 is a boundary. Just as for 11.2, we can assume M is simply connected. Then $\pi_2 M = H_2(M ; Z)$ and we can kill any element of the kernel of $w_2 : H_2(M ; Z) \to Z_2$, by surgery on 2-spheres in M. Killing the entire kernel we arrange that w_2 is injective.

We have $0 = (\beta\Delta) w_2[M] = w_2[D\beta\Delta]$. So the Poincaré dual $D\beta\Delta$ of $\beta\Delta$ is zero as w_2 is injective.

Now $\beta\Delta = 0$ means Δ is reduced integral ; indeed β is the Bockstein

$$\delta : H^4(M ; Z_2) \to H^5(M , Z)$$

followed by reduction mod 2. But

$$H^5(M ; Z) \cong H^5(M ; Z_2), \text{ since } H_2(M ; Z) \cong H_2(M ; Z_2)$$

(both isomorphisms by reduction). Thus $\beta\Delta = 0$ implies $\delta\Delta = 0$, which means Δ is reduced integral. Hence $D\Delta$ is reduced integral. Since the Hurewicz map $\pi_3 M \to H_3(M ; Z)$ is onto, $D\Delta$ is represented by an embedded 3-sphere S. Following the argument for dimension 6 and recalling $\pi_2 O = 0$, we can do surgery on S to obtain a smoothable manifold.

12. Unoriented cobordism (*).

Recalling calculations of Ω_i^{O} and Ω_i^{SO} from Thom [91] we get the following table

i	4	5	6	7
Ω_i^{SO}	Z	Z_2	0	0
$\Omega_i^{STOP}/\Omega_i^{SO} = R_i$	$\leqslant Z_2$	0	Z_2	$R_7 \leqslant Z_2$
Ω_i^{O}	$Z_2 \oplus Z_2$	Z_2	$Z_2 \oplus Z_2 \oplus Z_2$	Z_2
$\Omega_i^{TOP}/\Omega_i^{O}$	$\leqslant R_4 \leqslant Z_2$	Z_2	$Z_2 \oplus Z_2$	$Z_2 \oplus Z_2 \oplus R_7$

The only non zero-entry for $0 < i < 4$ would be $\Omega_2^{O} = Z_2$.

To deduce the last row from the first three, use the related long exact sequences (from Dold [16])

$$
(12.1)\qquad
\begin{array}{ccccccccc}
\ldots \to & \Omega_i^{SO} & \to & \Omega_i^{O} & \overset{(\partial,d)}{\to} & \Omega_{i-1}^{SO} \oplus \Omega_{i-2}^{O} & \overset{j}{\to} & \Omega_{i-1}^{SO} & \to \ldots \\
 & \downarrow & & \downarrow & & \downarrow & & \downarrow & \\
\ldots \to & \Omega_i^{STOP} & \to & \Omega_i^{TOP} & \overset{(\partial,d)}{\to} & \Omega_{i-1}^{STOP} \oplus \Omega_{i-2}^{TOP} & \overset{j}{\to} & \Omega_{i-1}^{STOP} & \to \ldots
\end{array}
$$

If transversality fails $\pi_4(M\ ?)$ should replace $\Omega_4^{?}$ in the TOP sequence. (See [93, §6], [3] for explanation).

All the maps are forgetful maps except those marked j and (∂, d). The map j kills the second summand, and is multiplication by 2 on the first summand (which is also the target of j).

At the level of representatives, ∂ maps M^i to a submanifold M^{i-1} dual to $w_1(M^i)$, and d maps M^i to $M^{i-2} \subset M^{i-1}$ dual to $w_1(M^i)|M^{i-1}$

The map d is onto with left inverse φ defined by associating to M^{i-2} the RP^2 bundle associated to $\lambda \oplus \epsilon^2$ over M^{i-2}, where λ is the line bundle with

$$w_1(\lambda) = w_1(M^{i-2})$$

and ϵ^2 is trivial.

The diagram (12.1) gives us the following generators for $S_i \equiv \Omega_i^{TOP}/\Omega_i^{O}$.

$S_4 \leqslant Z_2$: Any M^4 with $\Delta(M) \neq 0$ - - - if it exists.

$S_5 = Z_2$: Any M^5 detected by Δw_1.

$S_6 = Z_2 \oplus Z_2$: Non-smoothable $M_1^6 \simeq CP_3$, detected by Δw_2 ;

$$M_2^6 \simeq RP_2 \times (Q \cup \infty)$$

(*) Added in proof : A complete calculation of Ω_*^{TOP} has just been announced by Brumfiel, Madsen and Milgram (Bull. AMS to appear).

detected by Δw_1^2. $M_2^6 \times R$ can be $RP_2 \times \widetilde{X}^5$, where $\widetilde{X}^5$ is the universal covering of a manifold constructed in [80, § 5].

$S_7 = 2Z_2 \oplus R_7 : M_1^7 = \varphi M_1^5$ detected by Δw_1^3; M_2^7 is detected by $\Delta w_2 w_1$, $M_2^7 = T(\rho)$, the mapping torus of an orientation reversing homeomorphism of $M_1^6 \simeq CP_3$ homotopic to complex conjugation in CP_3. Such a ρ exists because conjugation doesn't shift the normal invariant for $M_1^6 \simeq CP_3$ in $[CP_3, G/TOP]$. Finally M_3^7 a generator of R_7 detected by $(\beta\Delta) w_2$ (if it exists).

13. **Spin cobordism.**

The stable classifying space $B_{SPINTOP}$ is the fiber of $w_2 : B_{STOP} \to K(Z_2, 2)$. So $\pi_i B_{SPINTOP}$ is 0 for $i \leqslant 3$ and equals $\pi_i B_{TOP}$ for $i \geqslant 3$. Topological spin cobordism is defined like smooth spin cobordism Ω_*^{SPIN} but using TOP manifolds. Thus $\Omega_*^{SPINTOP}$ is the cobordism ring for compact TOP manifolds M equipped with a spin structure – i.e. a lifting to $B_{SPINTOP}$ of a classifying map $M \to B_{TOP}$ for $\tau(M)$ – or equivalently for the normal bundle $\nu(M)$.

THEOREM 13.1. – *For* $n \leqslant 7$, $\Omega_n^{SPINTOP}$ *is isomorphic to* Ω_n^{SPIN}, *which for* $n = 0, 1, \ldots, 8$ *has the values* $Z, Z_2, Z_2, 0, Z, 0, 0, 0, Z \oplus Z$ [59] [86]. *The image of the forgetful map* $Z = \Omega_4^{SPIN} \to \Omega_4^{SPINTOP} = Z$ *is the kernel of the stable triangulation obstruction* $\Delta : \Omega_4^{SPINTOP} \to Z_2$.

The question whether Δ is zero or not is the question whether or not Rohlin's congruence for signature $\sigma(M^4) \equiv 0 \bmod 16$ holds for all topological spin manifolds M^4. Indeed $\sigma(M^4) \equiv 8\Delta(M^4) \bmod 16$, $\Delta(M^4)$ being 0 or 1.

Proof of 13.1. – The isomorphism $\Omega_n^{SPINTOP} \cong \Omega_n^{SPIN}$ for $n \leqslant 3$ comes from smoothing.

Postponing dimension 4 to the last, we next show $\Omega_n^{SPINTOP}/\Omega_n^{SPIN} = 0$ for $n = 5, 6, 7$. Note first that a smoothing and a topological spin structure determine a unique smooth spin structure. The argument of § 11 shows that the only obstruction to performing oriented surgery on M^n to obtain a smooth manifold is a characteristic number, viz. 0, Δw_2, $(\beta\Delta) w_2$ for $n = 5, 6, 7$ respectively. But $w_2(M^n) = 0$ for any spin topological manifold. It remains to show that the surgeries can be performed so that each one, say from M to M', thought of as an elementary cobordism $(W^{n+1}; M^n, M'^n)$, can be given a topological spin structure extending that of M. The only obstruction to this occurs in $H^2(W, M; Z_2)$, which is zero except if the surgery is on a 1-sphere. And in that case we can obviously find a possibly different surgery on it (by spinning the normal bundle !) for which the obstruction is zero.

Finally we deal with dimension 4. If $\Delta(M^4) = 0$ for any spin 4-manifold, then M^4 is spin cobordant to a smooth spin manifold by the proof of 11.1. Next suppose M^4 is a topological spin manifold such that the characteristic number $\Delta(M^4)$ is not zero. If we can show that $\sigma(M^4) \equiv 8\Delta(M^4) \bmod 16$ the rest of 13.1 will follow, including the fact that $\Omega_4^{SPINTOP} \cong Z$ rather than $Z \oplus Z_2$. We can assume M^4 connected (by surgery).

LEMMA 13.2. — *For any closed connected topological spin 4-manifold M^4, there exists a (stable)* TOP *bundle ξ over S^4 and a degree* 1 *map $M \to S^4$ covered by a* TOP *bundle map $\nu(M) \to \xi$. This ξ is necessarily fiber homotopically trivial. A similar result (similarly proved) holds for smooth spin manifolds.*

Proof of 13.2. — Since any map $M_0 \equiv M -$ (point) $\to B_{\mathrm{SPINTOP}}$ is contractible, $\nu(M)|M_0$ is trivial, and so $\nu(M) \to \xi$ exists as claimed. Now ξ is fiber homotopically trivial since it is – like $\nu(M)$ – reducible, hence a Spivak normal bundle for S^4. (Cf. proof in [40].)

LEMMA 13.3. — *A fiber homotopically trivialized* TOP *bundle ξ over the 4-sphere is (stably) a vector bundle iff $\frac{1}{3} p_1(\xi)[S^4] \equiv 0$ mod. 16.*

Proof of 13.3. — Consider the homomorphism $\frac{1}{3} p_1 : \pi_4 G/\mathrm{TOP} \to Z$ given by associating the integer $\frac{1}{3} p_1(\xi)[S^4]$ to a such a bundle ξ over S^4. The composed map $\frac{1}{3} p_1 : \pi_4 G/0 \to Z$ sends a generator η to $\pm 16 \in Z$. Indeed, by Lemma 13.2 , DIFF transversality, and the Hirzebruch index theorem, $\frac{1}{3} p_1(\eta)$ is the least index of a closed smooth spin 4-manifold, which is $\pm$ 16 by Rohlin's theorem [40]. The lemma follows if we grant that $\pi_4 G/\mathrm{TOP} = Z$ (not $Z \oplus Z_2$).

Now we complete 13.1. In $Z/16Z$ we have

$$\sigma(M^4) = \frac{1}{3} p_1(\tau M)[M] = \frac{1}{3} p_1(\xi)[S^4] = 8\Delta(\xi)[S^4] = 8\Delta(\tau(M))[M^4] = 8\Delta(M^4)$$

the third equality coming from the last lemma.

π_4G/TOP $= Z$ is used in 13.3 and in all following sections. So we prove it as

PROPOSITION 13.4. – *The forgetful map π_4 G/O $\to \pi_4$ G/TOP is $Z \overset{\times 2}{\to} Z$.*

Proof of 13.4. – (cf. naïve proof in [46A]). Since the cokernel is π_3TOP/O $=$ π_3TOP/PL $= Z_2$, it suffices to show that $\frac{1}{3} p_1$: π_4G/TOP $\to Z$ in the proof of 13.3 sends some element ζ to $\pm 8 \in Z$.

Such a ζ is constructed as follows. In §2, we constructed a closed TOP manifold X^{4+n} with $w_1(X) = w_2(X) = 0$ and a homotopy equivalence $f : N^4 \times T^n \simeq X^{4+n}$, where N^4 is a certain homology manifold (with one singularity) having $\sigma(N^4) = \pm 8$. Imitating the proof of 13.2 with N^4 and $\nu' = f^*\nu(X^{4+n})|\,N^4$ in place of M^4 and $\nu(M^4)$ we construct ξ over S^4 and $\nu' \to \xi$ over the degree 1 map $N^4 \to S^4$. This ξ is fiber homotopically trivial because $\nu(X), f^*\nu(X)$ and ν' are Spivak normal bundles. Let ξ represent ζ in π_4G/TOP.

It remains to show $\frac{1}{3} p_1(\zeta) = \pm 8$. First we reduce n to 1 in $f : N^4 \times T^n \simeq X^{4+n}$ by using repeatedly a splitting principle valid in dimension $\geqslant 6$. (eg. use the TOP version of [76], or just the PL or DIFF version as in the latter part of 5.4 (a)

in [80]). Consider the infinite cyclic covering

$$\bar{f} : N^4 \times R \simeq \bar{X}^5 \quad \text{of} \quad f : N^4 \times T^1 \simeq X^5 .$$

Splitting as above, we find that $CP_2 \times \bar{X}^5 \approx Y^8 \times R$ for some 8-manifold Y^8. Thus using the index theorem 10.7, and the multiplicativity of index and L-classes we have

$$\pm 8 = \sigma(N^4) = \sigma(CP_2 \times N) = \sigma(Y^8) = L_2(Y^8) = L_1(CP_2)\, L_1(\bar{X}^5)\,[CP_2 \times N] =$$

$$L_1(\bar{X}^5)\,[N^4] = -\, L_1(\nu')\,[N^4] = -\frac{1}{3}p_1(\nu')\,[N^4] = -\frac{1}{3}p_1(\xi)\,[S^4] = -\frac{1}{3}p_1(\zeta).$$

(We have suppressed some natural (co)homology isomorphisms).

14. The periodicity of Casson and Sullivan.

A geometric construction of a "periodicity" map

$$\pi : G/PL \to \Omega^4 G/PL$$

was discovered by A.J. Casson in early 1967 (unpublished)(*).

He showed that the fiber of π is $K(Z_2, 3)$, and used this fact with the ideas of Novikov's proof of topological invariance of the rational Pontrjagin classes to establish the Hauptvermutung for closed simply connected PL manifolds $M^m, m \geqslant 5$, with $H^3(M^m, Z_2) = 0$. (Sullivan had a slightly stronger result [88]).

Now precisely the same construction produces a periodicity map π' in a homotopy commutative square

$$\text{(14.1)} \qquad \begin{array}{ccc} G/PL & \xrightarrow{\pi} & \Omega^4 G/PL \\ {\scriptstyle\varphi}\downarrow & & \downarrow{\scriptstyle\Omega^4\varphi} \\ G/TOP & \xrightarrow{\pi'} & \Omega^4 G/TOP \end{array}$$

The construction uses TOP versions of simply connected surgery and transversality. Recalling that the fiber of φ is $K(Z_2, 3)$ we see that $\Omega^4\varphi$ is a homotopy equivalence. Hence π' must be a homotopy equivalence. Thus $(\pi')^{-1} \circ (\Omega^4\varphi)$ gives a homotopy identification of π to φ ; and an identification of the fiber of π to the fiber TOP/PL of φ. Thus TOP/PL had been found (but not identified) in 1967 !

The perfect periodicity $\pi' : G/TOP \simeq \Omega^4(G/TOP)$ is surely an attractive feature of TOP. It suggests that topological manifolds bear the simplest possible relation to their underlying homotopy types. This is a broad statement worth testing.

(*) Essentially the same construction was developed by Sullivan and Rourke later in 1967-68, see [72]. The "periodicity" π is implicit in Sullivan's analysis of G/PL as a fiber produc of $(G/PL)_{(2)}$ and $B_O \otimes Z[\frac{1}{2}]$ over $B_O \otimes Q$, [88] [89].

15. Hauptvermutung and triangulation for normal invariants ; Sullivan's thesis(*).

Since $\mathrm{TOP/PL} \xrightarrow{j} \mathrm{G/PL} \xrightarrow{\varphi} \mathrm{G/TOP} \xrightarrow{\Delta} K(Z_2, 4)$ is a fibration sequence of H-spaces see 15.5 we have an exact sequence for any complex X

$$H^3(X\,;Z_2) = [X\,, \mathrm{TOP/PL}] \to [X\,, \mathrm{G/PL}] \xrightarrow{\varphi_*} [X\,, \mathrm{G/TOP}] \xrightarrow{\Delta_*} H^4(X\,;Z_2)$$

Examining the kernel and cokernel of φ using Sullivan's analysis of $\mathrm{G/PL})_{(2)}$(**) we will obtain

THEOREM 15.1. — *For any countable finite dimensional complex X there is an exact sequence of abelian groups :*

$$H^3(X\,;Z_2)/\mathrm{Image}\, H^3(X\,;Z) \xrightarrow{j_*} [X\,, \mathrm{G/PL}] \xrightarrow{\varphi_*} [X\,, \mathrm{G/TOP}] \xrightarrow{\Delta_*} \{\mathrm{Image}\,(H^4(X\,;Z) + \mathrm{Sq}^2 H^2(X\,;Z_2))\}$$

The right hand member is a subgroup of $H^4(X\,;Z_2)$, and j^* comes from

$$K(Z_2, 3) \simeq \mathrm{TOP/PL} \xrightarrow{j} \mathrm{G/PL}$$

In 1966-67, Sullivan showed that φ_* is injective provided that the left hand group vanishes. Geometrically interpreted, this implies that a homeomorphism $h : M' \to M$ of closed simply connected PL manifolds of dimension $\geqslant 5$ is homotopic to a PL homeomorphism if $H^3(M\,;Z_2)/\mathrm{Image}\,H^3(M\,;Z) = 0$, or equivalently if $H^4(M\,;Z)$ has no 2-torsion [88]. Here $[M\,, \mathrm{G/PL}]$ is geometrically interpreted as a group of *normal invariants*, represented by suitably equipped degree 1 maps $f : M' \to M$ of PL manifolds to M, cf. [95]. The relevant theorem of Sullivan is :

(15.2) *The Postnikov K-invariants of* G/PL, *except for the first, are all odd ; hence*

$$(\mathrm{G/PL})_{(2)} = \{K(Z_2\,;2) \times_{\delta \mathrm{Sq}^2} K(Z_{(2)}, 4)\} \times K(Z_2, 6) \times K(Z_{(2)}, 8) \times K(Z_2, 10) \times K(Z_{(2)}, 12) \times \dots ,$$

where $Z_{(2)} = Z\left[\frac{1}{3}, \frac{1}{5}, \frac{1}{7}, \dots\right]$ *is Z with* $\frac{1}{p}$ *for all odd primes p adjoined.* This is one of the chief results of Sullivan's thesis 1966 [87]. For expositions of it see [72] [13] [74] [89].

(*) Section 15 (indeed §§10-16) discusses corollaries of $\pi_3(\mathrm{TOP/PL}) = Z_2$ collected in spring 1969. For further information along these lines, the reader should see work of Hollingsworth and Morgan (1970) and S. Morita (1971) (added in proof).

(**) The localisation at 2, $A_{(2)} = A \otimes Z_{(2)}$ of a space A will occur below, only for countable H-spaces A such that, for countable finite dimensional complexes X, $[X\,, A]$ is an abelian group (usually a group of some sort of stable bundles under Whitney sum). Thus E.H. Brown's representation theorem offers a space $A_{(2)}$ and map $A \to A_{(2)}$ so that $[X\,, A] \otimes Z_{(2)} = [X\,, A_{(2)}]$. For a more comprehensive treatment of localisation see [89]. The space $A \otimes Q$ is defined similarly.

Sullivan's argument adapts to prove

(15.3) *The Postnikov K-invariants of* G/TOP *are all odd ; Hence*

$$(\mathrm{G/TOP})_{(2)} = K(Z_2, 2) \times K(Z_{(2)}, 4) \times K(Z_2, 6) \times K(Z_{(2)}, 8) \times \ldots$$

Indeed his argument needs only the facts that (1) TOP surgery works, (2) the signature map $Z = \pi_{4k}(\mathrm{G/TOP}) \to Z$ is $\times 8$ (even for $k = 1$, by 13.4), and (3) the Arf invariant map $Z_2 = \pi_{4k+2}(\mathrm{G/TOP}) \to Z_2$ is an isomorphism.

Alternatively (15.2) $\Rightarrow$ (15.3) if we use $\Omega^4(\mathrm{G/PL}) \simeq \mathrm{G/TOP}$ from §14.

Remark 15.4.

It is easy to see directly that the 4-stage of G/TOP must be $K(Z_2, 2) \times K(Z, 4)$. For the only other possibility is the 4 stage of G/PL with K-invariant $\delta \mathrm{Sq}^2$ in $H^5(K(Z_2, 2), Z) = Z_2$. Then the fibration $K(Z_2, 3) = \mathrm{TOP/PL} \to \mathrm{GL/PL} \to \mathrm{G/TOP}$ would be impossible. (Hint : Look at the induced map of 4 stages and consider the transgression onto $\delta \mathrm{Sq}^2$). This remark suffices for many calculations in dimension $\leqslant 6$. On the other hand it is not clear to me that (15.2) $\Rightarrow$ (15.3) without geometry in TOP.

Proof that Kernel $\varphi \cong H^3(X ; Z_2)/\mathrm{image}\ H^3(X ; Z)$.

This amounts to showing that for the natural fibration

$$\Omega \mathrm{G/TOP} \to \mathrm{TOP/PL} \to \mathrm{G/PL}$$

the image of $[X, \Omega \mathrm{G/TOP}]$ in $[X, \mathrm{TOP/PL}] = H^3(X ; Z_2)$, consists of the reduced integral cohomology classes. Clearly this is the image of $[X, \Omega(\mathrm{G/TOP})_{(2)}]$ under $\Omega(\mathrm{G/TOP})_{(2)} \xrightarrow{j_{(2)}} (\mathrm{TOP/PL})_{(2)} = \mathrm{TOP/PL}$. Now $j_{(2)}$ is integral reduction on the factor $K(Z_{(2)}, 3)$ of $\Omega(\mathrm{G/TOP})_{(2)}$ because $\pi_4(\mathrm{G/TOP}) \to \pi_3(\mathrm{TOP/PL})$ is onto, and it is clearly zero on other factors. The result follows. The argument comes from [13] [72].

Proof that $\mathrm{Coker}(\varphi) = \{\mathrm{Image}\ H^4(X ; Z) + \mathrm{Sq}^2 H^2(X ; Z_2)\}$.

The following lemma is needed. Its proof is postponed to the end.

LEMMA 15.5. — *The triangulation obstruction* $\Delta : B_{\mathrm{TOP}} \to K(Z_2, 4)$ *is an H-map.*

Write $\varphi : A \to B$ for $\varphi : \mathrm{G/PL} \to \mathrm{G/TOP}$ and let $\varphi_4 : A_4 \to B_4$ be the induced map of Postnikov 4-stages, which have inherited H-space structure. Consider the fibration $A_4 \xrightarrow{\varphi_4} B_4 \xrightarrow{\Delta_4} K(Z_2, 4)$.

Assertion (1). — $(\Delta_4)_*[X, B_4] = \{\mathrm{Image}\ H^4(X ; Z) + \mathrm{Sq}^2 H^2(X ; Z_2)\}$.

Proof of (1). — Since $B_4 = K(Z_2, 2) \times K(Z, 4)$ and

$$[X, B_4] = H^2(X ; Z_2) \oplus H^4(X ; Z)$$

what we have to show is that the class of Δ_4 in

$$[B_4, K(Z_2, 4)] = H^4(K(Z_2, 2) \times K(Z, 4) ; Z_2) =$$

$$= H^4(K(Z_2, 2), Z_2) \oplus H^4(K(Z, 4) ; Z_2)$$

is (Sq^2, ρ) where ρ is reduction mod 2.

The second component of Δ_4 is $\Delta_4 | K(Z, 4)$ which is indeed ρ since

$$Z = \pi_4 \mathrm{G/TOP} \twoheadrightarrow Z_2 = \pi_4 K(Z_2, 4) \quad .$$

The first component $\Delta_4 \mid K(Z_2, 2)$ can be Sq^2 or 0 a priori, but it cannot be 0 as that would imply $A_4 \simeq K(Z_2, 2) \times K(Z, 4)$. This establishes Assertion (1).

Assertion (2). — $(\Delta_4)_* [X, B_4] = \Delta_* [X, B]$ by the projection $B \to B_4$.

Proof of (2). — In view of 15.5, localising B_4 and B at 2 does not change the left and right hand sides. But after localization, we have equality since $B_{(2)}$ is the product (15.3).

The theorem follows quickly

$$[X, A]/\varphi_* [X, A] = \Delta_* [X, B] = (\Delta_4)_* [X, B_4] = \{\mathrm{Image}\, H^4(X, Z) + \mathrm{Sq}^2 H^2(X; Z_2)\}$$

The three equalities come from Lemma 15.5 and (1) and (2) respectively.

It remains now to give

Proof of Lemma 15.5 (S. Morita's, replacing something more geometrical).

We must establish homotopy commutativity of the square

$$\begin{array}{ccc} B_{\mathrm{TOP}} \times B_{\mathrm{TOP}} & \xrightarrow{\sigma} & B_{\mathrm{TOP}} \\ {\scriptstyle \Delta \times \Delta} \downarrow & & \downarrow {\scriptstyle \Delta} \\ K(Z_2, 4) \times K(Z_2, 4) & \xrightarrow{\alpha} & K(Z_2, 4) \end{array}$$

where σ represents Whitney sum and α represents addition in cohomology.

Now $\alpha \circ (\Delta \times \Delta)$ represents $\Delta \times 1 + 1 \times \Delta$ in $H^4(B_{\mathrm{TOP}} \times B_{\mathrm{TOP}}\ ;\ Z_2)$. Also $\Delta \circ \sigma$ certainly represents something of the form $\Delta \times 1 + 1 \times \Delta + \Sigma$, where Σ is a sum of products $x \times y$ with x, y each in one of $H^i(B_{\mathrm{TOP}}\ ; Z_2) = H^i(B_{\mathrm{PL}}\ ;\ Z_2)$, for $i = 1,2$ or 3. Since $\Delta \circ \sigma$ restricted to $B_{\mathrm{PL}} \times B_{\mathrm{PL}}$ is zero, Σ must be zero.

Theorem 15.1 is very convenient for calculations. Let M be a closed PL manifold, m-manifold $m \geqslant 5$, and write $\mathcal{S}_{\mathrm{CAT}}(M)$, CAT = PL or TOP, for the set of h-cobordism classes of closed CAT m-manifolds M' equipped with a homotopy equivalence $f : M' \to M$. (See [95] for details).

There is an exact sequence of pointed sets (extending to the left) :

$$\ldots \to [\Sigma M, \mathrm{G/CAT}] \to L_{m+1}(\pi, w_1) \to \mathcal{S}_{\mathrm{CAT}}(M) \xrightarrow{\nu} [M, \mathrm{G/CAT}] \to L_m(\pi, w_1) \quad .$$

It is due to Sullivan and Wall [95]. The map ν equips each $f : M' \to M$(above) as a CAT normal invariant. Exactness at $\mathcal{S}_{\mathrm{CAT}}(M)$ is relative to an action of $L_{m+1}(\pi, w_1)$ on it. Here $L_k(\pi, w_1)$ is the surgery group of Wall in dimension k for fundamental group $\pi = \pi_1 M$ and for orientation map $w_1 = w_1(M)\ ;\ \pi \to Z_2$. There is a generalisation for manifolds with boundary. Since the PL sequence maps naturally to the TOP sequence, our knowledge of the kernel and cokernel of

$$[M, \mathrm{G/PL}] \to [M, \mathrm{G/TOP}]$$

will give a lot of information about $\mathcal{S}_{PL}(M) \to \mathcal{S}_{TOP}(M)$. Roughly speaking failure of triangulability in $\mathcal{S}_{TOP}(M)$ is detected by non triangulability of the TOP normal invariant ; and failure of Hauptvermutung in $\mathcal{S}_{PL}(M)$ cannot be less than its failure for the corresponding PL normal invariants.

In case $\pi_1 M = 0$, one has $\mathcal{S}_{CAT}(M) \cong \mathcal{S}_{CAT}(M_0) \cong [M_0, G/CAT]$ where M_0 is M with an open m-simplex deleted, and so Theorem 15.1 here gives complete information.

Example 15.6. — The exotic PL structure Σ on $S^3 \times S^n$, $n \geqslant 2$, from

$$1 \in H^3(S^3; Z_2) = Z_2$$

admits a PL isomorphism $(S^3 \times S^n)_\Sigma \cong S^3 \times S^n$ homotopic (not TOP isotopic) to the identity.

Example 15.7. — For $M = CP_n$ (= complex projective space), $n \geqslant 3$, the map $[M_0, G/PL] \to [M_0, G/TOP]$ is injective with cokernel $Z_2 = H^4(M_0.Z_2)$. This means that 'half' of all manifolds $M' \simeq CP_n$, $n \geqslant 3$, have PL structure. Such a PL structure is unique up to isotopy, since $H^3(CP_n, Z_2) = 0$.

16. Manifolds homotopy equivalent to real projective space P^n.

After sketching the general situation, we will have a look at an explicit example of failure of the Hauptvermutung in dimension 5.

From [54] [94] we recall that, for $n \geqslant 4$,

$$(16.1) \qquad [P^n, G/PL] = Z_4 \oplus \sum_{i=6}^{n} \pi_i(G/PL) \otimes Z_2 \quad .$$

This follows easily from (15.2). For G/TOP the calculation is only simpler. One gets

$$(16.2) \qquad [P^n, G/TOP] = \sum_{i=2}^{n} \pi_i(G/TOP) \otimes Z_2 \quad .$$

Calculation of $\mathcal{S}_{PL}(P^n) \equiv I_n$ is non-trivial [54] [94]. One gets (for $i \geqslant 1$)

$$(16.3) \qquad I_{4i+2} = I_{4i+1} = [P^{4i}, G/PL]\, ; I_{4i+3} = I_{4i+2} \oplus Z\, ; I_{4i+4} = I_{4i+2} \oplus Z_2 \, .$$

The result for $\mathcal{S}_{TOP}(P^n)$ is similar, when one uses TOP surgery. Then

$$\mathcal{S}_{PL}(P^n) \to \mathcal{S}_{TOP}(P^n)$$

is described as the direct sum of an isomorphism with the map

$$Z_4 = [P^4, G/PL] \to [P^4, G/TOP] = Z_2 \oplus Z_2 \quad ,$$

which sends Z_4 onto $Z_2 = \pi_2 G/TOP$.

Remark 16.4. — When two distinct elements of $\mathcal{S}_{PL}(P^n)$, $n \geqslant 5$, are topologically the same, we know already from 15.1 that their PL normal invariants are distinct since $H^3(P^n; Z_2)$ is not reduced integral. This facilitates detection of examples.

Consider the fixed point free involution T on the Brieskorn-Pham sphere in C^{m+1}

$$\Sigma_d^{2m-1} : z_0^d + z_1^2 + z_2^2 + \cdots z_m^2 = 0 \quad , \quad |z| = 1 \quad ,$$

given by $T(z_0, z_1, \ldots, z_m) = (z_0, -z_1, \ldots, -z_m)$. Here d and m must be odd positive integers, $m \geqslant 3$, in order that Σ_d^{2m-1} really be topologically a sphere [61].

As T is a fixed point free involution the orbit space $\Pi_d^{2m-1} \equiv \Sigma_d^{2m-1}/T$ is a DIFF manifold. And using obstruction theory one finds there is just one oriented equivalence $\Pi_d^{2m-1} \to P^{2m-1}$ (Recall $P^\infty = K(Z_2, 1)$). Its class in $\mathcal{S}_{CAT}(P^{2m-1})$ clearly determines the involution up to equivariant CAT isomorphism and conversely.

THEOREM 16.5. — *The manifolds Π_d^5, d odd, fall into four diffeomorphism classes according as $d \equiv 1, 3, 5, 7$ mod 8, and into two homeomorphism classes according as $d = \pm 1, \pm 3$ mod 8. Π_1^5 is diffeomorphic to P^5.*

Remark 16.6. — With Whitehead C^1 triangulations, the manifolds Π_d^5 have a PL isomorphism classification that coincides with the DIFF classification (§5, [9] [64]). Hence we have here rather explicit counterexamples to the Hauptvermutung. One can check that they don't depend on Sullivan's complete analysis of $(G/PL)_{(2)}$. The easily calculated 4-stage suffices. Nor do they depend on topological surgery.

PROBLEM. — Give an explicit homeomorphism $P^5 \approx \Pi_7^5$.

Remark 16.7. — Giffen states [23] that (with Whitehead C^1 triangulations) the manifolds Π_d^{2m-1}, $m = 5, 7, 9, \ldots$ fall into just four PL isomorphism classes $d \equiv 1, 3, 5, 7$ mod 8. In view of theorem 16.4., these classes are already distinguished by the restriction of the normal invariant to P^5 (which is that of Π_d^5). So Giffen's statement implies that the homeomorphism classification is $d \equiv \pm 1$, ± 3 mod 8.

Proof of 16.5. (**)

The first means of detecting exotic involutions on S^5, was found by Hirsch and Milnor 1963 [32]. They constructed explicit(*) involutions (M_{2r-1}^7, β_r), r an integer $\geqslant 0$, on Milnor's original homotopy 7-spheres, and found invariant spheres $M_{2r-1}^7 \supset M_{2r-1}^6 \supset M_{2r-1}^5$. They observed that the class of M_{2r-1}^7 in $\Gamma_{28}/2\Gamma_{28}$ is an invariant of the DIFF involution (M_{2r-1}^5, β_r) — (consider the suspension operation to retrieve (M_{2r-1}^7, β) and use $\Gamma_6 = 0$). Now the class of M_{2r-1}^7 in $Z_{28} = \Gamma_7$ is $r(r-1)/2$ according to Eells and Kuiper [18], which is odd iff $r \equiv 2$ or 3 mod 4. So this argument shows (M_{2r-1}^5, β_r) is an exotic involution if $r \equiv 2$ or 3 mod 4.

Fortunately the involution (M_{2r-1}^5, β_r) has been identified with the involution (Σ_{2r+1}^5, T).

(*) β_r is the antipodal map on the fibers of the orthogonal 3-sphere bundle M_{2r-1}^7.

(**) See major correction added on pg. 337.

There were two steps. In 1963 certain examples (X^5, α_r) of involutions were given by Bredon, which Yang [101] explicitly identified with (M^7_{2r-1}, β_r). Bredon's involutions extend to O(3) actions, α_r being the antipodal involution in O(3). And for any reflection α in O(3) , α has fixed point set diffeomorphic to $L^3(2r+1, 1) : z_0^{2r+1} + z_1^2 + z_2^2 = 0 ; |z| = 1$. This property is clearly shared by (Σ^5_{2r+1}, T), and Hirzebruch used this fact to identify (Σ^5_{2r+1}, T) to (X, α_r) [33, §4] [34]. The Hirsch-Milnor information now says that Π^5_d is DIFF exotic if $d \equiv 5,7 \bmod 8$.

Next we give a TOP invariant for Π_d in Z_2. Consider the normal invariant ν_d of Π_d in $[P^5, G/O] = Z_4$. Its restriction $\nu_d | P^2$ to P^2 is a TOP invariant because $[P^2, G/O] = [P^2, G/TOP] = Z_2$.

Now Giffen [22] shows that $\nu_d | P^2$ is the Arf invariant in Z_2 of the framed fiber of the torus knot $z_0^d + z_1^2 = 0, |z_0|^2 + |z_1|^2 = 1$ in $S^3 \subset C^2$. This turns out to be 0 for $d \equiv \pm 1 \bmod 8$ and 1 for $d \equiv \pm 3 \bmod 8$, (Levine [53], cf. [61, § 8]).

We have now shown that the diffeomorphism and homeomorphism classifications of the manifolds Π^5_d are *at least* as fine as asserted. But there can be at most the four diffeomorphism classes named, in view of 16.3. (Recall that the PL and DIFF classifications coincide since $\Gamma_i = \pi_i(PL/O) = 0, i \leqslant 5$). Hence, by Remark 16.4, there are exactly four – two in each homeomorphism class.

REFERENCES

[1] Alexander J.W. — On the deformation of an n-cell, *Proc. Nat. Acad. Sci. U.S.A.*, 9, 1923, p. 406-407.

[2] Alexander J.W. — Some problems in topology, *Proc. I.C.M.*, Zürich, 1932, Vol. 1, p. 249-257.

[3] Atiyah M. — Bordism and cobordism, *Proc. Camb. Phil. Soc.*, 57, 1961, p. 200-208.

[4] Browder W. and Hirsch M. — Surgery in piecewise linear manifolds and applications, *Bull. Amer. Math. Soc.*, 72, 1966, p. 959-964.

[5] Browder W., Liulevicius A. and Peterson F. — Cobordism theories, *Ann. of Math.*, 84, 1966, p. 91-101.

[6] Brown Edgar H. — Abstract homotopy theory, *Trans. Amer. Math. Soc.*, 119, 1965, p. 79-85.

[7] Brown M. — A proof of the generalized Shoenflies theorem, *Bull. Amer. Math. Soc.*, 66, 1960, p. 74-76.

[8] Brown M. and Gluck H. — Stable structures on manifolds, *Bull. Amer. Math. Soc.*, 69, 1963, p. 51-58.

[9] Cerf J. — Sur les difféomorphismes de la sphère de dimension trois ($\Gamma_4 = 0$), *Springer Lecture Notes in Math.*, No. 53, 1968.

[10] Černavskii A.V. — Local contractibility of the homeomorphism group of a manifold, *Soviet Math. Doklady*, 9, 1968, p. 1171-1174; *Dokl. Akad. Nauk. S.S.S.R.*, 182, No. 3, 1968,

[11] ČERNAVSKII A.V. — Local contractibility of the homeomorphism group of a manifold (Russian), *Math. Sbornik,* Vol. 79 (121), 1969, p. 307-356.

[12] CHEEGER J. and KISTER J. — Counting topological manifolds, *Topology,* 9, 1970, p. 149-151.

[13] COOKE G. — On the Hauptvermutung according to Sullivan, *Mimeo. I.A.S.,* Princeton, 1968.

[14] CONNELL E.H. — Approximating stable homeomorphisms by piecewise-linear ones, *Ann. of Math.,* 78, 1963, p. 326-338.

[15] DOLD A. — Halbexacte Homotopiefunktoren, *Springer Lecture Notes,* 12, 1966.

[16] DOLD A. — Structure de l'anneau de cobordisme Ω_*, d'après ROHLIN V.A. et WALL C.T.C., *Séminaire Bourbaki,* 1959-60, n° 186.

[17] EDWARDS R.D. and KIRBY R.C. — Deformations of spaces of imbeddings, *Ann. of Math.,* 93, 1971, p. 63-88.

[18] EELLS J. and KUIPER N. — An invariant for smooth manifolds, *Ann. Math. Pura Appl.,* 60, 1962, p. 93-110.

[19] FARRELL F.T. — These proceedings.

[20] FARRELL F.T. and HSIANG W.C. — *Manifolds with* $\pi_1 = G \times_\alpha T$ (to appear):

[21] FURCH R. — Zur Grundlegung der kombinatorischen Topologie, *Hamburger,* Abh. 3, 1924, p. 69-88.

[22] GIFFEN C.H. — Desuspendability of free involutions on Brieskorn spheres, *Bull. Amer. Math. Soc.,* 75, 1969, p. 426-429.

[23] GIFFEN C.H. — Smooth homotopy projective spaces, *Bull. Amer. Math. Soc.,* 75, 1969, p. 509-513.

[24] GLUCK H. — Embeddings in the trivial range, *Bull. Amer. Math. Soc.,* 69, 1963, p. 824-831; *Ann. of Math.,* 81, 1965, p. 195-210.

[25] GROMOV M.L. — Transversal maps of foliations, *Dokl. Akad. Nauk. S.S.S.R.,* 182, 1968, p. 225-258; *Soviet Math. Dokl.,* 9, 1968, p. 1126-1129. MR 38-6628.

[26] GROMOV M.L. — Thesis, *Izv. Akad. Nauk. S.S.S.R.,* 33, 1969, p. 707-734.

[27] HAEFLIGER A. and POENARU V. — Classification des immersions, *Publ. Math. Inst. Hautes Etudes Sci.,* 23, 1964, p. 75-91.

[28] HAEFLIGER A. — Feuilletages sur les variétés ouvertes, *Topology,* 9, 1970, p. 183-194.

[29] HAEFLIGER A. — Homotopy and integrability, *Manifolds Conf.,* Amsterdam, 1970, editor N. Kuiper, Springer Lecture Notes, 1971, No. 179.

[30] HIRSCH M. — On normal microbundles, *Topology,* 5, 1966, p. 229-240.

[31] HIRSCH M. — On nonlinear cell-bundles, *Ann. of Math.,* 84, 1966, p. 373-385.

[32] HIRSCH M. and MILNOR J. — Some curious involutions of spheres, *Bull. Amer. Math. Soc.,* 70, 1964, p. 372-377.

[33] HIRZEBRUCH F. — Singularities and exotic spheres, *Séminaire Bourbaki,* n° 314, 1966-67.

[34] HIRZEBRUCH F. and MAYER K.H. — O(n)-Mannifaltigkeiten, exotische Sphären und Singularitäten, *Lecture notes in Math.,* No. 57, 1968, Springer.

[35] HSIANG W.-C. and SHANESON J.L. — Fake tori, pages 18-51, in *Topology of Manifolds,* editors J.C. Cantrell and C.H. Edwards, Markham, Chicago 1970.

[36] HSIANG W.-C. and WALL C.T.C. — On homotopy tori II, *Bull. London Math. Soc.,* 1, 1969, p. 341-342. (Beware the homotopy invariance p. 342.)

[37] HUDSON J.F.P. — *Piecewise-linear topology,* W.A. Benjamin, Inc., New York, 1969.

[38] KERÉKJÁRTÓ B. v. — *Vorlesungen über Topologie, I, Flächentopologie,* Springer, 1923.

[39] KERVAIRE M. — Le théorème de Barden-Mazur-Stallings, *Comment. Math. Helv.,* 40, 1965, p. 31-42.

[40] KERVAIRE M. and MILNOR J. — Bernoulli numbers, homotopy groups and a theorem of Rohlin, *Proc. I.C.M.*, Edinborough 1958, p. 454-458.
[41] KERVAIRE M. and MILNOR J. — Groups of homotopy spheres, I, *Ann. of Math.*
[42] KIRBY R.C. — Stable homeomorphisms and the annulus conjecture, *Ann. of Math.*, 89, 1969, p. 575-582.
[43] KIRBY R.C. — Lectures on triangulation of manifolds, *Mimeo. U. of Calif.*, Los Angeles, 1969.
[44] KIRBY R.C. and SIEBENMANN L.C. — On the triangulation of manifolds and the Hauptermutung, *Bull. Amer. Math. Soc.*, 75, 1969, p. 742-749.
[44 A] KIRBY R.C. and SIEBENMANN L.C. — A straightening theorem and a Hauptvermutung for pairs, *Notices Amer. Math. Soc.*, 16, 1969, p. 582.
[44 B] KIRBY R.C. and SIEBENMANN L.C. — For manifolds the Hauptvermutung and the triangulation conjecture are false, *Notices Amer. Math. Soc.*, 16, 1969, p. 695.
[44 C] KIRBY R.C. and SIEBENMANN L.C. — Foundations of topology, *Notices Amer. Math. Soc.*, 16, 1969, 698.
[45] KIRBY R.C. and SIEBENMANN L.C. — Some theorems on topological manifolds, in *Manifolds*, editor N. Kuiper, Springer Lecture Notes, 1971.
[46] KIRBY R.C. and SIEBENMANN L.C. — Deformation of smooth and piecewise-linear manifold structures, *Ann. of Math.* (to appear).
[46 A] KIRBY R.C. and SIEBENMANN L.C. — To appear.
[47] KNESER H. — Die Topologie der Mannigfaltigkeiten, *Jahresbericht der D. Math. Ver.*, 34, 1925-26, p. 1-14.
[48] KUIPER N.H. — *These proceedings.*
[49] KUIPER N.H. and LASHOF R.K. — Microbundles and bundles I, *Invent. Math.*, 1, 1966, p. 1-17.
[50] LASHOF R. — The immersion approach to triagulation, pages 52-56 in *Topology of Manifolds*, editors J.C. Cantrell and C.H. Edwards, Markham, Chicago 1970, in extenso pages 282-355 in *Proc. Adv. Study Inst. Alg. Top.*, Aarhus, Denmark, 1970.
[51] LASHOF R. and ROTHENBERG M. — Triangulation of manifolds I, II, *Bull. Amer. Math. Soc.*, 75, 1969, p. 750-757.
[52] LEES J.A. — Immersions and surgeries of topological manifolds, *Bull. Amer. Soc.*, 75, 1969, p. 529-534.
[53] LEVINE J. — Polynomial invariants of knots of codimension two, *Ann. of Math.*, 84, 1966, p. 537-554.
[54] LOPEZ DE MEDRANO S. — Involutions on manifolds, *Ergebnisse Math.*, No. 59, Springer 1971.
[54 A] MILLER R.T. — *Trans. Amer. Math. Soc.* (to appear).
[55] MILNOR J. — On manifolds homeomorphic to the 7-sphere, *Ann. of Math.*, 64, 1956, p. 399-405.
[56] MILNOR J. — A proceedure for killing homotopy groups of differentiable manifolds, *Proc. Symposia in Pure Math.*, Vol. 3; Differentiable Geometry, *Amer. Math. Soc.*, 1961, p. 39-55.
[57] MILNOR J. — Two complexes that are homeomorphic but combinatorially distinct, *Ann. of Math.*, 74, 1961, p. 575-590.
[58] MILNOR J. — Topological manifolds and smooth manifolds, *Proc. Int. Congr. Math.*, Stockholm, 1962, p. 132-138.
[59] MILNOR J. — Spin structures on manifolds, *Enseignement Math.*, 9, 1963, p. 198-203.
[60] MILNOR J. — Microbundles, Part I, *Topology*, Vol. 3, 1964, Sup. 1, p. 53-80.

[61] MILNOR J. — Singular points of complex hypersurfaces, *Ann. of Math. Studies*, No. 61, Princeton U. Press, 1968.

[62] MOISE E. — Affine structures on 3-manifolds, *Ann. of Math.*, 56, 1952, p. 96-114.

[63] MORLET C. — Hauptvermutung et triangulation des variétés, *Sém. Bourbaki*, 1968-69, exposé 362.

[64] MUNKRES J.R. — Concordance of differentiable structures : two approaches, *Michigan Math. J.*, 14, 1967, p. 183-191.

[65] NEWMAN M.H.A. — The engulfing theorem for topological manifolds, *Ann. of Math.*, 84, 1966, p. 555-571.

[66] NÖEBLING G. — Zur Topologie der Manigfaltigkeiten, *Monatshefte für Math. und Phys.*, 42, 1935, p. 117-152.

[67] NOVIKOV S.P. — On manifolds with free abelian fundamental group and applications (Pontrjagin classes, smoothings, high dimensional knots), *Izvestia Akad. Nauk. S.S.S.R.*, 30, 1966, p. 208-246.

[68] POINCARÉ H. — *Dernières pensées,* E. Flammarion 1913, Dover (English translation), 1963.

[69] RADEMACHER H. — Partielle und totale Differentierbarkeit von Funktionen mehrer Variabeln, *Math. Annalen,* 79, 1919, p. 340-359.

[70] RADO T. — Ueber den Begriff der Riemannschen Fläche, *Acta Litt. Scient. Univ. Szegd,* 2, 1925, p. 101-121.

[71] ROHLIN V.A. — A new result in the theory of 4-dimensional manifolds, *Soviet Math. Doklady,* 8, 1952, p. 221-224.

[72] ROURKE C. — The Hauptvermutung according to Sullivan, Part I, *I.A.S. mimeo,* 1967-68.

[73] ROURKE C. and SANDERSON B. — On topological neighborhoods, *Compositio Math.*, 22, 1970, p. 387-424.

[74] ROURKE C. and SULLIVAN D. — *On the Kervaire obstruction* (to appear).

[75] SERRE J.P. — Groupes d'homotopie et classes de groupes abéliens, *Ann. of Math.*, 58, 1953, p. 258-294.

[76] SIEBENMANN L.C. — The obstruction to finding a boundary for an open manifold of dimension > 5, *thesis,* Princeton U., 1965.

[77] SIEBENMANN L.C. — A total Whitehead torsion obstruction, *Comment. Math. Helv.*, 45, 1970, p. 1-48.

[78] SIEBENMANN L.C. — Infinite simple homotopy types, *Indag. Math.*, 32, 1970, p. 479-495.

[79] SIEBENMANN L.C. — Are non-triangulable manifolds triangulable ? p. 77-84, *Topology of Manifolds,* ed. J.C. Cantrell and C.H. Edwards, Markham, Chicago 1970.

[80] SIEBENMANN L.C. — Disruption of low-dimensional handlebody theory by Rohlin's theorem, p. 57-76, *Topology of Manifolds,* ed. J.C. Cantrell and C.H. Edwards, Markham, Chicago 1970.

[81] SIEBENMANN L.C. — Approximating cellular maps by homeomorphisms, *Topology.*

[82] SIEBENMANN L.C. — Deformation of homoemorphisms on stratified sets, *Comment. Math. Helv.*

[83] SIEBENMANN L.C. — *Some locally triangulable compact metric spaces that are not simplicial complexes.*

[84] SMALE S. — On the structure of manifolds, *Amer. J. of Math.*, 84, 1962, p. 387-399.

[85] STEINITZ E. — Beiträge zur analysis situs, *Sitzungsberichte der Berliner Math. Gesellschaft,* 7, 1907, p. 29-49.

[86] STONG R.L.E. — Notes on cobordism theory, *Math. Notes,* Princeton U. Press, 1968.
[87] SULLIVAN D. — Triangulating homotopy equivalences, *Ph. D. thesis,* Princeton, 1966.
[88] SULLIVAN D. — On the Hauptvermutung for manifolds, *Bull. Amer. Math. Soc.* 73, 1967, 598-600.
[89] SULLIVAN D. — *Triangulation and smoothing homotopy equivalences and homeomorphisms* (seminar notes), Princeton U., 1967.
[90] SULLIVAN D. — Geometric Topology, Part I, *Mimeo., M.I.T.,* 1970.
[91] THOM R. — Quelques propriétés globales des variétés différentiables, *Comment. Math. Helv.,* 28, 1954, p. 17-86.
[92] THOM R. — Topological models in biology, *Topology,* 8, 1969, p. 313-336.
[93] WALL C.T.C. — Differential topology, Part V A, Cobordism : geometric theory, *Mineographed notes U. of Liverpool,* 1966.
[94] WALL C.T.C. — Free piecewise linear involutions on spheres, *Bull. Amer. Math. Soc.,* 74, 1968, p. 554-558.
[95] WALL C.T.C. — *Surgery on compact manifolds,* Academic Press, 1971.
[96] WEBER C. — Elimination des points doubles dans le cas combinatoire, *Comment. Math. Helv.,* 41, 1966-67, p. 179-182.
[99] WHITNEY H. — The self-intersections of a smooth n-manifold in $2\,n$-space, *Ann. of Math.,* 45, 1944, p. 220-246.
[97] WHITEHEAD J.H.C. — On C^1 complexes, *Ann. of Math.,* 41, 1940, p. 809-824.
[98] WHITEHEAD J.H.C. — Manifolds with transverse fields in euclidean space, *Ann. of Math.,* 73, 1961, 154-212.
[100] WILLIAMSON Jr. R.E. — Cobordism of combinatorial manifolds, *Ann. of Math.,* 83, 1966, p. 1-33.
[101] YANG C.T. — On involutions of the five sphere, *Topology,* 5, 1966, p. 17-19.

Mathématique
Univ. Paris-Sud
91-Orsay, France

Correction to proof of 16.5 : Glen Bredon has informed me that [101] is incorrect, and that in fact (X^5, α_r) can be identified to $(M^5_{2r+1}, \beta_{r+1})$. Thus, a different argument is required to show that the DIFF manifolds Π^5_d , $d \equiv 1, 3, 5, 7 \bmod 8$, respectively, occupy the four distinct diffeomorphism classes of DIFF 5-manifolds homotopy equivalent to P^5 . The only proof of this available in 1975 is the one provided by M. F. Atiyah in the note reproduced overleaf. So many mistakes, small and large, have been committed with these involutions that it would perhaps be wise to seek *several* proofs.

Note on involutions

by M. F. Atiyah

This is an addendum to Atiyah-Bott "A Lefschetz fixed-point formula for elliptic complexes: II".

Let X be a closed spin-manifold of dimension $4k+2$ and let $T: X \to X$ be an involution preserving orientation and spin-structure. Then T can be lifted to a map $\hat{T}$ of the principal spin bundle of X. The Lefschetz number $\mathrm{Spin}(\hat{T}, X) \in Z[i]$ is then defined. Assume now that T is of odd type, so that $\hat{T}^2 = -1$ (as explained in Atiyah-Bott p. 488, this happens if T has an isolated fixed-point). In this case the eigenvalues of $\hat{T}$ on the harmonic spinors are $\pm i$ (the eigenvalues ± 1 do not occur since $\hat{T}^2 = -1$).

Let a,b denote the multiplicity of $+i, -i$ respectively in $H^+ - H^-$ (the spaces of harmonic spinors). Then $\mathrm{Spin}(\hat{T}, X) = (a-b)i$, while the index of the Dirac operator $= a+b$. Since $\dim X = 4k+2$ this index $= \hat{A}(X) = 0$, so $a = -b$, hence $\mathrm{Spin}(\hat{T}, X) = 2ai$.

Applying this to the exotic Brieskorn involutions we pick up an extra factor of 2 in Theorem 9.8 [in fact this factor of 2 is already incorporated (by mistake) in the statement of 9.8 – the proof gives 2^{2m-1} (see the last line of paper on p. 490)].

COMBINED BIBLIOGRAPHY †
(in lexicographic order)

[Ad] J. F. **ADAMS**, A variant of E. H. Brown's representability theorem, Topology 10 (1971), 185-198.

[Al_1] **J. W. ALEXANDER**, On the deformation of an n-cell, Proc. Nat. Acad. Sci. U.S.A. 9 (1923), 406-407.

[Al_2] **J. W. ALEXANDER**, On the subdivision of 3-space by a polyhedron, Proc. Nat. Acad. Sci. U.S.A. 10 (1924), 6-8.

[Ak] **T. AKIBA**, Homotopy types of some PL complexes, Bull. Amer. Math. Soc. 77 (1971), 1060-2. (These results are unproved in 1974.)

[AnH] **D. R. ANDERSON and W. C. HSIANG**, Extending combinatorial PL structures on stratified spaces, Invent. Math. 32 (1976), 179-204.

[AnC] **J. J. ANDREWS and M. W. CURTIS**, n-space modulo an arc, Ann. of Math. 75 (1962), 1-7.

[Ar_1] **M. A. ARMSTRONG**, Collars and concordances of topological manifolds, Comment. Math. Helv. 45 (1970), 119-128.

[Ar_2] **M. A. ARMSTRONG**, Transversality for polyhedra, Ann. of Math. 86 (1967), 272-297 .

[AtB] **M. F. ATIYAH and R. BOTT**, A Lefshetz fixed point formula for elliptic complexes: II, Applications, Ann. of Math. 88 (1968), 451-491.

[Bar] **D. BARDEN**, A quick immersion, preprint, Cambridge 1970, (see [Rus]).

[Bas] **H. BASS**, Algebraic K-theory, W. A. Benjamin, New York, 1968.

[BHS] **H. BASS, A. HELLER and R. SWAN**, The Whitehead group of a polynomial extension, Publ. Inst. Hautes Etudes Sci. 22 (1964). See also [HS_2].

[Bl] **W. BLANKENSHIP**, Generalization of a construction of Antoine, Ann. of Math. 53 (1951), 176-297.

[BoV] **J. M. BOARDMAN and R. M. VOGT**, Homotopy invariant algebraic structures on topological spaces, Lecture Notes in Math. 347 (1973), Springer Vlg.

[BoS] **A. BOREL and J.-P. SERRE**, Corners and arithmetic groups. Avec un appendice: Arrondissement des variétés à coins, par A. Douady et L. Hérault, Comment. Math. Helv. 48 (1973), 436-491.

[Br_0] **W. BROWDER**, Manifolds with $\pi_1 = Z$, Bull. Amer. Math. 72 (1966), 238-244.

[Br_1] **W. BROWDER**, The Kervaire invariant, Ann. of Math. 90 (1969), 157-186.

[Br_2] **W. BROWDER**, Surgery on simply-connected manifolds, Ergebnisse der Math., Band 65, Springer 1972.

[BrH] **W. BROWDER and M. HIRSCH**, Surgery on piecewise linear manifolds and applications, Bull. Amer. Math. Soc. 72 (1966), 959-964.

[Brn] **E. H. BROWN**, Abstract homotopy theory, Trans. Amer. Math. Soc. 119 (1965), 79-85. See also [Ad] and [Vog] .

† Typing by Jeanne Polevoi and L.S., 1976 .

[Brn_1] M. BROWN, A proof of the generalized Shoenflies theorem, Bull. Amer. Math. Soc. 66 (1960), 74-76.

[Brn_2] M. BROWN, Locally flat embeddings of topological manifolds, Ann. of Math. 75 (1962), 331-342 ; also Topology of 3-manifolds, Prentice-Hall (1962). See also [Cny_1] .

[Brn_3] M. BROWN, Pushing graphs around, Topology of Manifolds, (Michigan Conference), editor J. G. Hocking, Prindle Weber and Schmidt, 1968, pp. 19-22.

[BrnG] M. BROWN and H. GLUCK, Stable structures on manifolds, Bull. Amer. Math. Soc. 69 (1963), 51-58.

[Bru] G. BRUMFIEL, On the homotopy groups of B_{PL} and PL/O , I, II, III, Ann. of Math. 88 (1968), 291-311; Topology 8 (1969), 305-311; Michigan Math. J. 17 (1970), 217-224.

[BruM] G. BRUMFIEL, I. MADSEN, and R. J. MILGRAM, PL characteristic classes and cobordism, Bull. Amer. Math. Soc 77 (1971), 1025-1030, and Ann. of Math. 97 (1973), 82-159.

[BruMo] G. BRUMFIEL and J. MORGAN, Homotopy theoretic consequences of N. Levitt's obstruction theory to transversality for spherical fibrations, to appear.

[Bry_1] J. L. BRYANT, On embedding of compacta in euclidean space, Proc. Amer. Math. Soc. 23 (1969), 46-51.

[Bry_2] J. L. BRYANT, Approximating embeddings of polyhedra in codimension $\geqslant 3$, Trans. Amer. Math. Soc. 170 (1972), 85-95.

[BS] J. L. BRYANT and C. L. SEEBECK, Locally nice embeddings in codimension three, Quart. J. Math. Oxford 21 (1970), 265-272.

[BuRS] S. BUONCHRISTIANO, C. P. ROURKE and B. J. SANDERSON, A geometric approach to homology theory, preprints, Univ. of Warwick, 1974.

[BuL] D. BURGHELIA and R. LASHOF , The homotopy type of diffeomorphisms, I, II, Trans. Amer. Math. Soc. 196 (1974), 1-50.

[BuLR] D. BURGHELIA, R. LASHOF, and M. ROTHENBERG, Groups of automorphisms of manifolds, Springer Lecture Notes 473 (1975).

[Ca] S. S. CAIRNS, The manifold smoothing problem, Bull. Amer. Math. Soc. 67 (1961), 237-238.

[Cap_1] S. CAPPELL, A splitting theorem for manifolds, Invent. Math., 1976.

[Cap_2] S. CAPPELL, Manifolds with fundamental group a generalized free product, (to appear). Cf. Springer lecture notes, volumes 341-343.

[Cap_3] S. CAPPELL, Splitting obstructions and manifolds with $Z_2 \subset \pi_1$, (to appear).

[CapLS] S. CAPPELL, R. LASHOF, and J. SHANESON, A splitting theorem and the structure of 5-manifolds, Ist. Naz. Alt. Mat. Symp. Mat. vol X (1972), 47-58.

[$CapS_1$] S. CAPPELL and J. SHANESON, On topological knots and knot cobordism, Topology 12 (1973), 33-40.

[$CapS_2$] S. CAPPELL and J. SHANESON, Construction of some new four-dimensional manifolds, Bull. Amer. Math. Soc. 82 (1976), 69-70.

[Ce_1] J. CERF, Groupes d'automorphismes et groupes de difféomorphismes des variétés compactes de dimension 3, Bull. Soc. Math. France 87 (1959), 319-329.

[Ce_2] J. CERF, Topologie de certains espaces de plongements, Bull. Soc. Math. de France 89 (1961), 227-380.

[Ce_3] **J. CERF**, Groupes d'homotopie locaux et groupes d'homotopie mixtes des espaces bitopologiques, définitions et propriétés, C. R. Acad. Sci. Paris, 252 (1961), 4093-4095; suite: Presque n-locale connexion, Applications, Ibid. 253 (1961), 363-365.

[Ce_4] **J. CERF**, Invariants des paires d'espaces, Applications à la topologie différentielle, Centro. Int. Mat. Estivo, Roma Ist. Mat. del Univ. (1962) polycopié. polycopié.

[Ce_5] **J. CERF**, Sur les difféomorphismes de la sphère de dimension trois ($\Gamma_4 = 0$), Springer Lecture Notes in Math. 53 (1968).

[Ce_6] **J. CERF**, Isotopie et pseudo-isotopie, Proc. Int. Congr. Math., Moscow, (1966), 429-437.

[Ce_7] **J. CERF**, La stratification naturelle des espaces de fonctions différentiables réelles et le théorème de la pseudo-isotopie, Publ. Math. IHES 39 (1970), 5-173.

[CeG] **J. CERF et A. GRAMAIN**, Le théorème du h-cobordisme (Smale), notes notes multigraphiées, Ecole Normale Supérieure, 45, rue d'Ulm, Paris (1968).

[$Če_1$] **A. V. ČERNAVSKII**, Local contractibility of the homeomorphism group of a manifold, Soviet Math. Doklady 9 (1968), 1171-1174; Dokl. Adad. Nauk. SSSR, Tom 182 (1968), No. 3, 510-513.

[$Če_2$] **A. V. ČERNAVSKII**, Local contractibility of the homeomorhpism group of a manifold (Russian) Math. Sbornik, Vol. 79 (121) 1969, 307-356; (English) Math. U.S.S.R. Sbornik 8 (1969), No. 3, 287-333.

[Che] **A. CHENCINER**, Pseudo-isotopies différentiables et pseudo-isotopies lineaires par morceaux, C.R. Acad. Sci. Paris, série A, 270 (1970), 1312-1315.

[Ch_1] **T. A. CHAPMAN**, Topological invariance of Whitehead torsion, Amer. J. of Math. 96 (1974), 488-497.

[Ch_2] **T. A. CHAPMAN**, Hilbert cube manifolds, Regional Conf. Series, Greensborough, Oct. 1975, Amer. Math. Soc. (to appear), preprint Univ. of Kentucky, Lexington 1976.

[Chn_1] **M. COHEN**, A general Theory of relative regular neighborhoods, Trans. Amer. Math. Soc. 136 (1969), 189-229.

[Chn_2] **M. COHEN**, A course in simple-homotopy theory, Springer Graduate Texts in Math. 10 (1973).

[Cnl_1] **E. H. CONNELL**, Approximating stable homeomorphisms by piecewise-linear ones, Ann. of Math. 78 (1963), 326-338.

[Cnl_2] **E. H. CONNELL**, A topological h-cobordism theorem, Illinois J. Math 11 (1967), 300-309.

[Cny_1] **R. CONNELLY Jr.**, A new proof of M. Brown's collaring theorem, Proc. Amer. Math. Soc. 27 (1971), 180-2.

[Cny_2] **R. CONNELLY Jr.**, Unknotting close embeddings of polyhedra in codimensions other than two, pages 384-389 in [Top].

[Da] **J. DANCIS**, General position maps for topological manifolds in the 2/3rds range, Trans. Amer. Math. Soc. 216 (1976), 249-266, cf. [Edw_5].

[Di] **T. tom DIECK**, Partitions of unity in the theory of fibrations, Compositio Math. 23 (1971), 159-167.

[Dol] **A. DOLD**, Lectures on algebraic topology, Grundlehren Band 200 (1972), Springer Vlg.

[Dow] **C. H. DOWKER**, Mapping theorems for non-compact spaces, Amer. J. Math. 69 (1947), 200-242.

[Dy_1] **A. DOUADY**, Variétés à bord anguleux, Sém. H. Cartan, 1961-2, exposé 1.

[Dy_2] **A. DOUADY**, Arrondissement des arêtes, Sém. H. Cartan, 1961-62, exposé 3, voir aussi [BoS].

[Edm] **A. L. EDMONDS**, Local contractability of spaces of group actions, Quart. J. Math Oxford 27 (1976), 71-84.

[Ed_1] **R. D. EDWARDS**, The equivalence of close piecewise-linear embeddings, Gen. Topology 5 (1975), 147-180, & errata to appear ibid.

[Ed_2] **R. D. EDWARDS**, On the topological invariance of simple homotopy type for polyhedra, Amer. J. Math. 1975.

[Ed_3] **R. D. EDWARDS**, Approximating codimension $\geqslant 3$ sigma-compacta with locally homotopically unknotted embeddings, (preprint, 1973).

[Ed_4] **R. D. EDWARDS**, Topological regular neighborhoods, polycopied manuscript, Univ. of Calif. at Los Angeles, 1973.

[Ed_5] **R. D. EDWARDS**, Demension theory, pages 195-211 in Geometric Topology, Proceedings Park City Utah Conf. 1974, Springer Lecture Notes 438 (1975).

[Ed_6] **R. D. EDWARDS**, The double suspension of a certain homology 3-sphere is S^5, (in preparation).

[Ed_7] **R. D. EDWARDS**, Locally compact ANR's are hilbert cube manifold factors, (in preparation).

[EK] **R. D. EDWARDS and R. C. KIRBY**, Deformations of spaces of imbeddings, Ann. of Math. 93 (1971), 63-88. (See also [Rus] .)

[EM] **R. D. EDWARDS and R. T. MILLER**, Cell-like closed 0-dimensional decompositions of R^3 are R^4 factors, Trans. Amer. Math. Soc. 215 (1976), 191-203.

[ES] **S. EILENBERG and N. STEENROD**, Foundations of Algebraic Topology, Princeton Univ. Press, 1952.

[EW] **S. EILENBERG and R. WILDER**, Uniform local connectedness and contractibility, Amer. J. Math. 64 (1942), 613-622.

[Fa] **F. T. FARRELL**, The obstruction to fibering a manifold over a circle, Proc. Int. Congress Math. Nice (1970), vol. 2, 69-72, Gauthier-Villars, 1971.

[FH_1] **F. T. FARRELL and W. C. HSIANG**, Manifolds with $\pi_1 = G \times T$, Amer. J. Math. 95 (1973), 813-848.

[FH_2] **F. T. FARRELL and W. C. HSIANG**, A formula for $K_1(R_\alpha[T])$, Proc. Symp. Pure Math. 17, Applications of Categorical Algebra, AMS, 1970.

[Fe] **S. FERRY**, An immersion of $T^n - D^n$ into R^n, Enseignement Math. 20 (1974), 177-178.

[FreK] **M. FREEDMAN and R. KIRBY**, A geometric proof of Rohlin's theorem, preprint, Berkeley, 1975 .

[Fr] **B. FRIBERG**, A topological proof of a theorem of Kneser, Proc. Amer. Math. Soc. 39 (1973), 421-426.

[$GalS_1$] **D. GALEWSKI and R. J. STERN**, The relationship between homology and topological manifolds via homology transversality, polycopied manuscript, 1975.

[$GalS_2$] **D. GALEWSKI and R. STERN**, Classification of simplicial triangulation of topological manifolds, Bull. Amer. Math. Soc. (to appear).

[Gla] **L. C. GLASER**, Geometrical Combinatorial Topology, I and II, Van Nostrand Math. Studies, 27, 28 (1970).

[Gl] **H. GLUCK**, Embeddings in the trivial range, Bull. Amer. Math. Soc. 69 (1963), 824-831; Ann. of Math. 81 (1965), 195-210.

[Gra] **A. GRAMAIN**, Construction explicite de certaines immersions de codimension 0 ou 1, Enseignement Math. 20 (1974), 333-337.

[Gr] **M. L. GROMOV**, Stable mappings of foliations into manifolds, Izv. Akad. Nauk. SSSR Mat. 33 (1969), 707-734; Trans. Math. USSR (Izvestia) 3 (1969), 671-693.

[Ha_0] **A. HAEFLIGER**, Knotted spheres and related geometric problems, Proc. ICM, Moscow 1966, pp. 337-345.

[Ha_1] **A. HAEFLIGER**, Lissage des immersions, I, Topology 6 (1967), 221-239, et II (polycopié, Univ. de Genève).

[Ha_2] **A. HAEFLIGER**, Lectures on the theorem of Gromov, pp. 118-142, in Liverpool Symposium II, ed. C. T. C. Wall, Springer Lecture Notes 209 (1971).

[HaP] **A. HAEFLIGER and V. POENARU**, Classification des immersions combinatoires, Publ. Math. IHES 23 (1964), 75-91.

[Hal] **A. J. S. HAMILTON**, The triangulation of 3-manifolds, Q. J. Math. Oxford 27 (1976), 63-70.

[Ham] **M. E. HAMSTROM**, Homotopy in homeomorphism spaces TOP and PL , Bull. Amer. Math. Soc. 80 (1974), 207-230.

[HamD] **M. E. HAMSTROM and E. DYER**, Regular mappings and the spaces of homeomorphisms on a 2-manifold, Duke Math. J. 25 (1958), 521-531.

[Hat_1] **A. HATCHER**, A K_2 invariant for pseudo-isotopies, thesis, Stanford Univ., 1971.

[Hat_2] **A. HATCHER**, The second obstruction for pseudo-isotopies, Astérisque 6 (1973), Soc. Math. de France. (Requires correction when the first k-invariant is non-zero, see Ann. of Math. 102 (1975), p. 133.)

[HaW] **A. E. HATCHER and J. B. WAGONER**, Pseudo-isotopies of compact manifolds, Astérisque 6 (1973), available from Offilib, 48, rue Gay Lussac 75240 Paris, France.

[HeM] **J. P. HEMPEL and D. R. McMILLAN Jr.**, Covering 3-manifolds with open cells, Fundamenta Math. 64 (1969), 99-104.

[Hi_1] **M. HIRSCH**, Immersions of manifolds, Trans. Amer. Math. Soc. 93 (1959), 242-276.

[Hi_2] **M. HIRSCH**, On combinatorial submanifolds of differentiable manifolds, Comment. Math. Helv. 36 (1962), 103-111.

[Hi_3] **M. HIRSCH**, Obstruction theories for smoothing manifolds and maps, Bull. Amer. Math. Soc. 69 (1963), 352-356.

[Hi_4] **M. HIRSCH**, Smoothings of piecewise-linear manifolds, Chap. I, mimeographed notes, Cambridge Univ., 1964.

[Hi_5] **M. HIRSCH**, On normal microbundles, Topology 5 (1966), 229-240.

[Hi_6] **M. HIRSCH**, On non-linear cell-bundles, Ann. of Math. 84 (1966), 373-385.

[HiM_1] **M. HIRSCH and B. MAZUR**, Smoothings of piecewise-linear manifolds, Chap. II, mimeographed notes, Cambridge Univ., 1964.

[HiM_2] **M. HIRSCH and B. MAZUR**, Smoothings of piecewise-linear manifolds, Annals of Math. Studies 80, Princeton Univ. Press, 1974.

[Hiz] **F. HIRZEBRUCH**, New topological methods in algebraic gemmetry, Grundlagen Series 134 (1966), Springer Vlg.

[Hod] **J. HODGSON**, Surgery on Poincaré complexes, preprint, Univ. of Pennsylvania 1973.

[Hol] **P. HOLM**, The microbundle representation theorem, Acta Math., Uppsala, 117 (1967), 191-213.

[HoM] **J. HOLLINGSWORTH and J. MORGAN**, Homotopy triangulations and topological manifolds, preprint, 1970.

[Hom] **T. HOMMA**, On the imbeddings of polyhedra into manifolds, Yokohama Math. J. 10 (1962), 5-10.

[HS_1] **W. C. HSIANG and J. L. SHANESON**, Fake tori, the annulus conjecture and the conjectures of Kirby. Proc. Nat. Acad. Sci. U.S.A. 62 (1969), 687-691.

[HS_2] **W. C. HSIANG and J. L. SHANESON**, Fake tori, pages 18-51 in [Top].

[Hu_1] **J. F. P. HUDSON**, Extending piecewise-linear isotopies, Proc. London Math. Soc. 16 (1966), 651-668.

[Hu_2] **J. F. P. HUDSON**, Piecewise-linear topology, Benjamin Inc., New York, 1969.

[Hu_3] **J. F. P. HUDSON**, On transversality, Proc. Camb. Phil. Soc. 66 (1969), 17-20.

[Hus] **D. HUSEMOLLER**, Fiber Bundles, McGraw-Hill, 1966.

[HuP] **L. S. HUSH and T. PRICE**, Finding a boundary for a 3-manifold, Ann. of Math. 91 (1970), 223-235.

[I], [II], [III], [IV], [V] These roman numerals refer to the essays I-V of this work.

[J] **L. JONES**, Patch spaces: a geometric representation for Poincaré duality spaces, Ann. of Math. 97 (1973), 306-343.

[Kap] **W. KAPLAN**, Regular curve families filling the plane I and II, Duke Math. J. 7 (1940), 154-185; ibidem 8 (1941), 11-48.

[Kar] **M. KAROUBI**, Périodicité de la K-théorie Hermitienne, Springer Lecture Notes 343, pages 381-383.

[KeaL] **C. KEARTON and W. B. R. LICKORISH**, Piecewise linear critical levels and collapsing, Trans. Amer. Math. Soc. 170 (1972), 415-424.

[Ke_1] **M. KERVAIRE**, Le théorème de Barden-Mazur-Stallings, Comment. Math. Helv. 40 (1965) 31-42.

[Ke_2] **M. KERVAIRE**, Lectures on the theorem of Browder and Novikov and Siebenmann's thesis, Tata Inst., Colaba, Bombay 5, India, 1969.

[KeM] **M. KERVAIRE and J. MILNOR**, Groups of homotopy spheres, I, Ann. of Math. 77 (1963), 504-537.

[Ki_1] **R. C. KIRBY**, Stable homeomorphisms and the annulus conjecture, Ann. of Math. 89 (1969), 575-582, reprinted in this volume.

[Ki_2] **R. C. KIRBY**, Lectures on triangulation of manifolds, Mimeo, Univ. of Calif. at Los Angeles, 1969.

[Ki_3] **R. C. KIRBY**, Locally flat codimension 2 imbeddings, pp. 416-23 in [Top].

[KiK] **R. C. KIRBY and J. KISTER**, revised and enlarged version of [Ki_2], to appear with Publish or Perish, P. O. Box 7108, Berkeley, CA94707.

[KS_1] **R. C. KIRBY and L.C. SIEBENMANN**, On the triangulation of manifolds and the Hauptvermutung, Bull. Amer. Math. Soc. 75 (1969), 742-749, reprinted in this volume.

[KS_2] **R. C. KIRBY and L. C. SIEBENMANN**, A straightening theorem and a Hauptvermutung for pairs, Notices Amer. Math. Soc. 16 (1969), p. 582.

[KS_3] **R. C. KIRBY and L. C. SIEBENMANN**, Remarks on topological manifolds, Notices Amer. Math. Soc. 16 (1969), 698 and 848.

[KS_4] **R. C. KIRBY and L. C. SIEBENMANN**, Some theorems on topological manifolds, in Manifolds, editor N. Kuiper, Springer Notes 97 (1971).

[KS_5] **R. C. KIRBY and L. C. SIEBENMANN**, Normal bundles for codimension 2 locally flat imbeddings, Proc. Topology Conf., Park City, Utah, Feb. 1974, Springer Lecture Notes in Math. 438 (1975).

[Kis] **J. M. KISTER**, Microbundles are fiber bundles, Ann. of Math. 80 (1964), 190-199.

[Kn] **H. KNESER**, Die Deformationssätze der einfach zuzammenhängende Flächen, Math. Z. 25 (1926), 362-372.

[KN] **S. KOBAYASHI and K. NOMIZU**, Foundations of Differential Geometry, Interscience, New York, 1963.

[KP] **W. KOCH and D. PUPPE**, Differenzierbare Structuren auf Manigfaltigkeiten ohne abzählbare Basis, Archiv der Math. 19 (1968), 95-102.

[KuL_1] **N. H. KUIPER and R. K. LASHOF**, Microbundles and bundles, I, Invent. Math. 1 (1966), 1-17.

[KuL_2] **N. H. KUIPER and R. K. LASHOF**, Microbundles and bundles, II, Semi-simplicial theory, Invent. Math. 1 (1966), 243-259.

[Lac] **C. LACHER**, Locally flat strings and half-strings, Proc. Amer. Math. Soc. 18 (1967), 299-304.

[Lam] **K. LAMOTKE**, Semi-simpliziale algebraische topologie, Grundlehren, Band 147 (1968), Springer Vlg.

[Lan] **S. LANG**, Introduction to differentiable manifolds, Interscience, New York, 1962.

[LLM] **J. LANNES, F. LATOUR et C. MORLET**, Géométrie des complexes de Poincaré et chirurgie, polycopié, IHES, 91-Bures sur Yvette, 1973.

[Las_1] **R. K. LASHOF**, Lees immersion theorem and the triangulation of manifolds, Bull. Amer. Math. Soc. 75 (1969), 535-538.

[Las_2] **R. K. LASHOF**, The immersion approach to triangulation, pages 52-56 in [Top]; in extenso, pages 282-355, Proc. Adv. Study Inst. Alg. Top. Aarhus, Denmark 1970.

[Las_3] **R. K. LASHOF**, Embedding spaces, Illinois J. Math. 20 (1976), 144-154.

[LR_1] **R. LASHOF and M. ROTHENBERG**, Microbundles and smoothing, Topology 3 (1965), 357-388.

[LR_2] **R. LASHOF and M. ROTHENBERG**, Triangulation of manifolds, I, II, Bull. Amer. Math. Soc. 75 (1969), 750-754, 755-757.

[LS] **R. LASHOF and J. SHANESON**, Smoothing four-manifolds, Invent. Math. 14 (1971), 197-210.

[Lau] **F. LAUDENBACH**, Formes différentielles de degré 1, Comment. Math. Helv. (à paraitre).

[Ls] **J. A. LEES**, Immersions and surgeries of topological manifolds, Bull. Amer. Math. Soc. 75 (1969), 529-534.

[Lt] **N. LEVITT**, Exotic singular structures on spheres, Trans. Amer. Math. Soc. 205 (1975), 371-388.

[Lick] **W. B. R. LICKORISCH**, A representation of orientable combinatorial 3-manifolds, Ann. of Math. 76 (1962), 531-540.

[Lick] **W. B. R. LICKORISCH,** A representation of orientable combinatorial 3-manifolds, Ann. of Math. 76 (1962), 531-540.

[Lu] **E. LUFT,** Covering manifolds with open discs, Ill. J. Math. 13 (1969), 321-326.

[Mad] **I. MADSEN,** Homology operations in G/TOP, to appear.

[MadM] **I. MADSEN and R. MILGRAM,** The universal smooth surgery class, Comment. Math. Helv. 50 (1975), 281-310.

[Mal] **B. MALGRANGE,** Ideals of differentiable functions, Oxford Univ. Press, 1966.

[Lusk] **E. LUSK,** Level preserving approximations ... and spaces of embeddings, Ill. J. Math. 18 (1974), 147-159.

[Mar] **A. MARIN,** La transversalité topologique, polycopié, Orsay, 1976.

[Mat] **Y. MATSUMOTO,** Topological t-regularity and Rohlin's theorem, J. Fac. Sci. Univ. Tokyo, Sect. I, 18 (1971), 97-108.

[Matu] **T. MATUMOTO,** Variétés simpliciales d'homologie et variétés topologiques topologiques métrisables, thèse, Univ. de Paris-Sud, 91405 Orsay, 1976.

[MatuM] **T. MATUMOTO and Y. MATSUMOTO,** The unstable difference between homology cobordism and piecewise linear block bundles, Tohoku Math J. 27 (1975), 57-68.

[Mau_1] **S. MAUMARY,** Whitehead torsion of Poincaré complexes ... , Preprint Institute for Advanced Study, June 1971.

[Mau_2] **S. MAUMARY,** Proper surgery groups for non-compact manifolds of finite dimension, preprint Berkeley 1972.

[Maun] **C. R. F. MAUNDER,** Surgery on homology manifolds II, ... An application to fake tori, J. London Math. Soc. 12 (1976), 169-175.

[May] **J. P. MAY,** Simplicial objects in algebraic topology, Van Nostrand Math. Studies No. 11.

[MazP] **B. MAZUR and V. POENARU,** Séminaire de topologie combinatoire et différentielle, IHES, Bures sur Yvette, 1962.

[Mc] **D. R. McMILLAN,** Cellular sets and a theorem of Armentrout, Notices Amer. Math. Soc. 17 (1970), p. 840.

[McR] **C. McCRORY,** Cone complexes and PL transversality, Trans. Amer. Math. Soc. 207 (1975), 269-291.

[Mic] **E. MICHAEL,** Continuous Selections, II, Ann. of Math. 64 (1956), 562-580.

[Mil_1] **R. T. MILLER,** Close isotopies on piecewise-linear manifolds, Trans. Amer. Math. Soc. 151 (1970), 597-628, see also Fiber-preserving equivalence, Trans. Amer. Math. Soc. 207 (1975), 241-268.

[Mil_2] **R. T. MILLER,** Approximating codimension $\geqslant 3$ embeddings, Ann. of Math. 95 (1972), 406-416.

[Mil_3] **R. T. MILLER,** Mapping cylinder neighborhoods of some ANR's, Bull. Amer. Math. Soc. 81 (1975), 187-8; and Ann. of Math. 103 (1976), 417-427.

[Milt] **K. C. MILLETT,** Piecewise linear concordances and isotopies, Memoir 153, Amer. Math. Soc. 1974.

[Mi_1] **J. MILNOR,** On manifolds homeomorphic to the 7-sphere, Ann. of Math. 64 (1956), 399-405.

[Mi_2] **J. MILNOR,** On spaces having the homotopy type of a CW-complex, Trans. Amer. Math. Soc. 90 (1959), 272-280.

[Mi_3] **J. MILNOR**, Microbundles and differentiable structures, mimeographed, Princeton Univ., September, 1961.

[Mi_4] **J. MILNOR**, Topological manifolds and smooth manifolds, Proc. Int. Congr. Math., Stockholm 1962, 132-138.

[Mi_5] **J. MILNOR**, Microbundles, Part I, Topology Vol. 3 (1964), Supplement 1, 53-80.

[Mi_6] **J. MILNOR**, Differential Topology, pp. 165-183 in Lectures on Modern Mathematics, Vol. 2, ed. Saaty, Wiley 1964.

[Mi_7] **J. MILNOR**, Topology from the differentiable viewpoint, Univ. Press of Virginia at Charlottesville, 1965.

[Mi_8] **J. MILNOR**, Lectures on the h-cobordism theorem, Princeton Mathematical Notes, 1965.

[Mi_9] **J. MILNOR**, Whitehead torsion, Bull. Amer. Math. Soc. 72 (1966), 358-426.

[Mi_{10}] **J. MILNOR**, On the 3-dimensional Brieskorn manifolds $M(p,q,r)$, pages 175-225 in: Knots, Groups and 3-Manifolds, ed. L. P. Neuwirth, Ann. of Math. Study 84, 1975.

[MiH] **J. MILNOR and J. HUSEMOLLER**, Symmetric Bilinear Forms, Springer 1973.

[MiK] **J. MILNOR and M. KERVAIRE**, Bernoulli numbers, homotopy groups and a theorem of Rohlin, Proc. Int. Congr. Math., Edinburgh, 1958, 454-458.

[Mo] **E. MOISE**, Affine structures on 3-manifolds, Ann. of Math. 56 (1952), 96-114.

[$Morg_1$] **J. MORGAN**, A product theorem for surgery obstructions, to appear.

[$Morg_2$] **J. MORGAN**, an article concerning Sullivan's characteristic varieties, in preparation.

[MorgL] **J. MORGAN and J. LEVITT**, Transversality structures and PL structures on spherical fibrations, Bull. Amer. Math. Soc. 78 (1972), 1064-1068. Cf. [BruMo].

[MorgS] **J. MORGAN and D. SULLIVAN**, Transversality characteristic class and linking cycles in surgery theory, Ann. of Math. 99 (1974), 463-544.

[$Mori_1$] **S. MORITA**, Some remarks on the Kirby-Siebenmann class, J. Fac. Sci. Univ. Tokyo, Sect. I, 18 (1972), 155-161.

[$Mori_2$] **S. MORITA**, Smoothability of PL manifolds is not topologically invariant, Manifolds-Tokyo 1973, pp. 51-56, Univ. of Tokyo Press, 1975.

[Mor_1] **C. MORLET**, Microfibrés et structures différentiables, Sém. Bourbaki (1963), Exposé no. 263.

[Mor_2] **C. MORLET**, Plongements et automorphismes des variétés, (notes multigraphiées), Cours Peccot, Collège de France, Paris, 1969.

[Mor_3] **C. MORLET**, Lissage des homéomorphismes, C. R. Acad. Paris, Série A, 268 (1969), 1323.

[Mor_4] **C. MORLET**, Hauptvermutung et triangulation des variétés, Sém. Bourbaki 1968-69, Exposé no. 362.

[Ms_1] **M. MORSE**, The critical points of a function of n variables, Trans. Amer. Math. Soc. 33 (1931), 71-91.

[Ms_2] **M. MORSE**, Topologically non-degenerate functions on a compact manifold, J. Analyse Math. 7 (1959), 189-208.

[Mu_1] J. MUNKRES, Elementary differential topology, Ann. of Math. Study Princeton Univ. Press.

[Mu_2] J. MUNKRES, Concordance is equivalent to smoothability, Topology 5 (1966), 371-389.

[Mu_3] J. MUNKRES, Concordance of differentiable structures, two approaches, Michigan Math. J. 14 (1967), 183-191.

[Na_1] J. NAGATA, Modern Dimension Theory, North-Holland 1965.

[Na_2] J. NAGATA, Modern General Topology, North-Holland 1968.

[Nar] R. NARASIMHAN, Analysis on real and complex manifolds, North-Holland 1968.

[Neu] L. NEUWIRTH, A topological classification of certain 3-manifolds, Bull. Amer. Math. Soc. 69 (1963), 372-375.

[Ne] M. H. A. NEWMAN, The engulfing theorem for topological manifolds, Ann. of Math. 84 (1966), 555-571.

[Ok] T. OKABE, The existence of topological Morse function and its application to topological h-cobordism theorem, J. Fac. Sci. Univ. Tokyo, Sect. I, 18 (1971), 23-35.

[P_1] R. S. PALAIS, Local triviality of the restriction map for embeddings, Comment. Math. Helv. 34 (1960), 305-312.

[P_2] R. S. PALAIS, On existence of slices for non-compact Lie group actions, Ann. of Math. 73 (1961), 295-323.

[Ped_1] E. K. PEDERSON, Spines of topological manifolds, Comment. Math. Hev. 50 (1975), 41-44.

[Ped_2] E. K. PEDERSON, Topological concordances, Bull. Amer. Math. Soc. 80 (1974), 658-660; see also Appendix 2 in [BuLR] , and article to appear.

[Ped_3] E. K. PEDERSON, Embeddings of topological manifolds, Ill. J. Math 19 (1975), 440-447.

[Q] D. G. QUILLEN, The geometrical realization of a Kan fibration is a Serre fibration, Proc. Amer. Math. Soc. 19 (1968), 1499-1500.

[Qn_1] F. QUINN, A geometric formulation of surgery, thesis, Princeton Univ. 1969; and in outline, pages 500-511 in [Top] .

[Qn_2] F. QUINN, $BTop_n$ and the surgery obstruction, Bull. Amer. Math. Soc. 77 (1971), 596-600.

[Qn_3] F. QUINN, Surgery on Poincaré and normal spaces, Bull. Amer. Math. Soc. 78 (1972), 262-267.

[Ran] A. RANICKI, The algebraic theory of surgery, preprint, Cambridge 1975.

[Roh_1] V. A. ROHLIN, A new result in the theory of 4-dimensional manifolds, Soviet. Math. Doklady 8 (1952), 221-224, (in Russian).

[Roh_2] V. A. ROHLIN, Proof of Gudkov's hypothesis, Funktsional'nyi Analiz i Ego Prilozheniya 6 (1971), 62-64 (Russian); English translation: Functional Analysis and Its Applications 6 (1972), 136-138.

[Rol] D. ROLFSEN, Knots and links, lecture notes, Univ. of British Columbia, Vancouver, 1975, to appear with Publish or Perish, ed. M. Spivak, P. O. Box 7108, Berkeley, CA94707.

[Ro_0] C. P. ROURKE, The Hauptvermutung according to Sullivan (and Casson), IAS mimeo 1967-8, reissued by Warwick Univ.

[Ro_1] **C. P. ROURKE**, On conjectures of Smale and others concerning the diffeomorphism group of S^n, preprint, Univ. of Warwick 1972.

[Ro_2] **C. P. ROURKE**, On structure theorems, preprint, Univ. of Warwick 1972.

[RS_0] **C. P. ROURKE and B. J. SANDERSON**, Block-bundles, I, II, III, Ann. of Math. 87 (1968).

[RS_1] **C. P. ROURKE and B. J. SANDERSON**, Δ-sets I, II, Q. J. Math. Oxford Ser. 2, 22 (1971), 321-338 and 465-485.

[RS_2] **C. P. ROURKE and B. J. SANDERSON**, On topological neighborhoods, Compositio Math. 22 (1970), p. 387-424.

[RS_3] **C. P. ROURKE and B. J. SANDERSON**, Introduction to Piecewise-linear Topology, Ergebnisse Math., Band 69, Springer Vlg., 1972.

[RS_4] **C. P. ROURKE and B. J. SANDERSON**, An imbedding without a normal bundle, Invent. Math. 3 (1967), 293-299.

[RS_5] **C. P. ROURKE and B. J. SANDERSON**, Decompositions and the relative tubular neighborhood conjecture, Topology 9 (1970), 225-229.

[RSu] **C. P. ROURKE and D. SULLIVAN**, On the Kervaire obstruction, Ann. of Math. 94 (1971), 397-413.

[Rud] **M. E. RUDIN**, Lectures on set theoretic topology, Regional Conf. Ser. no. 23, Amer. Math. Soc., 1975.

[Rus] **T. B. RUSHING**, Topological Embeddings, Academic Press, 1973.

[San] **B. J. SANDERSON**, The simplicial extension theorem, Proc. Camb. Phil. Soc. 77 (1975), 497-8.

[Sch_1] **M. SCHARLEMANN**, Constructing strange manifolds with the dodecahedral dodecahedral space, Duke Math. J. 43 (1976), 33-40.

[Sch_2] **M. SCHARLEMANN**, Transversality theories at dimension 4, Inventiones Math. 1976.

[SchS] **M. SCHARLEMANN and L. SIEBENMANN**, The Hauptvermutung for C^∞ homeomorphisms, Manifolds Tokyo 1973, Univ. of Tokyo Press, 1975, 85-91 and sequel Compositio Math. 3 (1974), 253-264.

[Sco] **G. P. SCOTT**, The space of homeomorphisms of a 2-manifold, Topology 9 (1970), 97-109. See corrections by M. E. Hamstrom, Math Reviews 41, no. 9267.

[See] **R. T. SEELEY**, Extension of C^∞ functions defined in a half-space, Proc. Amer. Math. Soc. 15 (1964), 625-626.

[Seg] **G. SEGAL**, Categories and cohomology theories, Topology 13 (1974), 293-312.

[Ser_1] **J. P. SERRE**, Extensions de groupes localement compacts, d'après H. Iwasawa et A. Gleason, Sém. Bourbaki 2 (1949-50), no. 27.

[Ser_2] **J. P. SERRE**, Groupes d'homotopie et classes de groupes abéliens, Ann. Math. 58 (1953), 258-294.

[Ser_3] **J. P. SERRE**, Cours d'Arithmétique, Presses Univ. de France, 1970.

[Sh] **J. SHANESON**, Wall's surgery obstruction groups for $Z \times G$, Ann. of Math. 90 (1969), 296-334.

[SGH_1] **L. SIEBENMANN, L. GUILLOU et H. HAHL**, Les voisinages ouverts réguliers, Annales Ecole Normale Sup. 6 (1973); 253-293.

[SGH_2] **L. SIEBENMANN, L. GUILLOU et H. HAHL**, Les voisinages ouverts réguliers: critères d'existence, Annales Ecole Normale Sup. 7 (1974), 431-462. Voir correction à la dernière phrase dans Acta Math. 1976-77.

[Si_1] L. C. SIEBENMANN, The obstruction to finding a boundary for an open manifold of dimension $\geqslant 5$, thesis, Princeton 1965, cf. [Ke_2] .

[Si_2] L. C. SIEBENMANN, Le fibré tangent, notes de cours, Orsay 1966-67, rédaction, de Chenciner, Herman, et Laudenbach, polycopié, Ecole Polytechnique, Paris.

[Si_3] L. C. SIEBENMANN, On the homotopy type of compact topological manifolds, Bull. Amer. Math. Soc. 74 (1968), 738-742; Z'blatt. 105, p. 567.

[Si_4] L. C. SIEBENMANN, Pseudo-annuli and invertible cobordisms, Archiv der Math. 19 (1968), 28-35.

[Si_5] L. C. SIEBENMANN, On detecting euclidean space homotopically among topological manifolds, Invent. Math. 6 (1968), 245-261.

[Si_6] L. C. SIEBENMANN, A total Whitehead torsion obstruction, Comment. Math. Helv. 45 (1970), 1-48.

[Si_7] L. C. SIEBENMANN, Infinite simple homotopy types, Indag. Math. 32 (1970), 479-495.

[Si_8] L. C. SIEBENMANN, Disruption of low-dimensional handlebody theory by Rohlin's theorem, p. 57-76 in [Top] .

[Si_9] L. C. SIEBENMANN, Are non-triangulable manifolds triangulable? p. 77-84 in [Top] .

[Si_{10}] L. C. SIEBENMANN, Topological manifolds, Proc. I.C.M. Nice (1970), Vol. 2, 143-163, Gauthier-Villars, 1971, reprinted in this volume.

[Si_{11}] L. C. SIEBENMANN, Approximating cellular maps by homeomorphisms, Notices Amer. Math. Soc. 17 (1970), p. 532, and Topology 11 (1973), 271-294.

[Si_{12}] L. C. SIEBENMANN, Deformation of homeomorphisms on stratified sets, Comment. Math. Helv. 47 (1972), 123-163. Note errata in §6.25, p. 158: on line 10, N becomes B ; on lines 23, 24, 25 $\mathfrak{F}_2'$ becomes $\mathfrak{F}_2^t$.

[Si_{13}] L. C. SIEBENMANN, Regular open neighborhoods, General Topology 3 (1973), 51-61, Compare [$SGH_{1,2}$] .

[Si_{14}] L. C. SIEBENMANN, L'invariance topologique du type simple d'homotopie (d'après T. Chapman et R. D. Edwards), Sém. Bourbaki, 25 (1973), No. 428, Springer Lecture Notes Vol. 383 (1974).

[Si_{15}] L. C. SIEBENMANN, (A note explaining the basic properties of k-smooth manifolds, in preparation).

[Sm_1] S. SMALE, Diffeomorphisms of the 2-sphere, Proc. Amer. Math. Soc. 10 (1959), 621-626.

[Sm_2] S. SMALE, On structure of manifolds, Amer. J. of Math. 84 (1962), 387-399.

[Sp] E. H. SPANIER, Algebraic Topology, McGraw-Hill 1966.

[St_1] J. STALLINGS, On fibering certain 3-manifolds, in Topology of 3-manifolds, Prentice Hall, 1962, pp. 95-100.

[St_2] J. STALLINGS, Infinite processes, pp. 245-253 in Differential and Combinatorial Topology, ed. S. Cairns, Princeton Univ. Press, 1965.

[St_3] J. STALLINGS, Lectures on polyhedral Topology, Notes by G. Ananda Swarup, Tata Institute of Fundamental Research, Colaba, Bombay (1967).

[$\check{S}t_1$] M. A. ŠTANKO, Embedding of compacta in euclidean space, Mat. Sbornik Tom 83 (1970), 234-255, (Russian); English translation: Math. USSR Sbornik 12 (1970), 234-255.

[Št$_2$] **M. A. ŠTANKO**, Approximation of imbeddings of compacta in codimension greater than two, Dokl. Akad. Nauk. USSR 198 (1971), 783-786 (Russian) English translation, Soviet Math. Dokl. 12 (1971), 906-909; In extenso: Mat. Sbornik 625-636, (Russian); English translation Math. USSR 190 (132) 1973, Sbornik 19 (1973), 615-626 .

[Ste$_1$] **R. STERN**, On topological vector fields, Topology 14 (1975), 257-270.

[Ste$_2$] **R. STERN**, Classification theorems for structures respecting a submanifold, Ill. J. Math. (to appear).

[Sto$_1$] **A. H. STONE**, Paracompactness and product spaces, Bull. Amer. Math. Soc. Soc. 54 (1948), 977-982.

[Sto$_2$] **D. A. STONE**, Stratified polyhedra, Springer Lecture Notes in Math. No. 252, 1972.

[Sull$_1$] **D. SULLIVAN**, Geometric periodicity and the invariants of manifolds, Manifolds Amsterdam 1970, pages 44-75, Springer Lecture Notes,197(1971).

[Sull$_2$] **D. SULLIVAN**, Geometric Topology, Part I, Mimeographed notes, Mass. Inst. of Tech. 1970.

[Su] **R. SUMMERHILL**, General position for compact subsets of euclidean space, Notices Amer. Math. Soc. 19 (1972) A-276, Abstract 72T-G179; General Topology 3 (1973), 339-345.

[Th$_1$] **R. THOM**, Quelques propriétés globales des variétés différentiables, Comment. Math. Helv. 28 (1954), 17-86.

[Th$_2$] **R. THOM**, Des variétés combinatoires aux variétés différentiables, Proc. Int. Congr. Math., Edinburgh, 1958 , Cambridge Univ. Press 1960.

[Toda] **H. TODA**, Composition methods in homotopy groups of spheres, Ann. of Math. Study 49, Princeton Univ. Press, 1962.

[Top] Topology of Manifolds, edited by J. C. Cantrell and C. H. Edwards, (Athens Georgia Conf. 1969), Markham, Chicago, 1970.

[V] Refers to Essay V of this work.

[Vog] **P. VOGEL**, le classifiant d'une classe de variétés, polycopié à Nantes, 1974, à paraître.

[Vol$_1$] **I. A. VOLODIN**, Algebraic K-theory, Uspehi Mat. Nauk. 27 (1972), No. 4 (166), 207-208, (Russian).

[Vol$_2$] **I. A. VOLODIN**, Generalized Whitehead groups and pseudo-isotopy, Uspehi Mat. Nauk.

[Wald] **F. WALDHAUSEN**, Algebraic K-theory of generalized free products, (to appear). Cf. Proc. Conf. Alg. K-theory, Springer Lecture Notes, vol. 341-343.

[Wa$_0$] **C. T. C. WALL**, Finiteness conditions for CW complexes, II, Proc. Royal Soc. 295 (1966), 129-139.

[Wa$_1$] **C. T. C. WALL**, Surgery on Compact Manifolds, Academic Press, 1971.

[Wa$_2$] **C. T. C. WALL**, Homeomorphism and diffeomorphism classification of manifolds, Proc. Int. Congr. Math., Moscow, 1966, pages 450-459.

[Wa$_3$] **C. T. C. WALL**, On homotopy tori and the annulus theorem, Bull. London Math. Soc. 1 (1969), 95-97.

[We] **C. WEBER**, Elimination des points doubles dans le cas combinatoire, Comment. Math. Helv. 41 (1966-67), 179-182.

[Webs] **D. E. WEBSTER**, Handles , semi-handles and destabilization of isotopies, Can. J. Math. 27 (1975), 439-445.

[Wes] **R. W. WEST**, Representability of Milnor's functor k_{PL}, J. Math. and Mech. 17 (1967), 299-314.

[West] **J. WEST**, Compact ANR's have finite type, Bull. Amer. Math. Soc. 187 (1975), 163-5 , and Ann. of Math. 1976.

[Wh] **D. WHITE**, Smoothing of embeddings and classifying spaces, thèse Univ. de Genève, 1970.

[Wh_1] **J. H. C. WHITEHEAD**, On C^1 complexes, Ann. of Math. 41 (1940), 804-824.

[Wh_2] **J. H. C. WHITEHEAD**, A certain exact sequence, Ann. of Math. 52 (1950), 51-110.

[Wh_3] **J. H. C. WHITEHEAD**, Manifolds with transverse fields in Euclidean space, Ann. of Math. 73 (1961) 154-212.

[Whn_1] **H. WHITNEY**, Analytic extension of differentiable functions defined on closed sets, Trans. Amer. Math. Soc. 36 (1934), 63-89.

[Whn_2] **H. WHITNEY**, The self-intersections of a smooth n-manifold in 2n-space, Ann. of Math. 45 (1944), 220-246.

[Whole] Whole Earth Catalog (access to tools), Portola Inst. & Random House, 1971, 1973.

[Wi] **R. E. WILLIAMSON Jr.**, Cobordism of combinatorial manifolds, Ann. of Math. 83 (1966), 1-33.

[Ze_1] **E. C. ZEEMAN**, Seminar on combinatorial topology, mimeo., IHES, Bures sur Yvette, and Univ. of Warwick, 1963.

[Ze_2] **E. C. ZEEMAN**, Relative simplicial approximation, Proc. Camb. Phil. Soc. 60 (1964), 39-43.

INDEX†

Many definitions of very standard concepts (manifolds, imbeddings, etc.) are given in Essay I, §2. They are not all indicated here.

† J.-C. Hausmann, J. Polevoi, and L. S., 1976.

ANNALS OF MATHEMATICS STUDIES

Edited by Wu-chung Hsiang, John Milnor, and Elias M. Stein

88. Foundational Essays on Topological Manifolds, Smoothings, and Triangulations, *by* ROBION C. KIRBY and LAURENCE C. SIEBENMANN
87. Scattering Theory for Automorphic Functions, *by* PETER D. LAX *and* RALPH S. PHILLIPS
86. 3-Manifolds, *by* JOHN HEMPEL
85. Entire Holomorphic Mappings in One and Several Complex Variables, *by* PHILLIP A. GRIFFITHS
84. Knots, Groups, and 3-Manifolds, *edited by* L. P. NEUWIRTH
83. Automorphic Forms on Adele Groups, *by* STEPHEN S. GELBART
82. Braids, Links, and Mapping Class Groups, *by* JOAN S. BIRMAN
81. The Discrete Series of GL_n over a Finite Field, *by* GEORGE LUSZTIG
80. Smoothings of Piecewise Linear Manifolds, *by* M. W. HIRSCH *and* B. MAZUR
79. Discontinuous Groups and Riemann Surfaces, *edited by* LEON GREENBERG
78. Strong Rigidity of Locally Symmetric Spaces, *by* G. D. MOSTOW
77. Stable and Random Motions in Dynamical Systems: With Special Emphasis on Celestial Mechanics, *by* JÜRGEN MOSER
76. Characteristic Classes, *by* JOHN W. MILNOR *and* JAMES D. STASHEFF
75. The Neumann Problem for the Cauchy-Riemann Complex, *by* G. B. FOLLAND *and* J. J. KOHN
74. Lectures on *p*-Adic *L*-Functions *by* KENKICHI IWASAWA
73. Lie Equations *by* ANTONIO KUMPERA *and* DONALD SPENCER
72. Introduction to Algebraic *K*-Theory, *by* JOHN MILNOR
71. Normal Two-Dimensional Singularities, *by* HENRY B. LAUFER
70. Prospects in Mathematics, *by* F. HIRZEBRUCH, LARS HÖRMANDER, JOHN MILNOR, JEAN-PIERRE SERRE, *and* I. M. SINGER
69. Symposium on Infinite Dimensional Topology, *edited by* R. D. ANDERSON
68. On Group-Theoretic Decision Problems and Their Classification, *by* CHARLES F. MILLER, III
67. Profinite Groups, Arithmetic, and Geometry, *by* STEPHEN S. SHATZ
66. Advances in the Theory of Riemann Surfaces, *edited by* L. V. AHLFORS, L. BERS, H. M. FARKAS, R. C. GUNNING, I. KRA, *and* H. E. RAUCH
65. Lectures on Boundary Theory for Markov Chains, *by* KAI LAI CHUNG
64. The Equidistribution Theory of Holomorphic Curves, *by* HUNG-HSI WU
63. Topics in Harmonic Analysis Related to the Littlewood-Paley Theory, *by* ELIAS M. STEIN
62. Generalized Feynman Amplitudes, *by* E. R. SPEER
61. Singular Points of Complex Hypersurfaces, *by* JOHN MILNOR
60. Topology Seminar, Wisconsin, 1965, *edited by* R. H. BING *and* R. J. BEAN
59. Lectures on Curves on an Algebraic Surface, *by* DAVID MUMFORD
58. Continuous Model Theory, *by* C. C. CHANG *and* H. J. KEISLER
56. Knot Groups, *by* L. P. NEUWIRTH
55. Degrees of Unsolvability, *by* G. E. SACKS
54. Elementary Differential Topology (Rev. edn., 1966), *by* J. R. MUNKRES
52. Advances in Game Theory, *edited by* M. DRESHER, L. SHAPLEY, *and* A. W. TUCKER
51. Morse Theory, *by* J. W. MILNOR
50. Cohomology Operations, *lectures by* N. E. STEENROD, *written and revised by* D.B.A. EPSTEIN
48. Lectures on Modular Forms, *by* R. C. GUNNING
47. Theory of Formal Systems, *by* R. SMULLYAN

A complete catalogue of Princeton mathematics and science books, with prices, is available upon request.

PRINCETON UNIVERSITY PRESS

PRINCETON, NEW JERSEY 08540

LIBRARY OF CONGRESS CATALOGING IN PUBLICATION DATA

Kirby, Robion C 1938–
Foundational essays on topological manifolds, smoothings, and triangulations.

(Annals of mathematics studies; no. 88)
Bibliography: p.
Includes index.
1. Manifolds (Mathematics)—Addresses, essays, lectures. 2. Piecewise linear topology—Addresses, essays, lectures. 3. Triangulating manifolds—Addresses, essays, lectures. I. Siebenmann, L., joint author. II. Title. III. Series.
QA613.K57 514′.7 76–45918
ISBN 0–691–08190–5
ISBN 0–691–08191–3 pbk.